Endangered Species and Fragile Ecosystems in the South China Sea

Alfredo C. Robles Jr.

Endangered Species and Fragile Ecosystems in the South China Sea

The Philippines v. China Arbitration

palgrave
macmillan

Alfredo C. Robles Jr.
De La Salle University
Manila, Philippines

ISBN 978-981-13-9812-4   ISBN 978-981-13-9813-1   (eBook)
https://doi.org/10.1007/978-981-13-9813-1

Jointly published with De La Salle University Publishing House
The print edition is not for sale in the Philippines. Customers from the Philippines please order the print book from: De La Salle University Publishing House.
ISBN of the Co-Publisher's edition: 978-971-555-675-0

This Palgrave Macmillan imprint is published by the registered company Springer Nature Singapore Pte Ltd.
The registered company address is: 152 Beach Road, #21-01/04 Gateway East, Singapore 189721, Singapore

PREFACE

The *South China Sea Arbitration* has attracted considerable scholarly attention since the issuance of the Award on the merits on 12 July 2016. The attention is understandable, as the Award decided a number of controversial issues over which States Parties to the United Nations Convention on the Law of the Sea in general and littoral States in the South China Sea in particular are divided. In the first category, one finds the distinctions between low-tide elevations and islands, on the one hand, and between islands and rocks which cannot sustain human habitation or economic life of their own. In the second category is China's claim to "historic rights" in the South China Sea, embodied in the "nine-dash line." The Tribunal's decisions on the Philippine Submissions No. 11 and 12(b), concerning China's toleration of the harvesting of endangered species and dynamite and cyanide fishing by Chinese nationals and the environmental consequences of China's construction activities in the South China Sea, have not been the object of the same outpouring of scholarly attention. Indeed, it seems that no Filipino scholar has undertaken an in-depth study of the Tribunal's decisions on the environmental claims of the Philippines.

The present work aims to fill this gap. One feature of the book that distinguishes it from the works that have been published so far on the topic, apart from its length, is the effort to present a brief scientific background of the Philippine claims. Unlike these other works, this book does not dwell at length on the significant role played by experts in the adjudication of Philippine Submissions No. 11 and 12(b), on which I am preparing a separate paper.

The analysis presented in the book draws on the scholarly literature in English and other languages on general international law, the international law of the environment, and the law of the sea. I have exploited the dozens of official documents submitted to the Tribunal by the Philippines, many of which were confidential and would have never been accessible to the public had it not been for the arbitration. I have made systematic use of the nearly thousand pages of transcripts of the Hearing on Jurisdiction and Admissibility and the Hearing on the Merits. Finally, I have attempted to take into account the critique of the Award of 12 July 2016 published by the Chinese Society of International Law in 2018.

This book complements the summaries of the two Awards that were published in a previous volume by the De La Salle University Publishing House and Sussex Academic Press. Its contents do not overlap with those of the short essays published in the same volume, which were primarily responses to China's campaign of defamation aimed at the Philippines, the Tribunal, and the Awards.

Writing this book saddened and horrified me. It saddened me, as I became aware of the threat to endangered species caused by the illegal activities of Chinese nationals. It horrified me, as I realized that the damage caused to coral reefs in the South China Sea by China's construction activities is irreversible. I hope that this book will be considered not just as an analysis of the Tribunal's decisions on the environmental submissions of the Philippines in the *South China Sea Arbitration* but also as an appeal to States Parties to the United Nations Convention on the Law of the Sea to hold to account States that glibly proclaim their commitment to preserve and protect the marine environment while acting in ways that belie any such commitment.

Manila, Philippines   Alfredo C. Robles Jr.

Acknowledgements

I am happy to acknowledge the assistance that I received from a number of people while I was researching and writing this book. At the top of the list is Dr. Leslie E. Bauzon, former chair of the Department of History and former Dean of the College of Social Sciences and Philosophy of the University of the Philippines. As the then chair of the Social Sciences Division (Division VIII) of the National Research Council of the Philippines (NRCP), he urged members of the Division at a regular meeting in December 2016 to prepare abstracts of papers for presentation to the June 2017 meeting of the Science Council of Asia (SCA) in Manila. My decision to focus on the environmental issues raised in the Arbitration was inspired by the hope that it would be attractive to an audience of natural scientists. Without Dr. Bauzon's urging, it would never have occurred to me to undertake this research. This is not the first intellectual debt that I owe Dr. Bauzon. Many years ago, I had the good fortune as a history major to have him as a professor in several courses at the University of the Philippines. I am glad to have another reason to express publicly my gratitude to him.

Dr. Maria Thaemar Tana helped me obtain a large number of materials in various languages, without which it would have been literally impossible to carry out this research.

Kareff May Rafisura gave me one of the most wonderful presents that I have ever received, a copy of the commentary of the United Nations Convention on the Law of the Sea edited by Alexander Proelss.

Darren Mangado obtained for me materials from the US Library of Congress in Washington, DC, at a time when I needed them most.

Dr. Wilfredo Licuanan, Director of the Br. Alfred Shields FSC Ocean Research Center and Biology Department, De La Salle University, very kindly read an earlier draft of the section on coral reefs and provided me with copies of his recent work and that of some of his colleagues.

Two anonymous reviewers gave me valuable feedback that guided the revision of the manuscript.

My sister-in-law Raissa Robles and my student Daryll Saclag gave me favorable feedback on the title of the book.

De La Salle University provided me with a grant that supported the research for the book between February and May 2017.

The Science Council of Asia accepted for a poster exhibition an earlier version of Chapters 2 and 3 of the book.

Dr. David Jonathan Bayot of the De La Salle University Publishing House steered the book through the review process in the Philippines and contacted Palgrave Macmillan for co-publication of the book. This is the second time that Dr. Bayot has been able to find a co-publisher outside the Philippines for my research on the *South China Sea Arbitration*. I cannot thank him enough for his efforts on my behalf.

Last, but not least, the production teams at Palgrave Macmillan and De La Salle University Publishing House deserve thanks for their painstaking work in preparing the book for publication.

It goes without saying that any remaining errors are my own responsibility.

CONTENTS

About the Author

Alfredo C. Robles Jr. is a University Fellow at De La Salle University. He holds doctorate degrees from the Université Paris 1 (Panthéon-Sorbonne) and Syracuse University. He is the author of *French Theories of Regulation and Conceptions of the International Division of Labour* (Macmillan, 1994), *The Political Economy of Interregional Relations: ASEAN and the EU* (Ashgate, 2004), *The Asia-Europe Meeting: The Theory and Practice of Interregionalism* (Routledge, 2008), and *The South China Sea Arbitration: Understanding the Awards and Debating with China* (De La Salle University Publishing House and Sussex Academic Press, 2018).

ABBREVIATIONS, ACRONYMS, AND TERMS

AFDI	*Annuaire Français de Droit International* (French Yearbook of International Law)
ASEAN	Association of Southeast Asian Nations
BRP	*Barko ng Republika ng Pilipinas* (Ship of the Republic of the Philippines)
CBD	Convention on Biological Diversity, signed at Rio de Janeiro on 5 June 1992, entered into force on 29 December 1993
CITES	Convention on International Trade in Endangered Species of Wildlife and Fauna, signed at Washington, DC, on 3 March 1973, amended at Bonn, on 22 June 1979, amended at Gaborone, on 30 April 1983
CLCS	Commission on the Limits of the Continental Shelf
CMS	China Marine Surveillance
Convention	United Nations Convention on the Law of the Sea, signed at Montego Bay on 10 December 1982, entered into force on 16 November 1994
COP	Conference of the Parties
CSIL	Chinese Society of International Law
CTI-CFF	Coral Triangle Initiative on Coral Reefs Fisheries and Food Security
DOC	Declaration on the Conduct of the Parties in the South China Sea, 4 November 2002
EAFM	Ecosystems Approach to Fisheries Management (Philippines)

EEZ	Exclusive Economic Zone
EIA	Environmental Impact Assessment
Espoo Convention	Convention on Environmental Impact Assessment in a Transboundary Context, done at Espoo (Finland) on 25 February 1991 and entered into force on 10 September 1997
Ferse Report	Dr. rer. Nat. Sebastian C.A. Ferse, Professor Peter Mumby, Ph.D. and Dr. Selina Ward, Ph.D., *Assessment of the Potential Environmental Consequences of Construction Activities on Seven Reefs in the Spratly Islands in the South China Sea* (26 April 2016)
First Carpenter Report	Professor Kent E. Carpenter, *Eastern South China Sea Environmental Disturbances and Irresponsible Fishing Practices and Their Effects on Coral Reefs and Fisheries* (22 March 2014)
ICJ	International Court of Justice
ITLOS	International Tribunal for the Law of the Sea
IUCN	International Union for Conservation of Nature
McManus Report	Professor John W. McManus, *Offshore Coral Reef Damage, Overfishing and Paths to Peace in the South China Sea* (rev. ed., 21 April 2016)
MEA	Multilateral Environmental Agreement
MLE	Maritime Law Enforcement (China)
MP	*Memorial of the Philippines*, 30 March 2014
MPA	Marine Protected Area
NBSAP	National Biodiversity Strategy and Action Plan
NCM	Non-Compliance Mechanism
OSCE	Organization for Security and Co-operation in Europe
PCA	Permanent Court of Arbitration
PLAN	People's Liberation Army Navy (China)
Ramsar Convention	Convention on Wetlands of International Importance especially as Waterfowl Habitat, 2 February 1971, as amended by the Protocol of 3 December 1982 and the Amendments of 28 May 1987
SCREMP	Sustainable Coral Reef Ecosystem Management Program (Philippines)
SDP	*Supplemental Documents of the Philippines*, 19 November 2015

Second Carpenter Report	Professor Kent E. Carpenter and Professor Loke Ming Chou, *Environmental Consequences of Land Reclamation Activities on Various Reefs in the South China Sea* (14 November 2015)
SOA	State Oceanic Administration (China)
SOA Report	Feng Aiping and Wang Yongzhi, First Ocean Research Institution of State Oceanic Administration, "Construction Activities at Nansha Reefs Did Not Affect the Coral Reef Ecosystem" (10 June 2015)
SOA Statement	State Oceanic Administration of China, "Construction Work at Nansha Reefs Will Not Harm Oceanic Ecosystems" (18 June 2015)
SWSP	*Supplemental Written Submissions of the Philippines*, 16 March 2015
Third Carpenter Report	Declaration of Professor K.E. Carpenter (24 April 2016)
TIHPA	Turtle Islands Heritage Protected Area
TIWS	Turtle Islands Wildlife Sanctuary
UNCLOS III	Third United Nations Conference on the Law of the Sea
UNEP	United Nations Environment Programme

LIST OF TABLES

CHAPTER 1

Introduction

The South China Sea is a large marine ecosystem characterized by a high degree of biodiversity, but its living resources have been overexploited for decades now. Overexploitation has been accompanied by an increasing number of fishing incidents between China and the Philippines since the mid-1990s, involving Chinese fishermen engaged in the harvesting of endangered species in ways that also harmed fragile marine ecosystems, specifically coral reefs. Concluding that bilateral mechanisms had failed to resolve its various disputes with China relating to the South China Sea, the Philippines initiated arbitration under Annex VII of the United Nations Convention on the Law of the Sea ("the Convention") on 22 January 2013. Among the Philippine submissions were the claims that China had breached its obligation to protect and preserve the marine environment by tolerating the activities of Chinese nationals who harvested endangered species and engaged in cyanide and dynamite fishing and by undertaking construction activities on several maritime features in the South China Sea. The proceedings are remarkable for the efforts of the Tribunal to ensure that its decisions were well founded in fact and law and, somewhat surprisingly, by China's informal participation in the proceedings. Following this introduction, the book will explain the importance of marine biodiversity for the Philippines; compare the international framework for the conservation of biodiversity and that for the conservation of marine living resources; examine the jurisdiction of the Arbitral Tribunal over endangered species and fragile marine

© The Author(s) 2020
A. C. Robles Jr., *Endangered Species
and Fragile Ecosystems in the South China Sea,*
https://doi.org/10.1007/978-981-13-9813-1_1

ecosystems; and discuss China's violation of the obligation to protect and preserve the marine environment through toleration of harvesting of endangered marine species and fragile marine ecosystems and the use of destructive fishing techniques as well as through construction activities in the South China Sea.

I. AN OVEREXPLOITED LARGE MARINE ECOSYSTEM: THE SOUTH CHINA SEA

The significance for international trade of the South China Sea, a semi-enclosed sea spanning approximately 3.5 million square kilometers in the Western Pacific, is already well known to the public and experts alike.[1] Approximately $5.3 trillion worth of trade passes through it every year. Such trade includes up to half of the world's oil shipments, 80% of maritime trade with China, and a large part of the trade connecting Europe, Africa, and Asia with Japan, Hawaii, and North and South America. Armed conflict that delayed or interrupted shipping in the area would result in losses estimated at $14.5 billion per day.[2]

For natural scientists, the South China Sea is significant as a large marine ecosystem, which is defined as a relatively large area of ocean space, of approximately 200,000 square kilometers or greater, adjacent to the continents in coastal waters where primary productivity

---

[1] *South China Sea Arbitration*, Award on Jurisdiction and Admissibility, 29 October 2015, 1, para. 3 ("Award on Jurisdiction"), https://pcacases.com/web/sendAttach/1506, accessed 29 March 2019. An enclosed or semi-enclosed sea is "a gulf, basin or sea surrounded by two or more States and connected to another sea or the ocean by a narrow outlet or consisting entirely or primarily of the territorial seas and exclusive economic zones of two or more coastal States." Article 122, *United Nations Convention on the Law of the Sea*, concluded at Montego Bay on 10 December, entered into force on 16 November 1994, http://www.un.org/Depts/los/convention_agreements/texts/unclos/closindx.htm, accessed 21 March 2019.

[2] John W. McManus, "Offshore Coral Reef Damage, Overfishing and Paths to Peace in the South China Sea," *The International Journal of Marine and Coastal Law* 32 (2017): 201; also accessible as Annex 850 in *The Philippines Annexes Cited During the Merits Hearing (Annexes 820–59)* (30 November 2015), 578–608, https://files.pca-cpa.org/pca-docs/The%20Philippines%27%20Annexes%20cited%20during%20Merits%20Hearing%20%28Annexes%20820-859%29.pdf, accessed 26 March 2019. The documentary annexes of the Philippine submissions are not paginated, and the Tribunal's decisions do not even indicate the volume in which an annex is to be found. For the reader's convenience, volume and page numbers are provided in this and other notes.

is generally higher than in open ocean areas.[3] The South China Sea is characterized by an extremely high diversity of species and habitats. The known species include 571 species of stony corals, 3365 species of marine fishes, more than 1500 species of sponges (of which the large majority appears to be endemic, or confined, to the region), 982 species of echinoderms (12% of which are endemic to the region),[4] 45 mangrove species,[5] 20 seagrass species,[6] 7 species of giant clams, and 6 of

[3] Kenneth Sherman and Gotthilf Hempel, "Perspectives on Regional Seas and the Large Marine Ecosystem Approach," in Kenneth Sherman and Gotthilf Hempel (eds.), *The UNEP Large Marine Ecosystems Report: A Perspective on Changing Conditions in LMEs of the World's Regional Seas* (Nairobi: United Nations Environment Programme, 2009), 5, http://lme.edc.uri.edu/index.php/reports-and-workbooks/92-unep-lme-report, accessed 20 February 2019.

[4] Echinodermata is a phylum of radially symmetrical marine animals, having the body wall strengthened by calcareous plates; there is a complex coelom; locomotion is usually carried out by the tube feet, which is distensible finger-like protrusions of a part of the coelom known as the water vascular system; the larva is bilaterally symmetrical and shows traces of metamerism; starfish, sea urchins, brittle stars, sea cucumbers, and sea lilies. *Larousse Dictionary of Science and Technology* (Edinburgh: Larousse plc, 1995), 349.

[5] Mangrove swamps were defined by the Contracting Parties to the 1971 Ramsar Convention as

> forested intertidal ecosystems that occupy sediment-rich sheltered tropical coastal environments, occurring from about 32° N (Bermuda Island) to almost 39° S (Victoria, Australia).... Mangrove swamps are characterized by salt-tolerant woody plants with morphological, physiological, and reproductive adaptations that enable them to colonize littoral habitats. The term mangrove is used in at least two different ways: a) to refer to the ecosystem composed of these plants, associated flora, fauna and their physico-chemical environment; and b) to describe those plant species (of different families and genera) that have common adaptations which allow them to cope with salty and oxygen-depleted (anaerobic) substrates.

8th Meeting of the Conference of the Contracting Parties to the Convention on Wetlands (Ramsar, Iran, 1971), "Wetlands: Water, Life, and Culture," Valencia, Spain, 18–26 November 2002, Resolution VIII.11, Guidance for Identifying and Designating Peatlands, Wet Grasslands, Mangroves and Coral Reefs as Wetlands of International Importance, 10, para. 52, www.ramsar.org/sites/default/files/documents/pdf/res/key_res_viii_11_e.pdf, accessed 26 March 2019. *Convention on Wetlands of International Importance especially as Waterfowl Habitat*, signed at Ramsar, Iran, on 2 February 1971, as amended by the Protocol of 3 December 1982 and the Amendments of 28 May 1987, https://www.ramsar.org/sites/default/files/documents/library/current_convention_text_e.pdf, accessed 24 March 2019.

[6] "Seagrasses are a type of submerged aquatic vegetation (SAV) [that] have evolved from terrestrial plants and have become specialized to live in the marine environment. Like terrestrial plants, seagrasses have leaves, roots, conducting tissues, flowers and seeds, and manufacture their own food via photosynthesis. Unlike terrestrial plants, however, seagrasses

the world's 7 sea (marine) turtle species.[7] The habitats of many of these species are coral reefs, which occupy 12,000 square kilometers in the South China Sea.[8] Most coral reefs in the South China Sea are shelf-edge reefs, found at the edge of a continental shelf, relatively near the coast.[9] Offshore are oceanic coral atolls, the most well known of which are the Spratly Islands, the Paracel Islands, and the Pratas Islands.[10] The first two are significantly rich in biodiversity. Thus, it has been estimated that in the Spratly Islands alone there are 382 stony coral species, 776 benthic species,[11] 524 spices of marine fish, 262 species of algae and seagrass, 35 species of seabirds, and 20 species of marine mammals and sea turtles. On Itu Aba, the largest feature in the Spratly Islands, a 1994 Vietnamese survey reported, among others, 399 reef fish species, 190 coral species, 99 mollusk species, 27 crustacean species, 4 echinoderm species, and 91

do not possess the strong, supportive stems and trunks required to overcome the force of gravity on land. Rather, seagrass blades are supported by the natural buoyancy of water, remaining flexible when exposed to waves and currents." K. Hill, "Indian River Lagoon Species Inventory: Seagrass Habitats," *Smithsonian Marine Station at Fort Pierce* (2002), http://www.sms.si.edu/irlspec/Seagrass_Habitat.htm, accessed 11 March 2017.

[7] *South China Sea Arbitration*, Independent Expert Report, Assessment of the Potential Environmental Consequences of Construction Activities on Seven Reefs in the Spratly Islands in the South China Sea, by Sebastian C. A. Ferse, Peter Mumby, and Selina Ward, 26 April 2016, 12–15 ("Ferse Report"), https://pcacases.com/web/sendAttach/1809, accessed 24 March 2019.

[8] Danwei Huang et al., "Extraordinary Diversity of Reef Corals in the South China Sea," *Marine Biodiversity* 45 (2017): 161.

[9] In geology, a continental shelf is the gently sloping offshore zone, extending usually to about 200 meters in depth. *Larousse Dictionary of Science and Technology*, 242. Article 76(1) of the UN Convention on the Law of the Sea defines the continental shelf as

the sea-bed and subsoil of the submarine areas that extend beyond its territorial sea throughout the natural prolongation of its land territory to the outer edge of the continental margin, or to a distance of 200 nautical miles from the baselines from which the breadth of the territorial sea is measured where the outer edge of the continental margin does not extend up to that distance.

[10] T. Spencer and M. D. Spalding, "The Coral Reefs of Southeast of Southeast Asia: Controls, Patterns and Human Impacts," in A. Gupta (ed.), *Physical Geography of Southeast Asia* (Oxford: Oxford University Press, 2005), 417. An atoll is a coral reef usually forming a circular, elliptical, or irregular chain of islets around a shallow lagoon and surrounded by deep water of the open tropical sea. *Larousse Dictionary of Science and Technology*, 68.

[11] Benthic species live at the soil–water interface at the bottom of a sea or lake. *Larousse Dictionary of Science and Technology*, 100.

other invertebrate species.[12] It is not surprising that coral reefs provide 80% of the fish caught by coastal communities in the South China Sea.[13]

The extent of biodiversity explains the importance of the South China Sea as a fisheries area. The annual landing of fish was estimated to be 10.5 million tons in 2010.[14] Pelagic (occurring in the open sea), transnational stocks of yellowfin tuna, mackerel, billfishes, anchovies, sardines, and several shark species, as well as demersal species (found in the deep water or on the sea bottom), such as groupers, some sharks, penaeid shrimps, giant clams, and sea cucumbers, are important commercial groups.[15] The abundance and diversity of fishery resources provided the basis for significant expansion of the catch of six Southeast Asian countries (Indonesia, Malaysia, Myanmar, the Philippines, Thailand, and Vietnam) between 1956 and 2000.[16]

The predictable result is the overfishing of the waters of the South China Sea. A rapid increase from 1976 to 1983 in the capture of 12 small, commercially valuable pelagic species, such as mackerel, tuna, sardines, and anchovies, gave rise to a situation of full exploitation of most of the pelagic fisheries after 1987.[17] Between 1950 and 2000, fish stocks

[12] Ma. Carmen A. Ablan-Lagman, "The Spratly Islands," in Charles Sheppard (ed.), *World Seas: An Environmental Evaluation, Vol. II: The Indian Ocean to the Pacific* (London: Academic Press, 2019), 587.

[13] United Nations Environment Programme (UNEP), *Coral Reefs in the South China Sea* (Bangkok: UNEP, 2004), 6, http://www.ais.unwater.org/ais/aiscm/getprojectdoc.php?d-ocid=3527#page=1&zoom=auto,-99,792, accessed 15 March 2019.

[14] Louise S. L. Teh, et al., "What Is at Stake? Status and Threats to South China Sea Marine Fisheries," *AMBIO: A Journal of the Human Environment* 46 (2017): 61.

[15] Ibid. For a description of Penaeid shrimps, see Kent E. Carpenter, *The Living Marine Resources of the Western Atlantic, Vol. I: Introduction, Molluscs, Crustaceans, Hagfishes, Sharks, Batoid fishes and Chimaeras* (Rome: Food and Agriculture Organization, 2002), 263, http://www.fao.org/3/y4160e/y4160e18.pdf, accessed 10 March 2019.

[16] See Fig.4.1 in David Rosenberg, "Fisheries Management in the South China Sea," in Sam Bateman and Ralf Emmers (eds.), *Security and International Politics in the South China Sea: Towards a Co-operative Management Regime* (London: Routledge, 2009), 62.

[17] Liane Talaue-McManus, *Transboundary Diagnostic Analysis for the South China Sea* (Bangkok: United Nations Environment Programme, 2000), 40, http://www.unepscs. org/remository/Download/01_-_Project_Development/PDF-B_Phase/Transboundary_ Diagnostic_Analysis/Transboundary_Diagnostic_Analysis_for_the_South_China_Sea.html, accessed 5 March 2019; Hiroyuki Yanagawa, "Small Pelagic Resources in the South China Sea," in M. Devara and P. Marftusubroto (eds.), *Small Pelagic Resources and Their Fisheries in the Asia-Pacific Region. Proceedings of APFIC [Asia-Pacific Fisheries Commission] Working Party on Marine Fisheries, First Session, 13–16 May 1997* (Bangkok: FAO Regional

at trophic levels 3 and above, e.g., tuna, mackerel, jacks, and sharks, in the South China Sea, declined by more than 50%.[18] The coral reefs seem to have reached maximum harvestable potential even earlier, around the mid-1970s.[19] That being said, the state of fish stocks varies from one part of the South China Sea to another. The heaviest fishing, indicated by depleted or fully fished species groupings, is on the western side of the South China Sea, in the shallower shelf fisheries. Stocks are in better conditions around Brunei, where fishing is prohibited around the numerous oil rigs and interconnecting pipes, Sabah, Sarawak, and parts of the Philippines.[20]

Marine biologists suggest that the existence of dense groups of coral reefs in the Spratly Islands, which are potential source of larvae, could explain why the heavily fished species did not become extinct, despite overfishing.[21] Unfortunately, the coral reefs in the South China Sea are no less threatened than the fish stocks. The most serious threats are overfishing and destructive fishing: As fishery resources close to shore are depleted, fishermen extend their activity to the more remote coral reefs and use explosives and cyanide.[22] Other threats to coral reefs, particularly those close to shore, are coastal development, pollution, sedimentation, deforestation, tourism, and bleaching.[23] The reefs that have suffered

Office for Asia and the Pacific, 1997), 365–80, http://www.fao.org/3/a-an020e.pdf, accessed 5 March 2019.

[18] Trophic level is defined as "broad class of organisms within an ecosystem characterized by mode of food supply. The first trophic level comprises the green plants, the second is the herbivores, and the third is the carnivores, which eat the herbivores." *Larousse Dictionary of Science and Technology*, 1134.

[19] UNEP, 5.

[20] Simon Funge-Smith et al., *Regional Overview of Fisheries and Aquaculture in Asia and the Pacific 2012* (Bangkok: FAO Regional Office for Asia and the Pacific, 2012), 15, http://www.fao.org/3/i3185e/i3185e00.pdf, accessed 20 February 2019; McManus, 206.

[21] McManus, 208.

[22] Lauretta Burke et al., *Reefs at Risk Revisited* (Washington, DC: World Resources Institute, 2011), 54–55, https://wriorg.s3.amazonaws.com/s3fs-public/pdf/reefs_at_risk_revisited_hi-res.pdf?_ga=2.177507173.1364385348.1552820007-1064952833.1552820007, accessed 17 March 2019.

[23] UNEP, 4; Huang et al., 157; Si Tuan Vo et al., "Status and Trends in Coastal Habitats of the South China Sea," *Ocean & Coastal Management* 85 (2013): 157.

the least adverse impact are in parts of the Philippines and the offshore reefs and atolls of the Pratas Islands, the Paracel Islands, and the Spratly Islands.[24]

Fishermen respond to overfishing or outright depletion of fishery resources in their country's territorial waters by venturing further out to sea. In so doing, they contribute to further overfishing in areas of the South China Sea where stocks may not have suffered as much from overfishing. Such is the risk for large pelagic fishes (tuna, mackerel, and swordfish) in the eastern part of the South China Sea, the Philippines, Sarawak, Sabah, and Indonesia, where they have been only moderately fished or underfished, as well as for small pelagic fish (sardines and anchovies), which have been only moderately fished or are underfished in northern and central Philippines.[25]

Even worse, fishermen's responses to overfishing create new types of environmental threats. The activities of fishermen from Tanmen, a small fishing town in Hainan Province, deserve particular attention.[26] Tanmen's inhabitants have been fishing for centuries in the South China Sea, particularly in the waters around the Spratly Islands and Scarborough Shoal. Now China's domestic fishery sources have been declining as a result of overfishing and pollution. As in other countries bordering the South China Sea, small demersal fish and small pelagic fish have been fully fished or overfished in China; large demersal fish (e.g., sharks) have been depleted.[27] The sorry state of Chinese domestic fisheries is driving fishermen from Tanmen further into the South China Sea, but they are not necessarily or always harvesting "fish." They are in fact harvesting high-value species, particularly giant clams, sea turtles, and corals, which just happen to be endangered species. Giant clams, the stocks of which have been completely depleted in the Paracel Islands, under Chinese control since 1974, are at the center

---

[24] Ferse Report, 3.

[25] Funge-Smith et al., 15.

[26] Unless otherwise indicated, the following paragraph is summarized from Hongzhou Zhang, "Chinese Fishermen in Disputed Waters: Not Quite a 'People's War'," *Marine Policy* 68 (2016): 65–73, and Hongzhou Zhang, "Chinese Fishermen at Frontline of Maritime Disputes: An Alternative Explanation," *RSIS [Rajaratnam School of International Studies] Commentaries* 152 (2016), http://hdl.handle.net/10220/40776, accessed 13 February 2019.

[27] Funge-Smith et al., 15.

of a handicraft industry in which hundreds of workshops and retailers employing 100,000 people are now engaged. In their search for giant clams, Tanmen fishermen are expanding their operations in waters and reefs in the South China Sea that are claimed by China to be within the "nine-dash line" and are even entering the exclusive economic zones ("EEZs") and territorial waters of other littoral States. Chinese fishermen harvesting corals and sea turtles are engaged in similar operations. It is thus hardly the case that the widespread capture of threatened and endangered species is accidental, as suggested by one report.[28]

The activities of fishermen who are engaged in fishing pelagic species, demersal species, or high-value species far beyond the territorial waters of their State of origin and into disputed waters or waters under the jurisdiction of other littoral States have in the last three decades given rise to an increasing number of incidents among the States bordering the South China Sea. More recently, China's construction activities in the Spratly Islands have created fresh sources of tension with other littoral States in the South China Sea.

II. Fishing Incidents and China's Construction Activities in the South China Sea in Philippine–China Relations

In the typical scenario, fishermen accused of poaching in waters of States other than those of their States of origin are arrested, tried, imprisoned or fined, their catch, their equipment and sometimes their vessels confiscated, while the State of origin exerts pressure to secure their release. Disputes over sovereignty complicate efforts to settle the incidents, as States with competing territorial and/or maritime claims trade charges of violation of sovereignty or jurisdiction. The arresting State alleges that the fishermen have been poaching in waters under its sovereignty or jurisdiction, to which the State of origin of the fishermen responds that they could not have been poaching as they were fishing in waters that were under its sovereignty or jurisdiction. The tendency of some developing States bordering the South China Sea to rely on their navies

---

[28] S. Heileman, "South China Sea," in Kenneth Sherman and Gotthilf Hempel (eds.), *The UNEP Large Marine Ecosystems Report: A Perspective on Changing Conditions in LMEs of the World's Regional Seas* (Nairobi: United Nations Environment Programme, 2009), 303, http://lme.edc.uri.edu/index.php/reports-and-workbooks/92-unep-lme-report, accessed 20 February 2019.

for tasks of maritime law enforcement further aggravates the tension and increases the risk of resort to force.[29]

Comprehensive statistics are difficult to come by, making it hazardous to determine the extent to which fishermen from particular littoral States are involved in such incidents. It seems that from the mid-1980s to the mid-1990s, many of the fishermen arrested in the territorial waters of littoral States other than their own, mainly Myanmar, Malaysia, and Vietnam, were from Thailand.[30] This phenomenon may probably be explained by the fact that it was around this time that the overfishing of the commercially valuable species (large and small pelagic and large and small demersal) in the Gulf of Thailand was beginning to make itself felt.[31] Since the 1990s, fishermen from other Southeast Asian States have also been apprehended in other littoral States. For instance, in May 1992, Thai navy vessels captured a Vietnamese fishing vessel in an area where Thai and Vietnamese maritime claims overlapped. In 1993, an Indonesian patrol boat fired on a Philippine fishing vessel, critically wounding two fishermen.[32] Between 1995 and 2001, fishermen from Malaysia, Indonesia, Thailand, and Vietnam, in addition to fishermen from China, were among those detained in the Philippines for illegal fishing.[33] That having been said, one gets the impression that from the mid-1990s onward, the number of reports of Chinese fishermen being apprehended or otherwise involved in incidents with Southeast Asian littoral States was rising. At one point, four separate groups of Chinese fishermen were incarcerated at almost the same time for poaching in Malaysia, the Philippines, and Vietnam.[34]

Since 2010, there has been a significant increase in the number of fishing incidents throughout the South China Sea, following China's decision to regularly dispatch patrol vessels to protect Chinese fishing

---

[29] Daniel Yarrow Coulter, "South China Sea Fisheries: Countdown to Calamity," *Contemporary Southeast Asia* 17 (1996): 384.

[30] Ibid., 383.

[31] Funge-Smith et al., 15.

[32] Coulter, 384.

[33] Brian Morton, "Fishing for Diplomacy in China's Seas," *Marine Pollution Bulletin* 46 (2003): 795.

[34] Ibid.

vessels against what China sees as harassment by other littoral States.[35] From 2010 to 2018, there were 55 incidents in the South China Sea involving different types of vessels (navy, maritime law enforcement, maritime surveillance, fishing, and research/survey) from China, Taiwan, Indonesia, Malaysia, the Philippines, Thailand, and Vietnam. In four-fifths of the incidents, Chinese maritime law enforcement and navy vessels harassed and occasionally rammed fishing vessels from Indonesia, the Philippines, and Vietnam. The remaining incidents saw them face off navy, maritime law enforcement, and maritime surveillance vessels from these Southeast Asian countries.[36]

Closer examination of the nature and extent of fishing incidents in Philippine–Chinese relations will provide the background for understanding of the Philippine environmental claims against China in the *South China Sea Arbitration*. Based on the documentation submitted by the Philippines to the Arbitral Tribunal, one can count five incidents in the 1990s (March 1992, March 1995, January and March 1998) and 13 in the dozen years or so preceding the initiation of the arbitration in January 2013 (April 2000, January 2001, February, March, and September 2002, October 2004, December 2005, April 2006, July 2010, October and December 2011, and April and May 2012). Most of the incidents arose from the activities of Chinese fishermen in the waters around Scarborough Shoal. In each instance in which the catch of the Chinese fishermen was reported, it consisted mainly or exclusively of endangered species—giant clams, sea turtles, corals, and occasionally, sharks. It was only on three occasions (February 2002, December 2005, and April 2006) that seaweed or unnamed fresh fish was found together with the endangered species. This pattern of activity is in sharp contrast to that of Filipino fishermen engaged in fishing at Scarborough Shoal, who reported that they caught mainly *bonito*,[37] *talakitok*,[38]

---

[35] Nguyen Dang Tang, "Fisheries Co-operation in the South China Sea and the (Ir)relevance of the Sovereignty Question," *Asian Journal of International Law* 2 (2012): 65.

[36] Figures calculated by the author from Center for Strategic and International Studies (CSIS), "Are Maritime Law Enforcement Forces Destabilizing Asia?", *China Power Project* (2018), https://chinapower.csis.org/maritime-forces-destabilizing-asia/, accessed 26 October 2018.

[37] Scientific name *sarda sarda*, *Tuna Species Guide* (2019), http://www.atuna.com/index.php/en/tuna-info/tuna-species-guide#bonito, accessed 26 March 2019.

[38] Yellow Spotted Trevally, scientific name *Carangoides fulvogutatus*, *Fishing the Philippines: Fish Fishing Spots, Lures, Tactics and More*, https://fishingthephilippines.com/category/trevally-talakitok/, accessed 26 March 2019.

Tanguinge,[39] and other species found beneath or near rocks,[40] or that of the Vietnamese fishing vessel sighted by the Philippine Navy in 2006, which had only fresh and salted fish on board.[41] On five occasions between 1995 and 2002, explosives, cyanide, and other paraphernalia for illegal fishing were found on board the Chinese fishing vessels.

Until the beginning of the twenty-first century, the Philippines charged the Chinese fishermen with causing harm to the marine environment, in violation of Philippine laws and international agreements, particularly the Convention on International Trade in Endangered Species of Wild Fauna and Flora ("CITES").[42] On several occasions, the Philippines requested that China instruct Chinese fishermen to respect Philippine sovereignty and to prevent the destruction of the marine environment. In reply, China always rejected the charge that the fishermen had violated Philippine laws, because it had sovereignty over the waters in question and the fishermen had been fishing in those waters for hundreds of years, and sometimes denied that the Philippines even had the right to protest. Once or twice, China declared that it also prohibited the capture of endangered species and the use of cyanide and dynamite in fishing, and that if Chinese fishermen were indeed engaged in illegal activities, it was up to China to prosecute them and indeed, it would prosecute them.[43] On one—and only one—occasion, the Chinese government relayed to the Philippines the Chinese fishermen's explanation

[39] Spanish Mackerel, scientific name *Scomberomonurs comerson*, Fisheries Research and Development Corporation, *Australian Fish Names Standard* (2018), http://www.fishnames.com.au/, accessed 30 March 2018.

[40] *South China Sea Arbitration*, Award of 12 July 2016, 301, para. 763, https://pca-cases.com/web/sendAttach/2086, accessed 26 March 2019.

[41] *Memorial of the Philippines* (30 March 2014), Annex 57, Letter from George T. Uy, Rear Admiral, Armed Forces of the Philippines, to Assistant Secretary, Office of Asia and Pacific Affairs, Department of Foreign Affairs of the Republic of the Philippines (2006), vol. III, 466 (*"MP"*), https://files.pca-cpa.org/pcadocs/The%20Philippines%27%20Memorial%20-%20Volume%20III%20%28Annexes%201-60%29.pdf, accessed 26 March 2019.

[42] *Convention on International Trade in Endangered Species of Wild Fauna and Flora*, signed at Washington, DC, on 3 March 1973, amended at Bonn, on 22 June 1979, amended at Gaborone, on 30 April 1983, https://www.cites.org/eng/disc/text.php, accessed 26 March 2019.

[43] See, for example, *MP*, Annex 19, Memorandum from Erlinda F. Basilio, Acting Assistant Secretary, Office of Asian and Pacific Affairs, Department of Foreign Affairs, Republic of the Philippines, to the Secretary of Foreign Affairs of the Republic of the Philippines (29 March 1995), vol. III, 217, https://files.pca-cpa.org/pcadocs/The%20

that defense against pirates was the reason for their possession of explosives.[44] Invariably, China insisted on the immediate release of the fishermen, to which the Philippines would reply that the judiciary was an independent branch of government and that once the judicial proceedings had been initiated, the government could not intervene. In September 2002, what appeared to be an agreement with the visiting head of the Chinese National People's Congress to release two groups of Chinese fishermen following a trial and payment of a fine seemed to break down when the Chinese ambassador to the Philippines demanded unconditional release. The Philippine Justice Secretary, claiming that the ambassador shouted at him and pounded his table, requested that he be declared *persona non grata*. The Philippine government, fearful of a deterioration in bilateral relations, instructed him to refrain from making further statements to the media until he finally withdrew his request that the ambassador be expelled.[45]

Perhaps it was the desire to avoid this type of diplomatic incident that explains the tendency, perceptible since 2002, of Philippine authorities to "exercise restraint," by releasing Chinese fishing vessels or escorting them away from Scarborough Shoal after confiscating their catch of endangered species (and in February 2002, throwing a sea turtle back to the sea) and illegal paraphernalia, rather than towing the vessels to

---

Philippines%27%20Memorial%20-%20Volume%20III%20%28Annexes%201-60%29.pdf, accessed 26 March 2019.

[44] *MP*, Annex 20, Memorandum from Lauro L. Baja, Jr., Assistant Secretary, Office of Asian and Pacific Affairs, Department of Foreign Affairs, Republic of the Philippines, to the Secretary of Foreign Affairs of the Republic of the Philippines (7 April 1995), vol. III, 222, https://files.pca-cpa.org/pcadocs/The%20Philippines%27%20Memorial%20-%20 Volume%20III%20%28Annexes%201-60%29.pdf, accessed 26 March 2019.

[45] Agence France Presse, "RP Minister Seeks Expulsion of Mainland Envoy," *China Post*, 21 September 2002, https://chinapost.nownews.com/20020921-142615, accessed 18 February 2019; Efren Danao, "Perez Backed vs Sino Envoy," *The Philippine Star*, 22 September 2002, https://www.philstar.com/nation/2002/09/22/176917/perez-backed-vs-sino-envoy, accessed 18 February 2019; Delon Porcalla and Mayen Jaymalin, "Perez Gagged on Sino Poachers," *The Philippine Star*, 24 September 2002, https://www. philstar.com/headlines/2002/09/24/177232/perez-gagged-sino-poachers, accessed 18 February 2019; and Delon Porcalla, Mayen Jaymalin, "Perez, Chinese Envoy Bury Hatchet," *The Philippine Star*, 26 September 2002, https://www.philstar.com/headlines/2002/09/26/177460/perez-chinese-envoy-bury-hatchet, 18 February 2019.

a Philippine port and detaining the Chinese fishermen.[46] What the Philippines saw as "restraint" did not prevent Chinese expressions of dissatisfaction with Philippine deployment of military vessels and aircraft in the waters around Scarborough Shoal.[47] The risk of misunderstanding and miscommunication accompanying the deployment of military vessels and aircraft was highlighted on at least three occasions. In February 2001, the Chinese Embassy in Manila alleged that Chinese fishing vessels that were seeking shelter from a typhoon in Scarborough Shoal had been captured by the Philippine Navy. Subsequently, the Chinese Embassy corrected itself and admitted that the vessels had only been asked to leave Scarborough Shoal by the Philippine Navy.[48] In 2006, China alleged that a Philippine Navy vessel fired gunshots at a Chinese fishing vessel. The Philippines denied that any shots had been fired. China warned that it might be "forced to take necessary measures" if such Philippine actions continued.[49] In the third incident, China claimed that one of the small boats accompanying a Chinese fishing vessel had been shot at, injuring one person.[50]

Almost from the first moment that the incidents occurred, the Philippines and China sought ways of preventing their recurrence. Among the documents that the Philippines submitted, one finds

[46] *MP*, Annex 51, Memorandum from Josue L. Villa, Embassy of the Republic of the Philippines in Beijing, to the Secretary of Foreign Affairs of the Republic of the Philippines (19 August 2002), vol. III, 408, https://files.pca-cpa.org/pcadocs/The%20 Philippines%27%20Memorial%20-%20Volume%20III%20%28Annexes%201-60%29.pdf, accessed 26 March 2019.

[47] Ibid.

[48] *MP*, Annex 46, Office of Asian and Pacific Affairs, Department of Foreign Affairs, Republic of the Philippines, Apprehension of Four Chinese Fishing Vessels in the Scarborough Shoal (23 February 2001), vol. III, 371–72, https://files.pca-cpa.org/pca-docs/The%20Philippines%27%20Memorial%20-%20Volume%20III%20%28Annexes%20 1-60%29.pdf, accessed 26 March 2019.

[49] *MP*, Annex 58, Memorandum from the Secretary of Foreign Affairs of the Republic of the Philippines to the President of the Republic of the Philippines (11 January 2006), vol. III, 482, 484, https://files.pca-cpa.org/pcadocs/The%20Philippines%27%20Memorial%20 -%20Volume%20III%20%28Annexes%201-60%29.pdf, accessed 26 March 2019.

[50] *MP*, Annex 75, Memorandum from the Embassy of the Republic of the Philippines in Beijing to the Secretary of Foreign Affairs of the Republic of the Philippines, No. ZPE-121-2011-S (2 December 2011), vol. IV, 123, https://files.pca-cpa.org/pcadocs/The%20 Philippines%27%20Memorial%20-%20Volume%20IV%20%28Annexes%2061-102%29.pdf, accessed 26 March 2019.

discussions of the possibility of concluding some type of bilateral fishing agreement or code of conduct at meetings in April, July, and August 1995, December 1998, May 2001, August 2002, and January 2012. None of these efforts seem to have borne fruit or to have had any impact on the activities of Chinese fishermen, as only three months after this last proposal by the Philippines, in April 2012, Chinese fishermen at Scarborough Shoal were observed on two separate occasions to be loaded with giant clams.[51] Chinese official vessels intervened to prevent Philippine law enforcement from arresting Chinese fishermen who were harvesting endangered species.[52] Moreover, on two separate occasions in April and May 2012, as Philippine vessels were conducting maritime law enforcement activities following reports of harvesting of giant clams by Chinese fishing vessels, Chinese official vessels undertook dangerous maneuvers that could have resulted in collisions with them.[53] In May 2012, China warned the Philippines to discontinue sending Philippine vessels to Scarborough Shoal.[54] In order to avoid confrontation, the Philippines complied.[55]

---

[51] MP, Annex 78, Report from Commanding Officer, SARV-003, Philippine Coast Guard, to Commander, Coast Guard District, Northwestern Luzon, Philippine Coast Guard (28 April 2012), vol. IV, 147–59, https://files.pca-cpa.org/pcadocs/The%20 Philippines%27%20Memorial%20-%20Volume%20IV%20%28Annexes%2061-102%29.pdf, accessed 26 March 2019.

[52] *South China Sea Arbitration*, Hearing on the Merits and Remaining Issues of Jurisdiction and Admissibility, Transcript, Day 2 (25 November 2015), 181 ("Hearing on the Merits"), https://pcacases.com/web/sendAttach/1548, accessed 26 March 2019.

[53] MP, Annex 77, Memorandum from Col. Nathaniel Y. Casem, Philippine Navy, to Chief of Staff, Armed Forces of the Philippines, No. N2E-0412-008 (11 April 2012), vol. IV, 131–46, https://files.pca-cpa.org/pcadocs/The%20Philippines%27%20Memorial%20-%20 Volume%20IV%20%28Annexes%2061-102%29.pdf, accessed 26 March 2019; MP, Annex 78, 147–58; MP, Annex 80, Report from Relly B. Garcia et al., FRPLEU/QRT [Fishery Resource Protection and Law Enforcement/Quick Response Team] Officers, Bureau of Fisheries and Aquatic Resources, Republic of the Philippines (2 May 2012), MP, vol. IV, 163–95, https://files.pca-cpa.org/pcadocs/The%20Philippines%27%20Memorial%20-%20 Volume%20IV%20%28Annexes%2061-102%29.pdf, accessed 26 March 2019; MP, vol. I, 202–206, paras. 6.115–6.127; Award of 12 July 2016, 417–21, paras. 1046–1058.

[54] MP, Annex 211, *Note Verbale* from the Embassy of the People's Republic of China in Manila to the Department of Foreign Affairs of the Republic of the Philippines, No (12) PG-239 (25 May 2012), vol. VI, 404, https://files.pca-cpa.org/pcadocs/The%20 Philippines%27%20Memorial%20-%20Volume%20VI%20%28Annexes%20158-221%29.pdf, accessed 26 March 2019.

[55] MP, vol. I, 184, para. 6.62,

Throughout most of the period being considered, Filipino fishermen did not seem to be involved in any incidents occurring in waters under Chinese jurisdiction or control. In 1995, the Philippines expressed its serious concern over the detention of Filipino fishermen by Chinese "military elements" deployed on and around Mischief Reef, which the Philippines claimed to be its territory. China denied that its Navy had arrested and detained Filipino fishermen.[56] In the waters around Scarborough Shoal, the Philippines observed that China did nothing to disturb fishing by Filipinos for fifty years following China's declaration of a territorial sea in 1958.[57] Filipino fishermen reported the presence of an official Chinese vessel in the waters around Scarborough Shoal in 2008, but the vessel did not prevent them from fishing.[58] There is no record in the documentation submitted by the Philippines to the Arbitral Tribunal of any incidents involving Filipino fishermen until April-May 2012, when China deployed and anchored Chinese vessels to form a physical barrier that prevented Philippine fishing boats from entering Scarborough Shoal. China thus abruptly ended a "long, peaceful and uninterrupted tradition" of Filipino fishing in the area.[59]

To the disputes over the illegal harvesting of endangered species and the use of cyanide and dynamite by Chinese fishermen were added China's construction activities and island-building on several reefs in the South China Sea. Such activities began in the mid-1990s at Mischief Reef, which in the Philippine view was located in the Philippine EEZ and on its continental shelf and was illegally occupied by China in 1995. Records of diplomatic exchanges submitted by the Philippines indicated that Chinese construction on Mischief Reef was discussed by the two countries on several occasions (February, March, and August 1995, October and November 1998, and March 1999). The exchanges had no discernible impact on Chinese construction activities on Mischief Reef. By February 1999, a helipad, new communications equipment, and

---

[56] *MP*, Annex 17, Memorandum from the Undersecretary of Foreign Affairs of the Republic of the Philippines to the Ambassador of the People's Republic of China in Manila (6 February 1995), vol. III, 205–208, https://files.pca-cpa.org/pcadocs/The%20 Philippines%27%20Memorial%20-%20Volume%20III%20%28Annexes%201-60%29.pdf, accessed 26 March 2019.

[57] *MP*, vol. I, 173, para. 6.44.

[58] Hearing on the Merits, Transcript, Day 2, 181.

[59] Ibid.

wharves were present on the reef. Between 2004 and 2012, China added telecommunications equipment to the structures on both sites.[60]

Concluding that bilateral channels had failed to resolve the disputes with China, the Philippines decided to join the disputes over the harvesting of endangered species, the use of cyanide and dynamite in fishing, and the construction of an artificial island at Mischief Reef with numerous other disputes and to submit them in January 2013 to an arbitral tribunal constituted under Annex VII of the Convention.

III. Overview of the *South China Sea Arbitration*

The Philippine decision may surely be explained in part by a desire for efficiency and optimal use of resources. After all, disputes between States are multifaceted, such that environmental disputes are often inextricably linked with political, and other disputes. It would thus make sense to attempt to submit as many of these different aspects as possible to a single settlement process.

The dispute settlement procedures of the Convention, embodied in its Part XV, reflect compromises reached during the Third United Nations Conference on the Law of the Sea ("UNCLOS III," 1973–1982).[61] On the one hand, they accord States the flexibility to resolve disputes in the manner of their choosing, a principle that is enshrined in Article 279. On the other hand, the Convention provides for compulsory dispute settlement procedures that entail binding decisions and are subject to very specific limitations and exceptions that are set out in the Convention itself.

Upon signature, ratification, or accession to the Convention or at any time thereafter, a State may declare that it accepts one or more means for the compulsory settlement of disputes entailing binding decisions that are identified in Article 287(1) of the Convention. These are the International Tribunal for the Law of the Sea ("ITLOS"), established by Annex VI of the Convention, the International Court of Justice ("ICJ"), arbitration under Annex VII of the Convention, or special arbitration

---

[60] *MP*, vol. I, 193–97, paras. 6.92–6.99.

[61] Award on Jurisdiction, 7, para. 107. United Nations Office of Legal Affairs, Codification Division, *United Nations Diplomatic Conferences: Third United Nations Conference on the Law of the Sea, 1973–1982*, 2019, http://legal.un.org/diplomaticconferences/1973_los/, accessed 4 April 2019.

under Annex VIII of the Convention. A State that does not make such a declaration will not be exempt from the obligation to participate in a dispute settlement proceedings entailing binding decisions, as it will be deemed to have accepted arbitration under Annex VII.[62] If two States Parties to a dispute have not chosen the same means of dispute settlement, they will also be deemed to have accepted arbitration under Annex VII of the Convention.[63]

By ratifying the Convention, the Philippines and China both gave their consent to the Convention's dispute settlement mechanisms.[64] Neither has made a declaration accepting the compulsory jurisdiction of any of the international courts or tribunals listed in Article 287(1). It follows that both are deemed to have given their consent in advance to arbitration as the compulsory procedure entailing binding decisions for the settlement of their disputes. As the Tribunal succinctly put it in its Award on Jurisdiction and Admissibility, "the present dispute has therefore correctly been submitted to arbitration by a tribunal constituted under Annex VII."[65]

International arbitration is defined by the 1907 Convention for the Pacific Settlement of International Disputes as "the settlement of disputes between States by judges of their own choice and on the basis of respect for law."[66] Arbitration is ad hoc: An arbitral tribunal is created to hear a dispute and is dissolved once it has issued an award (judgment). This ad hoc character differentiates it from permanent international tribunals, such as the ICJ, the ITLOS, and the International Criminal Court (ICC). The tribunal's award is binding on the parties to the dispute. The binding character of the award differentiates arbitration from diplomatic forms of dispute settlement (negotiation, good offices, mediation, and

---

[62] Convention, Article 287(3).

[63] Ibid., Article 287(5).

[64] The Convention was ratified by the Philippines on 8 May 1984 and by China on 7 June 1996. United Nations Office of Legal Affairs, Treaty Section, Treaty Collection, *Multilateral Treaties Deposited with the Secretary-General, Chapter XXI: Law of the Sea*, https://treaties.un.org/Pages/ViewDetailsIII.aspx?src=TREATY&mtdsg_no=XXI-6 &chapter=21&Temp=mtdsg3&clang=_en, accessed 4 April 2019.

[65] Award on Jurisdiction, 38, para. 109.

[66] *Convention for the Pacific Settlement of International Disputes*, done at The Hague, 18 October 1907, Article 37, https://pca-cpa.org/wp-content/uploads/sites/6/2016/01/1907-Convention-for-the-Pacific-Settlement-of-International-Disputes. pdf, accessed 4 April 2019.

conciliation), whose results are not binding on the parties to the dispute. Arbitration under Annex VII of the Convention is noteworthy in that it is an integral part of a compulsory system of dispute settlement. In this system, consent to arbitration is implicitly given upon ratification of the Convention if the State has not made a choice among the different means of dispute settlement entailing binding decisions. The recourse to compulsory dispute settlement is subject only to the conditions set out in Sections 1 and 3 of Part XV of the Convention.

Arbitration under Annex VII of the Convention was initiated by the Philippines by a Notification and Statement of Claim dated 22 January 2013 and addressed to China.[67] The Philippines clarified that it was not seeking a determination of which of the two States enjoyed sovereignty over the islands that they claimed.[68] This clarification reflected an awareness that at UNCLOS III, no State considered that a long-standing-dispute over territorial sovereignty would ever be considered a dispute "concerning the interpretation or application of the Convention."[69] China returned the Notification and rejected the invitation to participate in the arbitration. The Arbitral Tribunal was constituted in June 2013 in accordance with Annex VII of the Convention and was composed of five members: Judge Thomas A. Mensah (president), member (1996–2005) and first President (1996–1999) of the ITLOS; Judge Jean-Pierre Cot, member (since 2002) and President of the Chamber for Marine Environment Disputes (2008–2011) of the ITLOS; Judge Stanisław Pawlak, member (since 2005) of the ITLOS; Professor Alfred A.H. Soons, President of the Arbitral Tribunal in *Duzgit Integrity Arbitration (Malta v. São Tomé and Príncipe)*, initiated in 2013; and Judge Rüdiger Wolfrum, member (since 1996), Vice-President (1996–1999), and President (2005–2008) of the ITLOS.[70] The Permanent Court of

[67] Notification and Statement of Claim of the Republic of the Philippines, 22 January 2013, http://www.philippineembassy-usa.org/uploads/pdfs/embassy/2013/2013-0122-Notification%20and%20Statement%20of%20Claim%20on%20West%20Philippine%20Sea.pdf, accessed 3 May 2019.

[68] Ibid., 3, para. 7.

[69] *Chagos Marine Protected Area Arbitration* (Mauritius v. UK), Award of 18 March 2015, 89, paras. 215–216, http://www.pcacases.com/pcadocs/MU-UK%2020150318%20Award.pdf, accessed 3 May 2019.

[70] International Tribunal for the Law of the Sea (ITLOS), *New Arbitrator and President Appointed in the Arbitral Proceedings Instituted by the Republic of the Philippines against the People's Republic of China*, Doc. ITLOS/Press 197 (24 June 2013), https://www.

Arbitration ("PCA"), with headquarters in The Hague, was designated as the Registry in the case. The Philippines submitted its *Memorial* on 30 March 2014.

In December 2014, a week before the date that was set for China's submission of its *Counter-Memorial*, China issued a Position Paper in which it challenged the jurisdiction of the Tribunal.[71] The latter treated the Position Paper as a preliminary objection and held a Hearing on Jurisdiction and Admissibility at The Hague in July 2015. The Tribunal issued an Award on Jurisdiction and Admissibility on 29 October 2015, in which it declared that it had jurisdiction over certain of the Philippine claims and decided to reserve consideration of its jurisdiction over other claims to the merits phase of the proceedings. A Hearing on the Merits and Remaining Issues of Admissibility was held at The Hague in November 2015. The Tribunal issued its Award on the merits on 12 July 2016.[72]

Toward the end of 2013, approximately six months after the constitution of the Arbitral Tribunal, China began a program of intensive construction of artificial islands on Mischief Reef and six other reefs in the South China Sea (Cuarteron Reef, Fiery Cross Reef, Gaven Reef (North), Johnson Reef, Hughes Reef, and Subi Reef). In 2014, the

itlos.org/fileadmin/itlos/documents/press_releases_english/PR_197_E.pdf, accessed 26 March 2019.

[71] *South China Sea Arbitration, Supplemental Written Submission of the Philippines* (16 March 2015), Annex 467, People's Republic of China, Position Paper of the Government of the People's Republic of China on the Matter of Jurisdiction in the South China Sea Arbitration Initiated by the Republic of the Philippines (7 December 2014), Vol. VIII *(Annexes 466-499)*, 19–33 *("SWSP")*, https://files.pca-cpa.org/pcadocs/The%20 Philippines%27%20Supplemental%20Written%20Submission%20-%20Volume%20VIII%20 %28Annexes%20466-499%29.pdf, accessed 4 April 2019.

[72] Award of 12 July 2016, https://pcacases.com/web/sendAttach/2086, accessed 26 March 2019. The procedural history of the arbitration is summarized in the Award, 11–28, paras. 26–111. For a summary of the Award on Jurisdiction, see the Press Release issued by the PCA, *Arbitration between the Republic of the Philippines and the People's Republic of China. Tribunal Renders Award on Jurisdiction and Admissibility; Will Hold Further Hearings*, 29 October 2015, http://www.pcacases.com/web/sendAttach/1503, accessed 26 March 2019. For a summary of the Award of 12 July 2016, see the Press Release issued by the PCA, *The South China Sea Arbitration (The Republic of the Philippines v. the People's Republic of China. The Tribunal Renders Its Award*, 12 July 2016, https://pcacases.com/ web/endAttach/1801, 26 March 2019. The documents relating to the case may be found on the PCA Web site, https://pca-cpa.org/en/cases/7/, accessed 26 March 2019.

Philippines periodically expressed to the Tribunal its concern about China's extensive land reclamation and construction in the Spratly Islands and their impact on the fragile marine environment.[73] In April 2015, the Philippines alleged that China's massive reclamation activities were causing irreversible and widespread damage to biodiversity and the ecological balance of the South China Sea, resulting in the destruction of 300 hectares of coral reef systems.[74] China's island-building and construction activities on these reefs were the object of Philippine protests in April, June, August, and October 2014 and February 2015. All were rejected by China.

The Philippines' original two submissions on environmental matters were amended, with the authorization of the Arbitral Tribunal, in order to take into account China's island-building in the Spratly Islands. Under Submission No. 11, the Philippines alleged that China had violated its obligations under the Convention to protect and preserve the marine environment not just at Scarborough Shoal and Second Thomas Shoal, but also at Cuarteron Reef, Fiery Cross Reef, Gaven Reef, Johnson Reef, Hughes Reef, and Subi Reef. This submission covered both China's toleration of illegal harvesting of endangered species and cyanide and dynamite fishing at Scarborough Shoal and Second Thomas Shoal as well as China's island-building and construction at the six reefs listed in the submission.[75] Under Submission No. 12(b), the Philippines alleged that China's occupation of and construction activities on yet another feature, Mischief Reef, had, among other consequences, violated "China's duties to protect and preserve the marine environment under the Convention."[76] China's interference with Philippine traditional fishing at Scarborough Shoal, the dangerous and unlawful conduct of Chinese vessels at Scarborough Shoal in April and May 2012, and China's aggravation of the dispute through its construction activities at Mischief Reef,

---

[73] Award on Jurisdiction, 320, para. 819.

[74] *Supplemental Documents of the Philippines* (19 November 2015), Annex 608, Department of Foreign Affairs—Republic of the Philippines. Statement on China's Reclamation Activities and their Impact on the Region's Marine Environment (13 April 2015), vol. I, 7–9 ("*SDP*"), https://files.pca-cpa.org/pcadocs/The%20Philippines%27%20Supplemental%20Documents%20-%20Volume%20I%20%28Annexes%20607-667%29.pdf, accessed 26 March 2019.

[75] Award of 12 July 2016, 355–56, 358, paras. 891, 894, 901.

[76] Ibid., 41–42, para. 112.

Cuarteron Reef, Fiery Cross Reef, Gaven Reef, Johnson Reef, Hughes Reef, and Subi Reef were the object of separate submissions by the Philippines and will not be further discussed in the present work, since they did not raise issues of the protection and preservation of the marine environment.[77]

China's refusal to appear before the Tribunal did not empower the Tribunal to render automatically a default judgment, in favor of the Philippines. Article 9 on "Default of Appearance" of Annex VII of the Convention requires that "[b]efore making its award, the Arbitral Tribunal must satisfy itself not only that it has jurisdiction over the dispute but also that the claim is well founded in fact and law." Following China's rejection of the Philippine Notification and Statement of Claim, and in anticipation of China's failure to appear before the Tribunal, the latter instituted a procedure through which it would request further submissions from or pose questions to the Philippines on matters that in Tribunal's view had not been canvassed, or had been inadequately canvassed, in the pleadings submitted by the Philippines. The Tribunal was also empowered to take any steps that it deemed necessary in order to provide each of the parties to the dispute a full opportunity to present its case.[78]

Despite the fact that the Philippine *Memorial*, a copy of which had been sent to China, elaborated on the Philippine claims alleging China's toleration of harvesting of endangered species and destructive fishing techniques as well as construction on Mischief Reef in ways that harmed the marine environment, the Position Paper did not refer to environmental issues at all. The Tribunal included in its "Request for Further

---

[77] According to Submission No. 10, "China has unlawfully prevented Philippine fishermen from pursuing their livelihoods by interfering with traditional fishing activities at Scarborough Shoal." According to Submission No. 13, "China has breached its obligations under the Convention by operating its law enforcement vessels in a dangerous manner causing serious risk of collision to Philippine vessels navigating in the vicinity of Scarborough Shoal." Under Submission No. 14(d) the Philippines claimed that since the commencement of [the] arbitration…China has unlawfully aggravated and extended the dispute by, among other things (d) conducting dredging, artificial island-building and construction activities at Mischief Reef, Cuarteron Reef, Fiery Cross Reef, Gaven Reef, Johnson Reef, Hughes Reef and Subi Reef." For the Tribunal's decisions concerning Submissions No. 10, 13, and 14(d), see Award on Jurisdiction, 145, paras. 407 and 410, and 146–147, para. 410; Award of 12 July 2016, 304–318, paras. 772–814; 421–35, paras. 1059–1109; and 456–464, paras. 1163–1181.

[78] *South China Sea Arbitration*, Rules of Procedure, 27 August 2013, Article 25(1), 12, https://pcacases.com/web/sendAttach/233, accessed 7 April 2019.

Written Argument by the Philippines Pursuant to Article 25(2) of the Rules of Procedure," issued on 16 December 2014, a Question 11, requesting that the Philippines present arguments on the relationship between the alleged violation of the Convention on Biological Diversity ("CBD") and of the Convention.[79] China did not respond by the date set for its response to the Philippine arguments, 16 June 2015.

The Arbitral Tribunal went to great lengths to ensure itself that the Philippine claims concerning China's toleration of the harvesting of endangered species and cyanide and dynamite fishing as well as the environmental impact of China's construction activities in the South China Sea were well founded in fact and law, as was required by Annex VII of the Convention and the Rules of Procedure of the Arbitration, at all stages of the proceedings.[80] Prior to the Hearing on Jurisdiction, the Tribunal wrote to the parties on 23 June 2015, setting out a list of issues for possible consideration by the Philippines during the Hearing. In the Tribunal's list were questions concerning the applicability of the limitations and exceptions to the Tribunal's jurisdiction under Articles 297 and 298 of the Convention.[81] During the Hearing on Jurisdiction, the Tribunal posed a number of questions to the Philippines, one of which referred to the requirement to conciliate under the CBD.[82]

The Tribunal's response to the Philippine presentation of its scientific and legal arguments concerning Submissions No. 11 and 12(b) during the Hearing on the Merits in November 2015 was noteworthy. The Tribunal posed 22 questions to Dr. Kent E. Carpenter, a coral reef expert who had been presented by the Philippines to give testimony on the impact on the marine environment of China's construction activities

---

[79] The text of the Tribunal's Request is not available. The reader can only deduce its contents by reading the Philippine submissions. *SWSP*, 55–58, paras. 11.1–11.13. *Convention on Biological Diversity*, signed at Rio de Janeiro on 5 June 1992, entered into force on 29 December 1993, https://www.cbd.int/doc/legal/cbd-en.pdf, accessed 26 March 2019.

[80] Annex VII, Article 9; Rules of Procedure, Article 25(1).

[81] The text of the Tribunal's list of issues is not available. The reader can only deduce its contents by reading the transcripts of the Hearing itself. *South China Sea Arbitration*, Hearing on Jurisdiction and Admissibility, Transcript, Day 2 (8 July 2015), 74–86, https://pcacases.com/web/sendAttach/1400, accessed 7 April 2019.

[82] The text of the Tribunal's questions is not available. The reader can only deduce its contents by reading the transcripts of the Hearing itself. *South China Sea Arbitration*, Hearing on Jurisdiction and Admissibility, Transcript, Day 3 (13 July 2015), 42 ("Hearing on Jurisdiction"), https://pcacases.com/web/sendAttach/1401, accessed 7 April 2019.

in the South China Sea.[83] Dr. Carpenter had to respond to all questions in the second round of presentation of arguments.[84] In addition, the Tribunal directed six questions to the legal counsel of the Philippines who had presented the arguments concerning China's alleged violations of the obligation to protect and preserve the marine environment.[85] One of the questions obliged the legal team of the Philippines to search over the weekend for any publications by Chinese experts regarding the environmental impact of China's activities or of its toleration of activities that are contrary to or different from those of the Philippines. The research turned up a 500-word statement from China's State Oceanic Administration ("SOA"), published on 18 June 2015.[86]

The Tribunal's efforts to obtain further information and arguments did not cease after the conclusion of the oral hearings. First, the Tribunal itself undertook research to verify China's statements implying that China had conducted Environmental Impact Assessments ("EIA") before undertaking the construction activities in the South China Sea. The Tribunal found nine documents issued by the SOA between 2005

---

[83] *South China Sea Arbitration*, Hearing on the Merits and Remaining Issues of Jurisdiction and Admissibility, Transcript, Day 3 (26 November 2015), 47–54, https://pcacases.com/web/sendAttach/1549, accessed 26 March 2019. *MP*, Annex 240, Eastern South China Sea Environmental Disturbances and Irresponsible Fishing Practices and their Effects on Coral Reefs and Fisheries (22 March 2014), by Kent E. Carpenter, Ph.D., vol. VII, 389–437, https://files.pca-cpa.org/pcadocs/The%20Philippines%27%20Memorial%20-%20Volume%20VII%20%28Annexes%20222-255%29.pdf, accessed 7 April 2019; *SDP*, Annex 699, Environmental Consequences of Land Reclamation Activities on Various Reefs in the South China Sea (14 November 2015), by K. E. Carpenter & L. M. Chou, vol, II, 235–92, https://files.pca-cpa.org/pcadocs/The%20Philippines%27%20Supplemental%20Documents%20-%20Volume%20II%20%28Annexes%20608-709%29.pdf, accessed 7 April 2019.

[84] The text of the Tribunal's questions is not available. The reader can only deduce its contents by reading the transcripts of the Hearing itself. Hearing on the Merits, Transcript, Day 4 (30 November 2015), 138–162, https://pcacases.com/web/sendAttach/1550, accessed 7 April 2019.

[85] The text of the Tribunal's questions is not available. The reader can only deduce its contents by reading the transcripts of the Hearing itself. Ibid., 166–87.

[86] Ibid., 181–82. *The Philippines' Annexes cited during Merits Hearing*, Annex 821, China State Oceanic Administration, "Construction Work at Nansha Reefs Will Not Harm Oceanic Ecosystems" (18 June 2015), http://www.soa.gov.cn/xw/hyyw_90/201506/t20150618_38598.html, https://files.pca-cpa.org/pcadocs/The%20Philippines%27%20Annexes%20cited%20during%20Merits%20Hearing%20%28Annexes%20820-859%29.pdf, accessed 7 April 2019.

and 2015 as well as statements made by Chinese officials in 2015 and 2016. The Tribunal transmitted all these documents to China and the Philippines in February 2016, together with a request for their comments on them.[87] The Philippines filed its response to the SOA publications in March 2016.[88] Next, the Tribunal appointed in February 2016 an independent coral reef expert Dr. Sebastian Ferse. In April 2016, the Tribunal decided to appoint two additional coral reef experts, Professor Peter Mumby and Dr. Selina Ward. At the same time, Dr. Ferse sought clarification from coral reef expert Professor John W. McManus, author of a scientific report cited by the Philippines during the Hearing on the Merits.[89] On 28 April, the Philippines filed a response to Dr. Ferse's request, including a letter and updated report from Professor McManus and a supplementary declaration from Dr. Carpenter.[90] On 29 April, the three coral reef experts submitted their report to the Tribunal and to the parties.[91] Third, the Tribunal communicated with China directly. In December 2015, it forwarded to China for comments BBC reports on the damage to coral reefs caused by the harvesting of giant clams by Chinese nationals; in April 2016, the Tribunal transmitted on behalf of the Tribunal's coral reef expert another article reporting the widespread removal of giant clams by propeller cutting.[92] In February 2016, the Tribunal asked China directly whether it had conducted an EIA, and if so, it requested that China provide a copy to the Tribunal. China did not respond to any of the Tribunal's communications.

The fact is China, which refused to appear before the Tribunal, did make known its views on Philippine Submissions No. 11 and 12(b), albeit informally. While the Position Paper did not make any comments

---

[87] The list of documents may be found in *South China Sea Arbitration, The Philippines' Written Responses* (11 March 2016), https://files.pca-cpa.org/pcadocs/The%20 Philippines%27%20Written%20Responses%20%2811%20March%202016%29%20 %28Annexes%20864-892%29.pdf, accessed 7 April 2016.

[88] *South China Sea Arbitration, Responses of the Philippines to the Tribunal's 4 February 2016 Request for Comments* (11 March 2016), https://pcacases.com/web/sendAttach/1849, accessed 7 April 2019.

[89] *The Philippines Annexes Cited During the Merits* Hearing, Annex 850.

[90] *South China Sea Arbitration, The Philippines' Response to Tribunal Enquiry on Reef Damage (26 April 2016) (with Prof. McManus Revised Report and Third Carpenter Report)*, https://pcacases.com/web/sendAttach/1917, accessed 7 April 2019.

[91] Award of 12 July 2016, 29, 31, 32, paras. 84, 89, 90, 93; Ferse Report.

[92] Award of 12 July 2016, 364, para. 915.

on the Philippine Submissions No. 11 and 12(b), China subsequently corrected this omission. On at least five occasions (April, May and June 2015, February and May 2016), China publicly contradicted Philippine claims.[93] Two years after the issuance of the Award, the Chinese Society of International Law (CSIL) published a critique of the Tribunal's decisions on jurisdiction and admissibility and on the merits relating to the Philippine Submissions No. 11 and 12(b) as part of an overall critique of the two Awards.[94]

The *South China Sea Arbitration* is a landmark case, in that it is the first case requiring the interpretation and application of Part XII (Protection and Preservation of the Marine Environment) of the Convention that has actually examined the merits of the case.[95] In several cases involving environmental claims by one of the parties to the dispute, other arbitral tribunals constituted under the Convention have confined themselves to reiterating the duty of the parties to the dispute to negotiate.[96] In contrast, the Award in the *South China Sea Arbitration* devotes no less than 178 paragraphs (out of 1203), extending over 79 pages (out of 479), to a detailed factual investigation and painstaking legal analysis of the Philippine claims. There is no doubt that the Award's interpretation and application of Part XII of the Convention deserve to be the object of scholarly inquiry that is at least as intensive as that which will be

---

[93] Ibid., 364, para. 915; 365, paras. 917–920; 369, para. 924.

[94] Chinese Society of International Law (CSIL), "The South China Sea Arbitration Awards: A Critical Study," *Chinese Journal of International Law* 17 (2018): 207–748. For initial responses to the Critical Study, see Douglas Guilfoyle, "A New Twist in the South China Sea Arbitration: The Chinese Society of International Law's Critical Study," *ejil.talk*, 25 May 2018, ejiltalk.org/a-new-twist-in-the-south-china-sea-arbitration-the-chinese-society-of-international-laws-critical-study/, accessed 7 February 2019 and Douglas Guilfoyle, "Taking the Party Line on the South China Sea Arbitration," *ejil.talk*, 28 May 2018, ejiltalk.org/taking-the-party-line-on-the-south-china-sea-arbitration/, accessed 7 February 2019.

[95] Hearing on the Merits, Transcript, Day 3, 10; Paul Reichler, counsel for the Philippines, quoted in Stuart Leavenworth, "In South China Sea Case, Ruling on Environment Hailed as Precedent," *Christian Science Monitor*, 20 July 2016, http://www.csmonitor.com/World/Asia-Pacific/2016/0720/In-South-China-Sea-case-ruling-on-environment-hailed-as-precedent, accessed 26 March 2019.

[96] Patricia Birnie, Alan Boyle and Catherine Redgewell, *International Law and the Environment* (3rd ed.) (Oxford: Oxford University Press, 2009), 226. Boyle was one of the counsels for the Philippines in the arbitration.

accorded to the more controversial issues that the Tribunal adjudicated, such as China's "Nine-dash line."[97]

IV. Overview of the Book

The rest of the book is organized as follows. Chapter 2 will explain the importance of marine biodiversity for the Philippines, focusing on the ecological roles and economic importance of one endangered species, sea turtles, and of one fragile ecosystem, coral reefs. Sea turtles

[97]Since 2016, commentaries on the Tribunal's decisions relating to Philippine Submissions No. 11 and 12(b) have been published by distinguished scholars. See Chie Kojima, "South China Sea Arbitration and the Protection of the Marine Environment: Evolution of UNCLOS Part XII Through Interpretation and the Duty to Cooperate," *Asian Yearbook of International Law* 21 (2015): 166–80; Makane Moïse Mbengue, "The South China Sea Arbitration. Innovation in Marine Environmental Fact-finding and Due Diligence Obligations," *ASIL Unbound* (12 December 2016): 285–89; Nilufer Oral, "The South China Sea Arbitral Award, Part XII of UNCLOS and the Protection and Preservation of the Marine Environment," in S. Jayakumar et al. (eds.), *The South China Sea Arbitration: The Legal Dimension* (Cheltenham, Gloucestershire: Edward Elgar, 2018), 223–46; Ilias Plakokefalos, "Environmental Law Aspects of the South China Sea Arbitration Award," Symposium on the South China Sea Award, Netherlands Institute for the Law of the Sea (NILOS), Utrecht Centre for Oceans Water and Sustainability Law, School of Law Utrecht University, 7 December 2016, https://papers.ssrn.com/sol3/papers.cfm?abstract_id=2880624, accessed 3 May 2019; Tim Stephens, "The Collateral Damage from China's 'Great Wall of Sand': The Environmental Dimensions of the South China Sea Case," *Australian Yearbook of International Law* 24 (2017): 41–56; and Yoshifumi Tanaka, "The South China Sea Arbitration: Environmental Obligations under the Law of the Sea Convention," *Review of European, Comparative and International Environmental Law* 27 (2018): 90–96. Commentaries in French and Spanish that analyze the Award of 12 July 2016 as a whole devote little attention to the environmental issues discussed in the Award. See Romain Le Bœuf, "Différend en Mer de Chine méridionale. Sentence arbitrale du 12 juillet 2016 [The South China Sea Dispute. Arbitral Award of 12 July 2016]," *Annuaire Français de Droit International* [French Yearbook of International Law] 52 (2016): 178; Jean-Paul Pancracio, "La sentence arbitrale sur la mer de Chine méridionale du 12 juillet 2016 [The South China Sea Arbitral Award of 12 July 2016]," *Annuaire Français de Relations Internationales* [French Yearbook of International Relations] 18 (2017): 649, http://www.afri-ct.org/wp-content/uploads/2018/06/Article-Pancracio.pdf, accessed 5 February 2019; and Elena Pineros Polo, "Arbitraje del mar del sur de China. la estrategia procesal de la República Popular de China [The South China Sea Arbitration. The Procedural Strategy of the People's Republic of China]," *Revista Electrónica de Estudios Internacionales* [Electronic Journal of International Studies] 35 (2018): 20, www.reei.org/index.php/revista/.../09_Nota_PINEROS_Elena.pdf, accessed 9 February 2019.

maintain habitat, maintain a balanced food web, cycle nutrients, and provide habitat. While the economic importance of sea turtles for the Philippines as a whole has declined, the collection of their eggs, rather than the harvesting of the turtles themselves, is vital for marginalized communities in the Turtle Islands; a small indigenous group known as the Tagbanua consumes turtle meat and engages in small-scale trade of the meat. Conscious that sea turtles are an endangered species, the Philippines has been making great efforts to preserve the sea turtles that nest in the country. These efforts are undermined by trade in sea turtles by and with Chinese nationals. Coral reefs seem to be equally valuable to the Philippines for ecological and economic reasons. They provide physical structure services, biotic services, biogeochemical services, and information services. They are sources of living resources and mining materials; they are major tourist attractions as well. Coral reefs sustain the livelihood of many coastal communities. Hence, it is crucial for the Philippines to preserve this fragile ecosystem. Philippine efforts to conserve coral reefs are threatened by the activities of Chinese fishermen and by China's construction activities, first at Mischief Reef and later at six other reefs in the Spratly Islands.

Chapter 3 begins by explaining that the Philippines, without abandoning its claims to sovereignty, invoked the Convention, the CBD, and other international environmental agreements as the bases for its claims against China, as a way of preempting China's rejections of its protests on the ground that China had sovereignty over the maritime areas in which Chinese nationals exercised their activities. The Philippine approach suggested its belief that the obligation to protect and preserve the marine environment was independent of the question of sovereignty. The Philippines found the concepts of the CBD more appropriate for describing the consequences of the activities of Chinese nationals and of China than those of the Convention. The Philippines also probably found the CBD as a more useful framework, to the extent that the nature and jurisdictional scope of the obligations to conserve biodiversity seemed to be broader than those of the obligation to protect and preserve marine living resources under the Convention. Nevertheless, the Philippines did not present a claim under the CBD. What made the Convention more attractive to the Philippines as a framework for settling the two sets of disputes with China was the existence of compulsory mechanisms of dispute settlement entailing binding decisions. The CBD, in contrast, offered at most compulsory conciliation, which did not entail binding decisions.

These circumstances would explain the formulation of the Philippine claims in terms of the more general obligation established by Article 192 of the Convention to protect and preserve the marine environment and the Philippine insistence that its environmental disputes with China were not disputes about the interpretation and application of the CBD, but disputes over the interpretation and application of the Convention.

The Philippines also faced the challenge of demonstrating that the Tribunal constituted under the Convention had jurisdiction over its two environmental submissions. If the CBD could serve as a guide to interpreting the Convention, it could also be considered, under Convention rules, as an agreement precluding the jurisdiction of the Tribunal. Other Convention rules that permitted automatic limitations and optional exceptions to jurisdiction also had to be considered by the Tribunal, particularly since China had made a declaration in 2006 activating some of the optional exceptions. Chapter 4 will examine the arguments marshalled by the Philippines with the aim of convincing the Arbitral Tribunal that the latter had jurisdiction under the Convention over the disputes with China concerning China's toleration of the harvesting of endangered species and the use of cyanide and dynamite as well as China's island-building causing harm to fragile ecosystems. First, the Philippines had to demonstrate that the preconditions to the Tribunal's jurisdiction had been fulfilled: The CBD was not an agreement that precluded the Tribunal's jurisdiction under Articles 281 and 282 of the Convention, and China and the Philippines had exchanged views on the means of settling the dispute, as required by Article 283 of the Convention, the Philippines had to prove that the limitations to the Tribunal's jurisdiction concerning the coastal States' sovereign rights in the exclusive economic zone (EEZ) under Article 297 of the Convention and the military activities exception and the law enforcement activities exception to the Tribunal's jurisdiction under Article 298 of the Convention were not applicable.

Chapter 5 will examine the merits of the Philippine claims relating to the harvesting of endangered species and dynamite and cyanide fishing and to island-building causing harm to fragile ecosystems. It will show how the Arbitral Tribunal, following the reasoning of the Philippines, interpreted Article 192 of the Convention in such a way as to encompass the protection of endangered species and fragile ecosystems. In evaluating China's conduct with respect to the harvesting of endangered species, the Tribunal drew on other provisions of Part XII, notably Article 194(5), and the provisions of the CBD and CITES, concluding that China had failed to comply with its due diligence obligation to prevent harvesting of endangered species on a significant scale and harvesting of giant clams in a manner that was severely destructive of the coral reef ecosystem by its nationals. Contrary to the Philippine submissions, the Tribunal did not find that China had failed to comply with its due diligence obligation to prevent dynamite and cyanide fishing. China's island-building in the Spratly Islands, which was part of an official program, was analyzed by the Tribunal in light of China's obligations under Articles 192 and 194(5) of the Convention to protect and preserve the marine environment; the duty to cooperate with other States bordering the South China Sea under Articles 197 and 123 of the Convention; and the obligation to monitor and assess the impact on the marine environment of its construction activities, notably by carrying out an EIA and communicating the results of such an assessment under Articles 205 and 206 of the Convention. The Tribunal concluded that through its land reclamation and construction of artificial islands, installations, and structures in the South China Sea, China had breached its obligations under the Convention to protect and preserve the marine environment. China had also failed to cooperate with other littoral States of the South China Sea and had failed to communicate the results of its assessment of the environmental impact of its construction activities.

Chapter 6 will summarize the book's main arguments and attempt to draw up a balance sheet of the contribution of the South China Sea Arbitration to the protection of endangered species and fragile marine ecosystems.

References

Ablan-Lagman, Ma. Carmen A. "The Spratly Islands." *World Seas: An Environmental Evaluation, Vol. II: The Indian Ocean to the Pacific*, 583–91. Ed. Charles Sheppard. London: Academic Press, 2019.

Agence France Presse. "RP Minister Seeks Expulsion of Mainland Envoy." *China Post*, 21 September 2002, https://chinapost.nownews.com/20020921-142615, accessed 18 February 2019.

Birnie, Patricia, Alan Boyle, and Catherine Redgewell. *International Law and the Environment*, 3rd ed. Oxford: Oxford University Press, 2009.

Burke, Lauretta et al. *Reefs at Risk Revisited*. Washington, DC: World Resources Institute, 2011, https://wriorg.s3.amazonaws.com/s3fs-public/pdf/reefs_at_risk_revisited_hi-res.pdf?_ga=2.177507173.1364385348.1552820007-1064952833.1552820007, accessed 17 March 2019.

Carpenter, Kent E. *The Living Marine Resources of the Western Atlantic, Vol. I: Introduction, Molluscs, Crustaceans, Hagfishes, Sharks, Batoid Fishes and Chimaeras*. Rome: Food and Agriculture Organization, 2002, http://www.fao.org/3/y4160e/y4160e18.pdf, accessed 10 March 2019.

Center for Strategic and International Studies (CSIS). "Are Maritime Law Enforcement Forces Destabilizing Asia?" *China Power Project* (2018), https://chinapower.csis.org/maritime-forces-destabilizing-asia/, accessed 26 October 2018.

Chagos Marine Protected Area Arbitration (Mauritius v. UK). Award of 18 March 2015, http://www.pcacases.com/pcadocs/MU-UK%2020150318%20Award.pdf, accessed 3 May 2019.

Chinese Society of International Law. "The South China Sea Arbitration Awards: A Critical Study." *Chinese Journal of International Law* 17 (2018): 207–748.

Convention on Biological Diversity, signed at Rio de Janeiro on 5 June 1992, entered into force on 29 December 1993, https://www.cbd.int/doc/legal/cbd-en.pdf, accessed 26 March 2019.

Convention on International Trade in Endangered Species of Wild Fauna and Flora, signed at Washington, DC on 3 March 1973, amended at Bonn on 22 June 1979, amended at Gaborone on 30 April 1983, https://www.cites.org/eng/disc/text.php, accessed 26 March 2019.

Coulter, Daniel Yarrow. "South China Sea Fisheries: Countdown to Calamity." *Contemporary Southeast Asia* 17 (1996): 371–88.

Danao, Efren. "Perez Backed vs Sino Envoy." *The Philippine Star*, 22 September 2002, https://www.philstar.com/nation/2002/09/22/176917/perez-backed-vs-sino-envoy, accessed 18 February 2019.

Fisheries Research and Development Corporation. *Australian Fish Names Standard* (2018), http://www.fishnames.com.au/, accessed 30 March 2018.

Fishing the Philippines: Fish Fishing Spots, Lures, Tactics and More, https://fishingthephilippines.com/category/trevally-talakitok/, accessed 26 March 2019.

Funge-Smith, Simon et al. *Regional Overview of Fisheries and Aquaculture in Asia and the Pacific 2012*. Bangkok: FAO Regional Office for Asia and the Pacific, 2012, http://www.fao.org/3/i3185e/i3185e00.pdf, accessed 20 February 2019.

Guilfoyle, Douglas. "A New Twist in the South China Sea Arbitration: The Chinese Society of International Law's Critical Study." *ejil.talk*, 25 May 2018, ejiltalk.org/a-new-twist-in-the-south-china-sea-arbitration-the-chinese-society-of-international-laws-critical-study/, accessed 7 February 2019.

———. "Taking the Party Line on the South China Sea Arbitration." *ejil.talk*, 28 May 2018, ejiltalk.org/taking-the-party-line-on-the-south-china-sea-arbitration/, accessed 7 February 2019.

Heileman, S. "South China Sea." *The UNEP Large Marine Ecosystems Report: A Perspective on Changing Conditions in LMEs of the World's Regional Seas*, 297–308. Eds. Kenneth Sherman and Gotthilf Hempel. Nairobi: United Nations Environment Programme, 2009, http://lme.edc.uri.edu/index.php/reports-and-workbooks/92-unep-lme-report, accessed 20 February 2019.

Hill, K. "Indian River Lagoon Species Inventory: Seagrass Habitats." *Smithsonian Marine Station at Fort Pierce* (2002), http://www.sms.si.edu/irlspec/Seagrass_Habitat.htm, accessed 11 March 2017.

Huang, Danwei et al. "Extraordinary Diversity of Reef Corals in the South China Sea." *Marine Biodiversity* 45 (2017): 157–68.

International Tribunal for the Law of the Sea. *New Arbitrator and President Appointed in the Arbitral Proceedings Instituted by the Republic of the Philippines against the People's Republic of China.* Doc. ITLOS/Press 197 (24 June 2013), https://www.itlos.org/fileadmin/itlos/documents/press_releases_english/PR_197_E.pdf, accessed 26 March 2019.

Kojima, Chie. "South China Sea Arbitration and the Protection of the Marine Environment: Evolution of UNCLOS Part XII Through Interpretation and the Duty to Cooperate." *Asian Yearbook of International Law* 21 (2015): 166–80.

Larousse Dictionary of Science and Technology. Edinburgh: Larousse plc, 1995.

Leavenworth, Stuart. "In South China Sea Case, Ruling on Environment Hailed as Precedent." *Christian Science Monitor*, 20 July 2016, http://www.csmonitor.com/World/Asia-Pacific/2016/0720/In-South-China-Sea-case-ruling-on-environment-hailed-as-precedent, accessed 26 March 2019.

Le Bœuf, Romain. "Différend en Mer de Chine méridionale. Sentence arbitrale du 12 juillet 2016 [The South China Sea Dispute. Arbitral Award of 12 July 2016]." *Annuaire Français de Droit International* [French Yearbook of International Law] 52 (2016): 159–81.

Mbengue, Makane Moïse. "The South China Sea Arbitration: Innovation in Marine Environmental Fact-Finding and Due Diligence Obligations." *ASIL Unbound* (12 December 2016): 285–89.

McManus, John W. "Offshore Coral Reef Damage, Overfishing and Paths to Peace in the South China Sea." *The International Journal of Marine and Coastal Law* 32 (2017): 199–237.

Morton, Brian. "Fishing for Diplomacy in China's Seas." *Marine Pollution Bulletin* 46 (2003): 795–96.

Nguyen Dang Thang. "Fisheries Co-operation in the South China Sea and the (Ir)relevance of the Sovereignty Question." *Asian Journal of International Law* 2 (2012): 59–88.

Oral, Nilufer. "The South China Sea Arbitral Award, Part XII of UNCLOS and the Protection and Preservation of the Marine Environment." *The South China Sea Arbitration: The Legal Dimension*, 223–46. Eds. S. Jayakumar et al. Cheltenham, Gloucestershire: Edward Elgar, 2018.

Pancracio, Jean-Paul. "La sentence arbitrale sur la mer de Chine méridionale du 12 juillet 2016 [The South China Sea Arbitral Award of 12 July 2016]." *Annuaire Français de Relations Internationales* [French Yearbook of International Relations] 18 (2017): 639–57, http://www.afri-ct.org/wp-content/uploads/2018/06/Article-Pancracio.pdf, accessed 5 February 2019.

Permanent Court of Arbitration. *Arbitration Between the Republic of the Philippines and the People's Republic of China. Tribunal Renders Award on Jurisdiction and Admissibility; Will Hold Further Hearings*, 29 October 2015, http://www.pcacases.com/web/sendAttach/1503, accessed 26 March 2019.

———. *The South China Sea Arbitration (The Republic of the Philippines v. the People's Republic of China: The Tribunal Renders Its Award)*, 12 July 2016, https://pcacases.com/web/endAttach/1801, accessed 26 March 2019.

Pineros Polo, Elena. "Arbitraje del mar del sur de China. la estrategia procesal de la República Popular de China [The South China Sea Arbitration. The Procedural Strategy of the People's Republic of China]." *Revista Electrónica de Estudios Internacionales* [Electronic Journal of International Studies] 35 (2018): 1–23, www.reei.org/index.php/revista/.../09_Nota_PINEROS_Elena.pdf, accessed 9 February 2019.

Plakokefalos, Ilias. "Environmental Law Aspects of the South China Sea Arbitration Award." Symposium on the South China Sea Award, Netherlands Institute for the Law of the Sea (NILOS), Utrecht Centre for Oceans Water and Sustainability Law, School of Law Utrecht University, 7 December 2016, https://papers.ssrn.com/sol3/papers.cfm?abstract_id=2880624, accessed 3 May 2019.

Porcalla, Delon, and Mayen Jaymalin. "Perez, Chinese Envoy Bury Hatchet." *The Philippine Star*, 26 September 2002, https://www.philstar.com/headlines/2002/09/26/177460/perez-chinese-envoy-bury-hatchet, accessed 18 February 2019.

———. "Perez Gagged on Sino Poachers." *The Philippine Star*, 24 September 2002, https://www.philstar.com/headlines/2002/09/24/177232/perez-gagged-sino-poachers, accessed 18 February 2019.

Ramsar Convention: Convention on Wetlands of International Importance especially as Waterfowl Habitat, signed at Ramsar, Iran, on 2 February 1971, as amended by the Protocol of 3 December 1982 and the Amendments of 28 May 1987, https://www.ramsar.org/sites/default/files/documents/library/current_convention_text_e.pdf, accessed 24 March 2019.

———. 8th Meeting of the Conference of the Contracting Parties to the Convention on Wetlands (Ramsar, Iran, 1971). "Wetlands: Water, Life, and Culture," Valencia, Spain, 18–26 November 2002. Resolution VIII.11. Guidance for Identifying and Designating Peatlands, Wet Grasslands, Mangroves and Coral Reefs as Wetlands of International Importance, www.ramsar.org/sites/default/files/documents/pdf/res/key_res_viii_11_e.pdf, accessed 26 March 2019.

Rosenberg, David. "Fisheries Management in the South China Sea." *Security and International Politics in the South China Sea: Towards a Co-operative Management Regime*, 61–79. Eds. Sam Bateman and Ralf Emmers. London: Routledge, 2009.

Sherman, Kenneth, and Gotthilf Hempel. "Perspectives on Regional Seas and the Large Marine Ecosystem Approach." *The UNEP Large Marine Ecosystems Report: A Perspective on Changing Conditions in LMEs of the World's Regional Seas*, 3–22. Eds. Kenneth Sherman and Gotthilf Hempel. Nairobi: United Nations Environment Programme, 2009, http://lme.edc.uri.edu/index.php/reports-and-workbooks/92-unep-lme-report, accessed 20 February 2019.

Si Tuan Vo et al. "Status and Trends in Coastal Habitats of the South China Sea." *Ocean & Coastal Management* 85 (2013): 153–63.

South China Sea Arbitration. Award of 12 July 2016, https://pcacases.com/web/sendAttach/2086, accessed 26 March 2019.

———. Award on Jurisdiction and Admissibility, 29 October 2015, https://pca-cases.com/web/sendAttach/1506, accessed 26 March 2019.

———. Hearing on the Merits and Remaining Issues of Jurisdiction and Admissibility. Transcript. Day 2 (25 November 2015), https://pcacases.com/web/sendAttach/1548, accessed 26 March 2019.

———. Transcript. Day 3 (26 November 2015), https://pcacases.com/web/sendAttach/1549, accessed 26 March 2019.

———. Independent Expert Report. Assessment of the Potential Environmental Consequences of Construction Activities on Seven Reefs in the Spratly Islands in the South China Sea, by Sebastian C. A. Ferse, Peter Mumby, and Selina Ward, 26 April 2016, https://pcacases.com/web/sendAttach/1809, accessed 24 March 2019.

———. *Memorial of the Philippines* (30 March 2014). Vol. I, https://files.pca-cpa.org/pcadocs/Memorial%20of%20the%20Philippines%20Volume%20I.pdf, accessed 26 March 2019.

———. Annex 17. Memorandum from the Undersecretary of Foreign Affairs of the Republic of the Philippines to the Ambassador of the People's Republic of China in Manila (6 February 1995). Vol. III, 205–208, https://files.pca-cpa.org/pcadocs/The%20Philippines%27%20Memorial%20-%20Volume%20III%20%28Annexes%201-60%29.pdf, accessed 26 March 2019.

———. Annex 19. Memorandum from Erlinda F. Basilio, Acting Assistant Secretary, Office of Asian and Pacific Affairs, Department of Foreign Affairs, Republic of the Philippines, to the Secretary of Foreign Affairs of the Republic of the Philippines (29 March 1995). Vol. III, 213–17, https://files.pca-cpa.org/pcadocs/The%20Philippines%27%20Memorial%20-%20Volume%20III%20%28Annexes%201-60%29.pdf, accessed 26 March 2019.

———. Annex 20. Memorandum from Lauro L. Baja, Jr., Assistant Secretary, Office of Asian and Pacific Affairs, Department of Foreign Affairs, Republic of the Philippines, to the Secretary of Foreign Affairs of the Republic of the Philippines (7 April 1995). Vol. III, 219–24, https://files.pca-cpa.org/pcadocs/The%20Philippines%27%20Memorial%20-%20Volume%20III%20%28Annexes%201-60%29.pdf, accessed 26 March 2019.

———. Annex 37. Memorandum from the Embassy of the Republic of the Philippines in Beijing to the Secretary of Foreign Affairs of the Republic of the Philippines, No. ZPE-85–98-S (4 December 1998). Vol. III, 315–19, https://files.pca-cpa.org/pcadocs/The%20Philippines%27%20Memorial%20-%20Volume%20III%20%28Annexes%201-60%29.pdf, accessed 26 March 2019.

———. Annex 46. Office of Asian and Pacific Affairs, Department of Foreign Affairs, Republic of the Philippines, Apprehension of Four Chinese Fishing Vessels in the Scarborough Shoal (23 February 2001). Vol. III, 369–74, https://files.pca-cpa.org/pcadocs/The%20Philippines%27%20Memorial%20-%20Volume%20III%20%28Annexes%201-60%29.pdf, accessed 26 March 2019.

———. Annex 51. Memorandum from Josue L. Villa, Embassy of the Republic of the Philippines in Beijing, to the Secretary of Foreign Affairs of the Republic of the Philippines (19 August 2002). Vol. III, 405–409, https://files.pca-cpa.org/pcadocs/The%20Philippines%27%20Memorial%20-%20Volume%20III%20%28Annexes%201-60%29.pdf, accessed 26 March 2019.

———. Annex 57. Letter from George T. Uy, Rear Admiral, Armed Forces of the Philippines, to Assistant Secretary, Office of Asia and Pacific Affairs, Department of Foreign Affairs of the Republic of the Philippines (2006). Vol. III, 461–80, https://files.pca-cpa.org/pcadocs/The%20Philippines%27%20Memorial%20-%20Volume%20III%20%28Annexes%201-60%29.pdf, accessed 26 March 2019.

———. Annex 58. Memorandum from the Secretary of Foreign Affairs of the Republic of the Philippines to the President of the Republic of the Philippines (11 January 2006). Vol. III, 481–84, https://files.pca-cpa.org/pcadocs/The%20Philippines%27%20Memorial%20-%20Volume%20III%20%28Annexes%201-60%29.pdf, accessed 26 March 2019.

———. Annex 75. Memorandum from the Embassy of the Republic of the Philippines in Beijing to the Secretary of Foreign Affairs of the Republic of the Philippines, No. ZPE-121–2011-S (2 December 2011), Vol. IV, 121–24, https://files.pca-cpa.org/pcadocs/The%20Philippines%27%20Memorial%20 -%20Volume%20IV%20%28Annexes%2061-102%29.pdf, accessed 26 March 2019.

———. Annex 77. Memorandum from Col. Nathaniel Y. Casem, Philippine Navy, to Chief of Staff, Armed Forces of the Philippines, No. N2E-0412- 008 (11 April 2012). Vol. IV, 131–46, https://files.pca-cpa.org/pca- docs/The%20Philippines%27%20Memorial%20-%20Volume%20IV%20 %28Annexes%2061-102%29.pdf, accessed 26 March 2019.

———. Annex 78. Report from Commanding Officer, SARV-003, Philippine Coast Guard, to Commander, Coast Guard District, Northwestern Luzon, Philippine Coast Guard (28 April 2012). Vol. IV, 147–59, https://files.pca- cpa.org/pcadocs/The%20Philippines%27%20Memorial%20-%20Volume%20 IV%20%28Annexes%2061-102%29.pdf, accessed 26 March 2019.

———. Annex 80. Report from Relly B. Garcia *et al.*, FRPLEU/QRT [Fishery Resource Protection and Law Enforcement/Quick Response Team] Officers, Bureau of Fisheries and Aquatic Resources, Republic of the Philippines (2 May 2012). Vol. IV, 163–95, https://files.pca-cpa.org/pcadocs/The%20 Philippines%27%20Memorial%20-%20Volume%20IV%20%28Annexes%20 61-102%29.pdf, accessed 26 March 2019.

———. Annex 211. Note Verbale from the Embassy of the People's Republic of China in Manila to the Department of Foreign Affairs of the Republic of the Philippines, No (12) PG-239 (25 May 2012). Vol. VI, 401–404, https:// files.pca-cpa.org/pcadocs/The%20Philippines%27%20Memorial%20-%20 Volume%20VI%20%28Annexes%20158-221%29.pdf, accessed 26 March 2019.

———. Annex 240. Eastern South China Sea Environmental Disturbances and Irresponsible Fishing Practices and their Effects on Coral Reefs and Fisheries (22 March 2014), by Kent E. Carpenter, Ph.D. Vol. VII, 389–437, https:// files.pca-cpa.org/pcadocs/The%20Philippines%27%20Memorial%20-%20 Volume%20VII%20%28Annexes%20222-255%29.pdf, accessed 7 April 2019.

———. *The Philippines' Annexes cited during Merits Hearing.* Annex 821. China State Oceanic Administration, "Construction Work at Nansha Reefs Will Not Harm Oceanic Ecosystems" (18 June 2015), http://www.soa.gov.cn/xw/ hyyw_90/201506/t20150618_38598.html, https://files.pca-cpa.org/pca- docs/The%20Philippines%27%20Annexes%20cited%20during%20Merits%20 Hearing%20%28Annexes%20820-859%29.pdf, accessed 7 April 2019.

———. Notification and Statement of Claim of the Republic of the Philippines, 22 January 2013, http://www.philippineembassy-usa.org/uploads/pdfs/ embassy/2013/2013-0122-Notification%20and%20Statement%20of%20 Claim%20on%20West%20Philippine%20Sea.pdf, accessed 3 May 2019.

————. *The Philippines Annexes cited during the Merits Hearing (Annexes 820-59)* (30 November 2015). Annex 850. "Offshore Coral Reef Damage, Overfishing and Paths to Peace in the South China Sea," by John W. McManus, 578–608, https://pcacases.com/web/view/7, accessed 26 March 2019.

————. *The Philippines' Response to Tribunal Enquiry on Reef Damage (26 April 2016) (with Prof. McManus Revised Report; and Third Carpenter Report)*, https://pcacases.com/web/sendAttach/1917, accessed 7 April 2019.

————. *Responses of the Philippines to the Tribunal's 4 February 2016 Request for Comments* (11 March 2016), https://pcacases.com/web/sendAttach/1849, accessed 7 April 2019.

————. *Supplemental Documents of the Philippines* (19 November 2015). Annex 608. Department of Foreign Affairs—Republic of the Philippines. Statement on China's Reclamation Activities and their Impact on the Region's Marine Environment (13 April 2015). Vol. I, 7–9, https://files.pca-cpa.org/pca-docs/The%20Philippines%27%20Supplemental%20Documents%20-%20Volume%20I%20%28Annexes%20607-667%29.pdf, accessed 26 March 2019.

————. Annex 699. Environmental Consequences of Land Reclamation Activities on Various Reefs in the South China Sea (14 November 2015), by K. E. Carpenter and L. M. Chou, Vol. II, 235–92, https://files.pca-cpa.org/pcadocs/The%20Philippines%27%20Supplemental%20Documents%20-%20Volume%20II%20%28Annexes%20608-709%29.pdf, accessed 7 April 2019.

Spencer, T., and M. D. Spalding. "The Coral Reefs of Southeast of Southeast Asia: Controls, Patterns and Human Impacts." *Physical Geography of Southeast Asia*, 402–28. Ed. A. Gupta. Oxford: Oxford University Press, 2005.

Stephens, Tim. "The Collateral Damage from China's 'Great Wall of Sand': The Environmental Dimensions of the South China Sea Case." *Australian Yearbook of International Law* 24 (2017): 41–56.

Talaue-McManus, Liane. *Transboundary Diagnostic Analysis for the South China Sea*. Bangkok: United Nations Environment Programme, 2000, http://www.unepscs.org/remository/Download/01_-_Project_Development/PDF-B_Phase/Transboundary_Diagnostic_Analysis/Transboundary_Diagnostic_Analysis_for_the_South_China_Sea.html, accessed 5 March 2019.

Tanaka, Yoshifumi. "The South China Sea Arbitration: Environmental Obligations under the Law of the Sea Convention." *Review of European, Comparative and International Environmental Law* 27 (2018): 90–96.

Teh, Louise S. L. et al. "What Is at Stake? Status and Threats to South China Sea Marine Fisheries." *AMBIO: A Journal of the Human Environment* 46 (2017): 57–72.

Tuna Species Guide, http://www.atuna.com/index.php/en/tuna-info/tuna-species-guide#bonito, accessed 26 March 2019.

United Nations Convention on the Law of the Sea, concluded at Montego Bay on 10 December, entered into force on 16 November 1994, http://www.un.org/Depts/los/convention_agreements/texts/unclos/closindx.htm, accessed 21 March 2019.

United Nations Environment Programme (UNEP). *Coral Reefs in the South China Sea*. Bangkok: UNEP, 2004, http://www.ais.unwater.org/ais/aiscm/getprojectdoc.php?docid=3527#page=1&zoom=auto,-99, accessed 15 March 2019.

Yanagawa, Hiroyuki. "Small Pelagic Resources in the South China Sea." *Small Pelagic Resources and Their Fisheries in the Asia-Pacific Region. Proceedings of APFIC [Asia-Pacific Fisheries Commission] Working Party on Marine Fisheries, First Session, 13–16 May 1997*, 365–80. Eds. M. Devara and P. Marftusubroto. Bangkok: FAO Regional Office for Asia and the Pacific, 1997, http://www.fao.org/3/a-an020e.pdf, accessed 5 March 2019.

Zhang, Hongzhou. "Chinese Fishermen at Frontline of Maritime Disputes: An Alternative Explanation." *RSIS [Rajaratnam School of International Studies] Commentaries* 152 (2016), http://hdl.handle.net/10220/40776, accessed 13 February 2019.

———. "Chinese Fishermen in Disputed Waters: Not Quite a 'People's War'." *Marine Policy* 68 (2016): 65–73.

CHAPTER 2

Endangered Species, Fragile Marine Ecosystems, and the Philippines

In the nearly two decades preceding the initiation of the *South China Sea Arbitration* in 2013, the Philippines repeatedly complained to China about the activities of Chinese fishermen in the South China Sea. At the outset, the Philippines, when charging the Chinese fishermen with violation of Philippine laws and international conventions such as the United Nations Convention on the Law of the Sea ("the Convention"), tended to frame its complaints in terms of "the conservation of the natural resources of the South China Sea."[1] This approach was an echo

---

[1] *South China Sea Arbitration, Memorial of the Philippines* (30 March 2014), Annex 181, Government of the Republic of the Philippines, Transcript of Proceedings Republic of the Philippines-People's Republic of China Bilateral Talks (10 August 1995), vol. VI, 203 ("*MP*"), https://files.pca-cpa.org/pcadocs/The%20Philippines%27%20Memorial%20-%20Volume%20VI%20%28Annexes%20158-221%29.pdf, accessed 26 March 2019; *MP*, Annex 19, Memorandum from Erlinda F. Basilio, Acting Assistant Secretary, Office of Asian and Pacific Affairs, Department of Foreign Affairs, Republic of the Philippines, to the Secretary of Foreign Affairs of the Republic of the Philippines (29 March 1995), vol. III, 215, https://files.pca-cpa.org/pcadocs/The%20Philippines%27%20Memorial%20-%20Volume%20III%20%28Annexes%201-60%29.pdf, accessed 26 March 2019; *MP*, Annex 20, Memorandum from Lauro L. Baja, Jr., Assistant Secretary, Office of Asian and Pacific Affairs, Department of Foreign Affairs, Republic of the Philippines, to the Secretary of Foreign Affairs of the Republic of the Philippines (7 April 1995), vol. III, 223, https://files.pca-cpa.org/pcadocs/The%20Philippines%27%20Memorial%20-%20Volume%20III%20%28Annexes%201-60%29.pdf, accessed 26 March 2019. *United Nations Convention on the Law of the Sea*, concluded at Montego Bay on 10 December, entered into force on 16 November 1994, http://www.un.org/Depts/los/convention_agreements/texts/unclos/closindx.htm, accessed 21 March 2019.

© The Author(s) 2020
A. C. Robles Jr., *Endangered Species and Fragile Ecosystems in the South China Sea*,
https://doi.org/10.1007/978-981-13-9813-1_2

of the approach of the Convention itself, which has been characterized as exploitation-oriented, emphasizing the protection of marine resources against overexploitation.[2] Yet even at this early stage, the Philippines already included references to the status of giant sea turtles as endangered species and to the threat caused by the use of destructive fishing techniques (cyanide and dynamite fishing) to the ecological balance of the Spratly Islands.[3]

As the number of incidents arising from fishing of endangered species and from dynamite and cyanide fishing by Chinese nationals multiplied, the Philippines increasingly described their conduct in terms of threats to endangered species, to ecosystems or habitats, and to the marine environment in general. The concepts used by the Philippines implied a recognition of the intrinsic value of biodiversity, going beyond the mere focus on marine resources harvested for human use.[4] For example, in 2000, the Philippines declared that the illegal activity of Chinese fishermen "disturbed the tranquility of ecosystems and habitat of important species of marine life."[5] A month later, the Philippines complained that Chinese fishermen were engaged in "illegal fishing of endangered marine life."[6] In 2001, the Philippines justified its confiscation of blasting caps, dynamite, and cyanide that were used to harvest endangered marine

---

[2] Rüdiger Wolfrum and Nele Matz, "The Interplay of the United Nations Convention on the Law of the Sea and the Convention on Biological Diversity," *Max Planck Yearbook of United Nations Law* 4 (2000): 448; Nele Matz, *Wege zur Koordinierung völkerrechtlicher Verträge. Völkervertragsrechtliche und institutionelle Ansätze* [Means to Co-ordinate International Treaties. International Treaty Law Approaches And Institutional Approaches] (Heidelberg: Springer Verlag, 2005), 136.

[3] *MP*, Annex 20, 224.

[4] Wolfrum and Matz, 464.

[5] *MP*, Annex 186, Note Verbale from the Department of Foreign Affairs of the Republic of the Philippines to the Embassy of the People's Republic of China in Manila No. 2000100 (14 January 2000), vol. VI, 259, https://files.pca-cpa.org/pcadocs/The%20Philippines%27%20Memorial%20-%20Volume%20VI%20%28Annexes%20158-221%29.pdf, accessed 26 March 2019.

[6] *MP*, Annex 45, Memorandum from Willy C. Gaa, Assistant Secretary of Foreign Affairs, Republic of the Philippines to Secretary of Foreign Affairs, Republic of the Philippines (14 February 2001), vol. III, 366, https://files.pca-cpa.org/pcadocs/The%20Philippines%27%20Memorial%20-%20Volume%20III%20%28Annexes%201-60%29.pdf, accessed 26 March 2019.

species, on the grounds of protection of the marine environment.[7] In 2006, the Philippines protested that Chinese fishermen continued to engage in "harmful fishing and rampant trading of endangered corals and marine species in the South China Sea."[8] In 2012, the Philippines informed the members of the Association of Southeast Asian Nations ("ASEAN") that the increase in the number of Chinese vessels fishing around Scarborough Shoal imperiled marine diversity in the Shoal and threatened the whole marine ecosystem in the South China Sea.[9] By the time of the submission of its *Memorial* to the Arbitral Tribunal in 2014, the Philippines, when explaining China's violations of Philippine rights under the Convention, alleged that China had violated the Convention on Biological Diversity ("CBD").[10]

The Philippine insistence on the value of biodiversity is the manifestation of a greater awareness of the problems of conserving biodiversity. The Philippines had become a party to the CBD on 6 January 1994 and submitted its first two reports on the implementation of the CBD in 1998 and 2002.[11] Certainly, the stress on biodiversity reflects the vital

[7] *MP*, Annex 48, Memorandum from Josue L. Villa, Embassy of the Republic of the Philippines in Beijing, to the Secretary of Foreign Affairs of the Republic of the Philippines (21 May 2001), vol. III, 390, https://files.pca-cpa.org/pcadocs/The%20Philippines%27%20Memorial%20-%20Volume%20III%20%28Annexes%201-60%29.pdf, accessed 26 March 2019.

[8] *MP*, Annex 205, Note Verbale from the Department of Foreign Affairs of the Republic of Philippines to the Embassy of the People's Republic of China in Manila, No. 12-0894 (11 April 2012), vol. VI, 377, https://files.pca-cpa.org/pcadocs/The%20Philippines%27%20Memorial%20-%20Volume%20VI%20%28Annexes%20158-221%29.pdf, accessed 26 March 2019.

[9] *MP*, Annex 210, Note Verbale from the Department of Foreign Affairs of the Philippines to the Embassies of ASEAN Member States in Manila, No. 12-1372 (21 May 2012), vol. VI, 400, https://files.pca-cpa.org/pcadocs/The%20Philippines%27%20Memorial%20-%20Volume%20VI%20%28Annexes%20158-221%29.pdf, accessed 26 March 2019. The other Member States of ASEAN (Association of Southeast Asian Nations) are Brunei Darussalam, Cambodia, Indonesia, Lao People's Democratic Republic, Malaysia, Myanmar, Singapore, Thailand, and Vietnam.

[10] *MP*, vol. I, 190–93, paras. 6.82–6.89, https://files.pca-cpa.org/pcadocs/Memorial%20of%20the%20Philippines%20Volume%20I.pdf, accessed 27 March 2019. *Convention on Biological Diversity* (CBD), signed at Rio de Janeiro, on 5 June 1992, entered into force on 29 December 1993, https://www.cbd.int/doc/legal/cbd-en.pdf, accessed 26 March 2019.

[11] Republic of the Philippines, *The First Philippine National Report to the Convention on Biological Diversity May 1998* [Quezon City: Department of Environment and Natural Resources—Protected Areas and Wildlife Bureau, 1998], https://www.cbd.int/

importance of coastal and marine diversity for the country. It has at least 4951 species of marine plants and animals, of which 15 are endemic (found only in the Philippines). Of the total, 142 species are under threat and 15 are endangered. In distribution among ecosystems, coral reefs are the most biodiverse, with 3967 species, covering an area of approximately 2500 square kilometers. Based on the distribution of marine taxa, the Philippines can be divided into two areas.[12] The area of higher biodiversity is the South China Sea in the west, while the area of lower biodiversity is in the Pacific Ocean area in the east.[13] It goes without saying that coastal and marine biodiversity is an essential resource for more than 60% of the country's population living in coastal areas.[14] For the entire country, more than half of the animal protein consumed comes from fish.[15]

Marine biodiversity, like biodiversity in general, has three components: diversity within species, species diversity, and ecosystem diversity.[16] It was concern for the second and third components of biodiversity that underlay Philippine protests to China, although these terms were not actually used in the diplomatic communications. Over the years, eels, sharks, corals, sea turtles, and giant clams were identified by the Philippines as endangered species, and coral reefs, as endangered habitats. Since it is not possible to examine here in any detail the Philippine interest in each of the endangered

reports/search/, accessed 3 May 2019; Republic of the Philippines. *The Second Philippine National Report to the Convention on Biological Diversity* [Quezon City: Department of Environment and Natural Resources—Protected Areas and Wildlife Bureau, 2002], https://www.cbd.int/reports/search/, accessed 3 May 2019.

[12]A taxon is defined as "any group of organisms to which any rank of taxonomic name is applied." *Larousse Dictionary of Science and Technology* (Edinburgh: Larousse plc, 1995), 1086.

[13]*The First Philippine National Report to the Convention on Biological Diversity*, 1–3.

[14]Republic of the Philippines, *The Fifth National Report to the Convention on Biological Diversity 2014* [Quezon City: Department of Environment and Natural Resources—Biodiversity Management Bureau, 2014], 4, https://www.cbd.int/reports/search/, accessed 25 March 2019.

[15]*MP*, Annex 240, "Eastern South China Sea Environmental Disturbances and Irresponsible Fishing Practices and Their Effects on Coral Reefs and Fisheries," by Kent E. Carpenter, vol. VII, 411 ("First Carpenter Report"), https://files.pca-cpa.org/pcadocs/The%20Philippines%27%20Memorial%20-%20Volume%20VII%20%28Annexes%20222-255%29.pdf, accessed 27 March 2019.

[16]CBD, Article 2.

species, attention will be limited to one of them, sea (marine) turtles. While giant clams were specifically named in the Arbitral Tribunal's findings, several reasons justify the focus on sea turtles. First, during the Hearing on the Merits, the Tribunal asked Dr. Kent E. Carpenter, an expert presented by the Philippines, questions relating to sea turtles.[17] Second, as we shall see below, while the arbitral proceedings were ongoing, illegal harvesting of the sea turtles continued unabated and may have even increased in scale. Third, sea turtles are not fished using dynamite and cyanide, but with the use of nets or spears, so that the mere harvesting of sea turtles, regardless of methods employed for this purpose, constitutes in itself a threat to these reptiles. The first part of this chapter will discuss the four ecological roles of sea turtles and their economic importance to specific groups, particularly marginalized communities and indigenous groups, in the Philippines.

The second part of the chapter considers coral reefs as constituting fragile marine ecosystems that should be protected and preserved. Mangroves and seagrass beds are also fragile marine ecosystems that are no less threatened by natural causes and human activity in the South China Sea and are equally deserving of protection and preservation. They will not be discussed here, for the simple reason that they were not the object of diplomatic exchanges between the Philippines and China and were not encompassed by Philippine Submissions No. 11 and 12(b) to the Arbitral Tribunal. If the focus on coral reefs needed further justification, three circumstances may be cited: corals and fish inhabiting the reefs were harvested by Chinese fishermen with the use of cyanide and dynamite; several coral reefs in the Spratly Islands were irreversibly damaged by Chinese construction activities; and the Arbitral Tribunal appointed its own experts to provide an independent opinion on the effects of China's island-building in the Spratly Islands.[18] The second section will briefly explain the ecological services performed by coral reefs and their economic value to the Philippines.

---

[17] *South China Sea Arbitration*, Hearing on the Merits and Remaining Issues of Jurisdiction and Admissibility, Transcript, Day 4 (30 November 2015), 145–46 ("Hearing on the Merits"), https://pcacases.com/web/sendAttach/1550, accessed 27 March 2019. The Tribunal also asked questions concerning the extent of shark fishing conducted by Chinese fishermen and the volume of shark fishing that Dr. Carpenter would consider to constitute over-exploitation.

[18] *South China Sea Arbitration*, Award of 12 July 2016, 29–31, paras. 84, 85, 90, https://pcacases.com/web/sendAttach/2086, accessed 26 March 2019.

I. THE ECOLOGICAL ROLES AND ECONOMIC IMPORTANCE OF ENDANGERED SPECIES: THE EXAMPLE OF SEA TURTLES

The diverse ecological roles of sea turtles cannot be underestimated, even if they are economically important mainly for small marginalized and indigenous communities in the Philippines.

A. The Ecological Roles of Sea Turtles

A few words are in order concerning the biology of sea turtles before their ecological roles are discussed.

1. The Biology of Sea Turtles

Sea turtles differ from land turtles (tortoises), which have legs, and freshwater turtles, which have webbed claws, in that they have long flippers for swimming.[19]

Sea turtles are solitary cold-blooded reptiles that need to surface to breathe.[20] They spend all their life cycle in the ocean. Females emerge on land to mate, while males and females occasionally emerge on land to bask. They migrate from distant feeding grounds to different nesting sites. All species use beach habitats for nesting. Sea turtles mate during one or two months, although individual females are receptive for only two to three weeks.[21] After mating, it takes the female two to four weeks to lay the first clutch of eggs, of which there may be 80 to 120 in each nest, depending on the species. The female may return two to eight more times in the same season to mate. The eggs, which take 45–65 days to hatch, invariably hatch after dark when the cool sand surface prevents

---

[19]Marine Wildlife Watch of the Philippines, *Philippine Aquatic Wildlife Rescue and Response Manual to Marine Turtle Incidents* (Taguig City: Marine Wildlife Watch of the Philippines, 2014), 2, http://www.mwwphilippines.org/downloads/rm-marineturtles.pdf, accessed 19 March 2019.

[20]The following paragraph is summarized from Emirates Wildlife Society (EWS)-World Wildlife Fund (WWF), *Marine Turtle Conservation Project. Final Scientific Report. Arabian Region* (Abu Dhabi: EWS-WWF, 2015), 15–16, https://www.cbd.int/doc/meetings/mar/ebsaws-2015-02/other/ebsaws-2015-02-ews-wwf-submission1-en.pdf, accessed 25 March 2019.

[21]Receptive means "capable of being effectively pollinated or fertilized." *Larousse Dictionary of Science and Technology*, 918.

hatchlings from suffering from the heat on emergence. Hatchlings dig through the sand for two to three days before emerging, then crawl down the beach and head in the offshore direction. For guidance and orientation, they first rely on light—they move away from the dark silhouetted shoreline toward the brighter ocean horizon. The presence of man-made light sources may disorient the hatchlings, causing them to head inland rather than offshore and increasing the risk of exposure to predators, entrapment in vegetation and debris or dehydration. As they reach offshore areas, they then rely for orientation on waves through nearshore waters and finally on magnetic fields. During their offshore migration, they face additional threats from currents, waves, and predators. They swim for one to two days to get as far offshore as possible. Once in the ocean, they float on the surface for several years until they recruit as small 20–40-centimeter juveniles from oceanic waters to nearshore shallow feeding areas. When the sea turtles have grown, the risk of predation decreases. Adults have very few natural predators other than killer whales, which prey on the green sea turtle and leatherbacks, and tiger sharks, which prey on green sea turtles. The sea turtles remain at one or more feeding grounds for one to ten years until they reach sexual maturity and undertake their first migration to nesting areas, generally returning to their natal beach, when the cycle recommences.

One might judge the importance that the Philippines attaches to sea turtles from the fact that it devoted a separate section to them in its *Fourth National Report* on the implementation of the CBD (2009).[22] Five out of seven sea turtle species (*Testudines*) can be found in the Philippines.[23] Green (*Chelonia mydas*) sea turtles are an endangered species, according to the criteria of the International Union

---

[22] Republic of the Philippines, *Assessing Progress Towards the 2010 Biodiversity Targets. The Fourth National Report to the Convention on Biological Diversity* [Quezon City: Department of Environment and Natural Resources—Protected Areas and Wildlife Bureau, 2009], 53–55 ("*Philippines Fourth National Report to the CBD*"), https://www.cbd.int/reports/, accessed 25 March 2019.

[23] Unless otherwise indicated, the following is summarized from Chloe Schauble, *Sea Turtles* (Hanoi: Marine and Coastal Program/IUCN Vietnam, 2007), 3–5, http://www.ioseaturtles.org/electronic_lib2.php?cat_id=3, accessed 10 March 2017. Photographs of each of the five species may be found on Ocean Ambassadors Track a Turtle, *Philippine Turtle Islands, Turtle Biology* [2001], http://www.oneocean.org/ambassadors/track_a_turtle/tihpa/pti.html, accessed 31 March 2019. See also Vicky Viray-Mendoza, "Marine Turtles and the Philippine Forest Turtle," *The Maritime Review. The Online Edition of the*

for the Conservation of Nature (IUCN).[24] The females average over 100 centimeters in carapace length and 120 kilograms in weight.[25] They stay near the coastline and around islands on seagrass beds. They are herbivores, feeding on seagrass and algae. Hawksbill sea turtles (*Eretmochelys imbricata*), which are critically endangered, are among the smaller sea turtles, with carapace length of 76 to 91 centimeters and can weigh 40 to 60 kilograms.[26] Hawksbill sea turtles are found around coastal coral reefs, rocky areas, and estuaries. The olive ridley sea turtles (*Lepidochelys olivacea*) can grow up to 60 centimeters in length and weigh up to 70 kilograms. They forage offshore in surface waters or dive to depths of 150 meters to feed on crabs, clams, mussels, shrimp, fish, sea urchins, squid, and jellyfish. These three species of sea turtles are widespread in the Philippines. High nesting aggregations of green turtles are found in the Turtle Islands Wildlife Sanctuary, on Bancauan Island in Mapun, other islands in Tawi-Tawi Province and Panikian island in Zamboanga del Sur Province, all in the southern region of Mindanao; a significant developmental area is found in Palawan Province in the western part of the country. Major aggregations of hawksbill turtles may be found in Romblon Island, Magsaysay in Misamis Oriental Province, Mindanao, and in the Davao Gulf in the south. Important nesting aggregations of the olive ridley may be found in the provinces of Zambales, Bataan, and Batangas, on the main island of Luzon.[27]

Leatherback sea turtles (*Dermochelys coriacea*), classified as critically endangered, grow the largest, reaching 1.2 to 2.4 meters in length and 294 to 589 kilograms in weight. They dive the deepest and travel the furthest. They are found mainly in the open ocean, where they feed almost exclusively on jellyfish that are borne along on ocean currents. The leatherbacks are the only sea turtles that lack a large shell. Their

---

Maritime League's Maritime Review Magazine, 24 May 2016, http://maritimereview. ph/2016/05/24/marine-turtles-and-the-philippine-forest-turtle/, accessed 25 March 2019.

[24] See Annex 2.1.

[25] Carapace is an exoskeleton covering part or all of the dorsal surface of an animal, e.g., the bony dorsal shield of a tortoise. *Larousse Dictionary of Science and Technology*, 161. See the IUCN criteria for identification of endangered species in Annex 2.1.

[26] See the IUCN criteria for identification of critically endangered species in Annex 2.2.

[27] Marine Wildlife Watch of the Philippines, 1.

elongated shell is composed of layers of thin, tough, rubber skin. The leatherback turtles nest mostly in Malaysia and Indonesia, but they often forage in the Philippines, around Palawan Province, in the Central Visayas and Bicol regions, and in the Davao Gulf in the south. The nesting of a leatherback turtle in the Bicol region was first documented in 2013.[28] The loggerhead sea turtles (*Caretta caretta*), which are classified as endangered, can have carapace length of 82 to 105 centimeters and weight of up to 158 kilograms. They feed in coastal bays and estuaries on shellfish living on the bottom of the ocean. The loggerhead turtles nest in Japan, but they come to the Philippines to forage in the Bicol region and in the waters of Basilan Province in the south.[29]

All species of sea turtles perform important ecological roles.

2. Sea Turtles and the Marine Environment
Sea turtles perform four main ecological roles: They maintain habitat, maintain a balanced food web, cycle nutrients, and provide habitat.[30]

Green sea turtles help to maintain habitat by feeding on seagrass just a few centimeters from the bottom of the blades, allowing the older, upper portions of the blades to float away. Without constant grazing by green sea turtles, seagrass beds would become overgrown, obstruct currents, shade the bottom, begin to decompose, and provide a habitat for the growth of slime molds. The hawksbill turtles feed on sponges, thus removing them from reefs and allowing other species, such as corals, to colonize and grow. The hawksbill sea turtles rip the sponges apart during feeding; in this way, they remove the physical and chemical defenses of sponges that prevent most fish and marine mammals from eating them. Sea turtles improve the beaches where they nest, as the unhatched eggs provide nutrients, such as nitrogen, phosphate, and potassium, to the

---

[28] Ibid.

[29] Ibid.

[30] The following is summarized from E. D. Wilson et al., *Why Healthy Oceans Need Sea Turtles. The Importance of Sea Turtles to Marine Ecosystems* (Washington, DC: Oceana, 2010), 5–15, http://www.oceana.org/sites/default/files/reports/Why_Healthy_Oceans_ Need_Sea_Turtles.pdf, accessed 25 March 2019. The references to the scientific literature are to be found in this publication. See also World Wildlife Fund-Philippines, *Turtle Islands: Resources and Livelihoods Under Threat. A Case Study on the Philippines* (Quezon City: WWF-Philippines, 2005), 6 ("WWF-Philippines, *Turtle Islands*"), https://wwf.org. ph/wp-content/uploads/2017/11/Turtle-Islands-2005.pdf, accessed 28 March 2019 and EWS-WWF, 18.

dune ecosystem. The nutrients allow continued growth of vegetation, which helps to stabilize dunes and provides food for plant-eating animals. The eggs themselves provide food for many predators, which redistribute them among the dunes through their feces.

Second, sea turtles help to maintain a balanced food web. The leatherbacks consume large quantities of jellyfish, which prey on fish eggs and larvae. Sea turtles provide a food source for fish and shrimp by carrying around epibionts, such as barnacles and algae.[31] The fish and shrimp that feed on the epibionts reduce drag and keep the skin and shells of the sea turtles clean. All sea turtle species are also prey for other animals, both on shore and at sea. Their eggs are sources of food for ants, crabs, rats, raccoons, foxes, coyotes, dogs, mongoose, vultures, and feral cats.[32] The hatchlings that emerge on the beaches as well as those that make it to the water are food for a variety of seabirds. Sea turtle hatchlings and juveniles are also preyed upon by reef fish, such as groupers and jacks.

Third, sea turtles play a role in nutrient cycling. Loggerheads feed on hard-shelled prey, such as crustaceans, the shells of which are reduced into fragments that are discarded on site or further away in the form of feces. In this manner, the loggerheads increase the rate of nutrient recycling for benthic (ocean bottom) ecosystems. Female sea turtles improve the beaches where they nest by introducing nutrients and energy from distant and dispersed foraging grounds into small and nutrient-poor beaches. The eggs that they lay may be consumed by predators. The eggs are penetrated by roots, enabling plants to absorb nutrients. When the eggs hatch, most of the nutrients return to the sea as hatchlings, but some nutrients are left on the beach in the form of eggshells and embryonic fluid. If the eggs fail to hatch, they allow the nutrients to enter the detrital food chain.[33] The processes of nesting and emergence of

---

[31] Epibiosis is a relationship in which one organism lives on the surface of another without causing it harm. Plant epibionts are epiphytes, animal epibionts, epizoites. *Larousse Dictionary of Science and Technology*, 380.

[32] Feral refers to a domesticated animal that has reverted to the wild. Ibid., 412.

[33] In a detrital food chain, dead organic matter of plants and animals is broken down by decomposers, e.g., bacteria and fungi, and moves to detritivores and then carnivores. In a grazing food chain, energy and nutrients move from plants to the herbivores consuming them, and to the carnivores or omnivores preying upon the herbivores. Dafeng Hui, "Food Web: Concept and Applications," *Nature Education Knowledge* 3 (2012): 6, https://www.nature.com/scitable/knowledge/library/food-web-concept-and-applications-84077181, accessed 28 March 2019.

hatchlings bring egg matter and nest organisms to the surface and thus help disperse nutrients to small organisms on beach sand.

Fourth, sea turtles provide habitat to many marine organisms. Loggerheads play host to the largest and most diverse communities of epibionts, more than 100 species of which have been identified. Sea turtles transport plants and crustaceans to and from reefs, seagrass beds, and the open ocean. Sea turtles can be used as a place to rest for seabirds. Olive ridleys, as they surface to bask in the ocean, expose the center of their shell and create platforms on which seabirds whose feathers are not extremely water-resistant can perch. Seabirds that would be vulnerable to attack from sharks find refuge by perching on sea turtles. Small baitfish use sea turtles for protection by forming tight schools beneath the sea turtle's body; these fish provide a food source for resting seabirds. Other seabirds sometimes feed on the epibionts inhabiting the sea turtle's shell.

The ecological roles played by sea turtles make conservation of these reptiles an imperative, even though they are economically valuable resources only for small marginalized and indigenous communities in the Philippines.

B. The Economic Importance of Sea Turtles

The consumption of and trade in sea turtles and their by-products are vital primarily for small groups in the Philippines. At present, a primary threat to their survival is the illegal harvesting of and trade in these reptiles and their products.

1. The Consumption of and Trade in Sea Turtles and Their Products in the Philippines

It is extremely difficult to assess the extent of consumption of and trade in turtles and turtle products in the Philippines prior to the second half of the twentieth century. On these issues as well as on scientific questions relating to sea turtles, there is a dearth of research in the country.[34] It is commonplace among commentators to assert and repeat, without citing Spanish and American sources, that exploitation and use of marine turtles, particularly green turtles, in the Philippines predates the Spanish

---

[34]Christopher N. S. Poonian et al., "Diversity, Habitat Distribution, and Indigenous Hunting of Marine Turtles in the Calamian Islands, Palawan, Republic of the Philippines," *Journal of Asia-Pacific Biodiversity* 9 (2016): 69.

period (1521–1898) and continues to the present.[35] Some even go so far as to say, without providing evidence, that during the colonial period and for three decades following the end of the Second World War, the volume of exports of sea turtle shells and derivative products made them among the country's primary foreign exchange-earners.[36] The available data indicate that toward the end of Spanish rule, for five years (1886–1890), the Philippines exported a total of 39,606 pounds of tortoise-shell, the destination of which was unknown; they were ranked 25th in terms of volume, hardly qualifying them to be major export earners.[37] During the early American colonial period, it was reported that approximately 8000 kilograms of tortoiseshell were gathered in the Philippines annually and that in 1914, 2296 kilograms of tortoiseshell originating from the southern regions of Mindanao and Sulu were exported, primarily to Japan.[38] A Filipino scientist who visited the Turtle Islands in 1950 reported that during the Second World War, the Japanese occupiers had butchered 20,000–25,000 females, but he did not explain whether they did so for their meat, their eggs, or their shells.[39]

[35] Karen L. Eckert, *The Biology and Population Status of Marine Turtles in the North Pacific Ocean*, NOAA Technical Memorandum NMFS, NOAA-TM-NMFS-SWFSC-186 (Washington, DC: U.S. Department of Commerce, National Oceanic and Atmospheric Administration, National Marine Fisheries Service, Southwest Fisheries Science Center, 1993), 21, https://repository.library.noaa.gov/view/noaa/6133/noaa_6133_DS1.pdf?, accessed 19 March 2019; Monnyeen Nida R. Alava and Jose Alfred B. Cantos, "Marine Protected Species in the Philippines," in Geronimo Silvestre et al. (eds.), *In Turbulent Seas: The Status of Philippine Marine Fisheries* (Cebu City: Department of Agriculture—Bureau of Fisheries and Aquatic Resources, 2004), 111, http://oneocean.org/download/db_files/fshprofl.pdf, accessed 25 March 2019.

[36] WWF-Philippines, *Turtle Islands*, 20.

[37] Frank Harris Hitchcock, *Trade of the Philippine Islands* (Washington, DC: Government Printing Office, 1898), 33, https://ia800200.us.archive.org/28/items/tradephilippine-00hitcgoog/tradephilippine00hitcgoog.pdf, accessed 20 March 2019. The exact figures (in pounds) are as follows: 1886—1623; 1887—3904; 1888—1974; 1889—4910; and 1890—27,194.

[38] Alvin Seale, "Sea Products of Mindanao and Sulu, III: Sponges, Tortoise Shell, Corals and Trepang," *Philippine Journal of Science D. General Biology, Ethnology and Anthropology* 13 (1917): 202–203, http://philjournalsci.dost.gov.ph/images/pdf_upload/pjs1917/PJS_Vol_12D_No1_Jan_1917.pdf, accessed 19 March 2019.

[39] Jose S. Domantay, "The Turtle Fisheries of the Turtle Islands," *Bulletin of the Fisheries Society of the Philippines* 3–4 (1953): 15.

In the 1970s and the 1980s, a series of turtle products became the object of substantial exports from the Philippines to Japan. The trade amounted to 32,921 kilograms of hawksbill shells between 1970 and 1986, which implied the hunting of over 47,000 hawksbill turtles; 26,610 kilograms of tortoiseshell from 1970 to 1986; 10,603 kilograms of stuffed hawksbills, which entailed the capture of approximately 8700 hawksbills; 59,771 kilograms of worked tortoiseshell between 1972 and 1983, equivalent to the taking of 24,800 sea turtles; and 46,706 kilograms of green sea turtle skin, representing the hunting of 8800 large sea turtles. The trade in these turtle by-products practically ceased following Philippine ratification of the Convention on International Trade in Endangered Species of Wild Fauna and Flora ("CITES") in 1981.[40]

At present, different species of sea turtles are sought after by consumers for different reasons. The green and the leatherback are hunted for their meat. The hawksbill is hunted for its carapace, used as raw material for souvenirs. The eggs of the loggerhead and the olive ridley are considered delicacies. In China, demand is high from the middle class for turtle meat and fat to cook turtle soup and for costly accessories, such as guitar picks and bags made of shell. Sea turtle bones are also in demand for the preparation of medicine. Stuffed whole turtles are considered as a status symbol for display in homes. The trade in shell products seems to be predominant in East Asia.[41]

Sea turtles are still hunted in some parts of the Philippines for consumption and trade of their meat, eggs, and other products.[42] Green

---

[40] Tom Milliken and Hideomi Tokunaga, *The Japanese Sea Turtle Trade 1970–1986* (Tokyo: TRAFFIC [Japan] [Wildlife Trade Monitoring Network], 1987), 107–10, https://www.traffic.org/site/assets/files/9659/japanese-sea-turtle-trade-1970-1986.pdf, accessed 19 March 2019. *Convention on International Trade in Endangered Species of Wild Fauna and Flora*, signed at Washington, DC, on 3 March 1973, amended at Bonn, on 22 June 1979, amended at Gaborone, on 30 April 1983, https://www.cites.org/eng/disc/text.php, accessed 26 March 2019.

[41] Memorandum of Understanding on the Conservation and Management of Marine Turtles and Their Habitats of the Indian Ocean and Southeast Asia, *Illegal Take and Trade of Marine Turtles in the IOSEA Region*, Doc. No. MT-IOSEA/SS.7/Doc. 10.1 (28 August 2014), 14, 18, 20, 21, paras. 34, 46, 54, 58 ("MT-IOSEA"), https://www.cms.int/sites/default/files/publication/MT_IO7_DOC10-1_Illegal_Take%26Trade-final.pdf, accessed 25 March 2019.

[42] Angelo L. Alcala, "Observations on the Ecology of the Pacific Hawksbill Turtle in the Central Visayas, Philippines," *Fisheries Resource Journal of the Philippines* 5 (1980) 50, http://scinet.dost.gov.ph/union/Downloads/Alcala%20AC%201980%20Observations%20 on%20the%20Ecology%20of%20the%20Pacific%20Hawksbill%20Turtle%20in%20the%20

turtle meat, fat, and cartilage are used in the preparation of turtle soup.[43] Turtle meat is consumed partly in the belief that it would endow the consumer with a long life.[44] Millions of turtle eggs have been harvested since the Second World War. One can get an idea of the scale of the harvesting from the fact that in the four years between 1948 and 1951, 2,654,255 eggs were collected in the Turtle Islands alone.[45] In the 1990s, turtle eggs were sold at PhP3.00 to PhP4.00 ($0.11 to $0.15), and in the 2000s, at the price of three eggs for PhP20.00 ($0.37).[46]

Ratification of CITES was only one of a series of measures adopted by the Philippines since the Second World War with the aim of protecting sea turtles. For three decades, they were protected as part of fisheries management, involving such measures as seasonal closure of fisheries and controls on the quantities, species, and sizes of the fisheries that were caught.[47] For example, it was required in 1974 that 100 eggs had to be retained in every nest.[48] One can understand that enforcement of this type of regulation would always raise problems, but in the Philippines enforcement is complicated by the fact that it is primarily marginalized and indigenous communities that rely on the collection of and trade in eggs and to a lesser extent meat for their livelihoods.

The Philippine Turtle Islands, consisting of six islands in Tawi-Tawi Province in the southern part of the country, have been described as the most important major green turtle rookery (breeding ground) in Southeast Asia.[49] The Turtle Islands produce up to two million eggs per year, making them the 11th major nesting site of sea turtles in the world.

Central%20Visayas_3988.pdf, accessed 19 March 2019; Jonathan L. Mayuga, "From Predators to Protectors of 'Pawikan [Sea turtles]'," *Business Mirror*, 26 November 2018 ("Mayuga, 'From Predators'"), https://businessmirror.com.ph/2018/11/26/from-predators-to-protectors-of-pawikan/, accessed 18 March 2019.

[43] MT-IOSEA, 8, paras. 13, 21, 59.

[44] Eckert, 22.

[45] Domantay, 18.

[46] Mayuga, "From Predators."

[47] Alava and Cantos, 113.

[48] Pejabat Perikanan Negeri and Siow Kuan Tow, "Observations on the Exploitation of Turtles in the Philippines," *Marine Turtle Newsletter* 3 (1977), http://www.seaturtle.org/mtn/PDF/MTN3.pdf, accessed 19 March 2019.

[49] The six islands are Taganak, Boan, Lihiman, Langaan, Bakkungan, and Baguan.

In 2009, sea turtles laid approximately one million eggs on the shores of Baguan Island alone.[50] Fringing reefs, which are relatively well-developed and in good condition, are the benthic communities (living in and on the bottom of the ocean floor) that surround the islands, particularly on the north and northeast coasts; the islands have a fair hard coral cover (28–46%) and high coral diversity (24–27 genera).[51] The Turtle Islands were first settled in 1949 by an ethno-linguistic clan known as Jama Mapun, whose members were Muslims.[52] As of 2010, the Turtle Islands have a population of 2734, of which 868 are gainfully employed, mostly in agriculture and fishing (717). The level of poverty of the islands is reflected in dismal socioeconomic indicators. No household was reported as having electricity as fuel for cooking; 535 out of a total of 656 households used a shared tubed/dipped deep well as a source of water supply for cooking.[53] A quarter of the population lacked any formal education, compared to 5% for the national average.[54]

In the late 1940s, residents of the Turtle Islands became aware of the demand for green turtle eggs in Taiwan and in other Asian countries. They then started collecting green turtle eggs, which they sold through Sandakan, Malaysia. As a result of the collection of eggs over several decades, green turtle egg production declined by 88% between 1951 and 1990.[55] In response, the government declared marine turtle sanctuaries in Tawi-Tawi Province in the late 1970s and early 1980s, prohibiting

---

[50] Dennis L. Maliwanag, "Turtle Islands: Hundreds of Nautical Miles Away from Justice," *Philippine Daily Inquirer*, 22 August 2010, https://www.pressreader.com/philippines/philippine-daily-inquirer-1109/20100822/281775625465558, accessed 29 March 2019.

[51] Ocean Ambassadors, Track a Turtle, "The Philippine Turtle Islands" [2001], http://www.oneocean.org/ambassadors/track_a_turtle/tihpa/pti.html, accessed 31 March 2019.

[52] Raul P. Lejano and Helen Ingram, "Place-Based Conservation: Lessons from the Turtle Islands," *Environment: Science and Policy for Sustainable Development* 49 (2007): 21.

[53] National Statistics Office, *Statistical Tables on Sample Variables from the Results of 2010 Census of Population and Housing—Tawi-Tawi*, 31 July 2014, https://psa.gov.ph/content/statistical-tables-sample-variables-results-2010-census-population-and-housing-tawi-tawi, accessed 18 March 2019.

[54] WWF-Philippines, *Turtle Islands*, 14–15.

[55] Romeo B. Trono, "Management and Conservation Program of a Protected Wildlife. Species," *Philippine Journal of Public Administration* 26 (1990): 91–92, http://lynchlibrary.pssc.org.ph:8081/bitstream/handle/0/4078/10_Management%20and%20Conservation%20Program.pdf?sequence=1, accessed 25 March 2019.

extractive activities within 250 meters from the shore at the lowest tidal line. Recognizing the importance of egg collection for the livelihood of Turtle Islands residents, a program was created that allowed them to apply for permits to collect the eggs in four of the islands; a fifth, Baguan Island, was off-limits to human beings.[56] The permit holders then sold the permits to middlemen, who took charge of the actual collection. Under the so-called 60/40 scheme, 60 of every 100 eggs were allocated to a permit holder, and 40 were set aside for conservation. Of the 40 eggs, 30 were transplanted to a hatchery and 10 were sold, with the proceeds going to a Marine Turtle Foundation.

The sale of the permits, while contributing only 23% of total household income, was a valuable income supplement for people who lived a hand-to-mouth existence. The proceeds from the sale of permits enabled them to purchase big-ticket items that they would not normally be able to afford (such as boats), to construct their houses, or to meet emergencies, such as illness of children.[57] Significantly, the program also encouraged egg conservation, which rose from a 50% conservation rate in the 1980s to 80% in the late 1990s.[58] It has been estimated that the program resulted in the protection of 10 million eggs.[59]

The eggs were not consumed locally but were traded. Since sea turtles are on the endangered species list of CITES, up to 70% of the eggs laid in the Turtle Islands was smuggled to neighboring countries.[60] In 2002, the latest year for which data are available, Sabah was said to be the hub for the illegal eggs trade. An egg that was sold at PhP3.50 ($0.06) in Taganak Island, one of the six Turtle Islands, was sold at approximately six times the price (RM10.00 or PhP23.50) in Kota Kinabalu, the capital of Sabah.[61] The distribution network was suspected to extend to Brunei Darussalam, Singapore, and Hong Kong.[62]

---

[56] The following discussion is based on WWF-Philippines, *Turtle Islands*, 20–24.

[57] Ibid., 24.

[58] Lejano and Ingram, 23.

[59] Dennis Atienza Maliwanag, "DOH Team Accused of Turtle Eggs Poaching in Tawi-Tawi," *Philippine Daily Inquirer*, 2 March 2017 ("Maliwanag, 'DOH Team'"), https://newsinfo.inquirer.net/876941/doh-team-accused-of-turtle-eggs-poaching?utm_expid=.XqNwTug2W6nwDVUSgFJXed.1, accessed 18 March 2019.

[60] MT-IOSEA, 10, para. 21.

[61] Maliwanag, "DOH Team."

[62] WWF-Philippines, *Turtle Islands*, 23.

In 2001, the Philippine Wildlife Act prohibited the harvest of wildlife and its by-products, including turtle eggs. In practice, a de facto moratorium on the implementation of the law was introduced, with the municipal government taking over the issuance of permits. The reasons for the moratorium were the important contribution of egg collection to household income, as well as the need to maintain the conservation rate of the eggs, which had fallen by half in one year following the implementation of the law.[63] It seems that as of 2016, collection of eggs in the Turtle Islands continues.[64]

The collection of turtle eggs and the hunting of turtles themselves are still permitted for a small indigenous group, known as Tagbanua and living on Calamanian Islands in Palawan Province, in the west of the country.[65] The Tagbanua number only 17,514 members, accounting for 2.3% of the population of a province that is itself not very densely populated (771,667 out of a total Philippine population of 92.34 million in 2010).[66] They are one of the few indigenous groups still practicing their traditional lifestyle, part of which involves hunting of turtles and collecting of turtle eggs according to traditional management practices. As hunting of sea turtles is part of their tradition, they are legally permitted to do so, yet they are aware that the sea turtles are in need of protection and that a series of local and national laws protect them and prohibit hunting. Turtle meat is considered a reliable and cheap source of protein and is served as a main dish at weddings, birthdays, and funerals. The Tagbanua also sell turtle meat to Chinese residents on a nearby island. A live subadult sells for $4.00; a kilogram of meat for $0.70, and a kilogram of boiled and dried scutes (scales) for $33.00 to $67.00.

---

[63] Lejano and Ingram, 24.

[64] Jonathan L. Mayuga, "Saving the Endangered 'Pawikan' [Sea Turtles]," *Business Mirror*, 17 January 2016, http://www.businessmirror.com.ph/saving-the-endangered-pawikan-2/, accessed 25 March 2019.

[65] Unless otherwise indicated, the following paragraph is summarized from Poonian et al., 69–73.

[66] Republic of the Philippines, National Statistics Office, *2010 Census of Population and Housing. Report No. 2A—Demographic and Housing Characteristics (Non-Sample Variables). Palawan* (Manila: National Statistics Office, 2013), 124, https://psa.gov.ph/sites/default/files/PALAWAN_FINAL%20PDF.pdf, accessed 18 March 2019. On the Tagbanua in general, see Celeste Lacuna-Richman, "Subsistence Strategies of an Indigenous Minority in the Philippines: Nonwood Forest Product Use by the Tagbanua of Narra, Palawan," *Economic Botany* 58 (2004): 266–85.

At the level of this indigenous group, this illegal trade raises the question how one is to distinguish between the legal indigenous practice and illegal commercial exploitation.[67] At other levels, the conflict is between conservation, on the one hand, and illegal harvesting of sea turtles by Chinese nationals and the trade in sea turtles with them, on the other.

2. The Illegal Harvesting of and Trade in Sea Turtles as a Threat to Their Survival

One can describe the present situation of sea turtles in the Philippines in terms of a paradox. On the one hand, a significant increase in conservation awareness since the 1990s manifests itself in the multiplication of initiatives at the local, national, and international levels. For example, in the province of Zambales, in the previous decade, the Zambales Turtle Conservation Program (2004–2006) paid PhP5.00 to PhP6.00 ($0.09 to $0.10) per egg to the program's hatcheries.[68] More recently, the provincial government recruited former hunters and gatherers of sea turtles and their eggs and made them advocates of sea turtle preservation, known as Sea Turtle Rangers.[69] In Batangas Province on the main island of Luzon, a Conservation Center was set up in 1999, while coastal communities recruited Sea Turtle Guards (*"Bantay Pawikan"*) to prevent poaching of eggs by collecting them and transferring them to hatcheries. These "guards" also used to be hunters and collectors of sea turtles and their eggs.[70] At the national level, the Philippine Turtle Islands were established as a protected area, the Turtle Islands Wildlife Sanctuary (TIWS), by virtue of Presidential Proclamation No. 171 of 26 August 1999, becoming a breeding or nesting area for endangered hawksbill and green sea turtles. At the international level, the Philippines and Malaysia

---

[67] Poonian et al., 72.

[68] "Zambales Turtle Conservation Program 2004–2005 Technical Turtle Seminar," *Environmental Protection of Asia*, 2006, http://www.environmentalprotectionofasia.com/ztcp/reports/2004_2005.htm, accessed 13 March 2017.

[69] Jonas Reyes, "215 'Pawikan' [Sea Turtles] Hatchlings Released in Subic," *Manila Bulletin*, 21 December 2017, https://news.mb.com.ph/2017/12/20/215-pawikan-hatchlings-released-in-subic/, accessed 29 March 2019.

[70] Mayuga, "From Predators," 2018.

established in 1996 a Turtle Islands Heritage Protected Area (TIHPA), consisting of the Philippine TIWS and Turtle Islands—Malaysia.[71]

On the other hand, local, national, and international initiatives are undermined by illegal harvesting by or illegal trade with Chinese nationals. Illegal harvesting, targeting mostly hawksbill and green turtles, is carried out mainly by Chinese fishermen from Hainan Province.[72] In January 2002, four Chinese vessels with 58 sea turtles, mostly green turtles, were caught in the Philippine Tubbataha Marine Park, a UNESCO Natural Heritage Park located in the Sulu Sea.[73] Between 2012, the year immediately preceding the initiation of the arbitration, and 2015, the year when the Award on Jurisdiction and Admissibility was issued, the Philippine Navy and the Philippine Coast Guard seized 200 sea turtles, most of which were already dead. According to reports, the operation was financed by Chinese nationals.[74] In May 2014, 354 sea turtles were found on a Chinese vessel. 234 of them were dead, many with their bodies removed from their shells, a sight described by a journalist as "heartbreaking."[75] In 2015, eleven Chinese fishermen were arrested and tried

---

[71] Republic of the Philippines, Department of Environment and Natural Resources, Biodiversity Management Bureau, *Philippines-Malaysia Partnership in Marine Turtle Conservation*, 2016, http://www.bmb.gov.ph/index.php/mainmenu-news-events/ mainmenu-news/396-philippine-malaysia-partnership-in-marine-turtle-conservation, accessed 29 March 2019. On the management of the Turtle Islands Heritage Protected Area, see Rolando C. Esteban, "The Turtle Island Heritage Protected Area: The Possibilities and Limits of Transborder Conservation," *Ostrom Workshop Indiana University Bloomington*, July 2008, http://dlc.dlib.indiana.edu/dlc/bitstream/handle/10535/993/ Esteban_223701.pdf?sequence=1, accessed 31 March 2019; Evangeline Miclat and Enrique Nunez, "The Philippine-Sabah Turtle Island Heritage Protected Area (TIHPA)," in Peter Mackelworth (ed.), *Marine Transboundary Conservation and Protected Areas* (New York: Routledge, 2016), 32–47.

[72] Glenda Cadigal of the Palawan Council on Sustainable Development, quoted in Public Radio International, "High Demand for Sea Turtles in China Sends Poachers Toward Philippines," *PRI's* [Public Radio International] *The World*, 9 February 2012, https:// www.pri.org/stories/2012-02-09/high-demand-sea-turtles-china-sends-poachers-toward-philippines, accessed 29 March 2019; MT-IOSEA, 7, para. 10.

[73] Cruz, 63.

[74] Mayuga, "Saving."

[75] John B. Virata, "Philippine Police Arrest Sea Turtle Poachers from China," *Reptiles Magazine*, 11 May 2014, http://www.reptilesmagazine.com/Turtles-Tortoises/ Information-News/Philippine-Police-Arrest-Chinese-Sea-Turtle-Poachers/, accessed 27 March 2019.

in Palawan for poaching more than 500 marine turtles.[76] Perhaps in order to avoid being apprehended, Chinese nationals are trading with Filipinos, who do the actual harvesting. For instance, in 2012, Half Moon Shoal in the South China Sea was reported to be a buying station for Chinese nationals stationed there to purchase turtles, mostly caught by Filipino fishermen around the "turtle corridor" of Balabac Strait.[77] Not without reason the former project leader of the government's conservation project in the Turtle Islands in the 1980s and 1990s declared that Chinese traders now constitute the greatest threat to sea turtles in the Philippines.[78]

As mentioned above, sea turtles feeding on sponges remove them from reefs and allow corals to colonize and grow. Attention should now be turned to another major component of biodiversity in the South China Sea, coral reefs.

II. The Ecological Roles and Economic Importance of Fragile Ecosystems: The Example of Coral Reefs

Coral reefs seem to be equally valuable to the Philippines for ecological and economic reasons.

A. The Ecological Roles of Coral Reefs

An overview of the nature of coral reefs as geological structures and living ecosystems will help us to understand the ecological roles that they perform.

1. Coral Reefs as Geological Structures and Living Ecosystems

A scientific definition of coral reef is that it is a tract of corals growing on massive, wave-resistant structures and associated sediments, substantially built by skeletons of successive generations of corals and other calcareous

---

[76] Cecil Morella, "Turtles' Vulnerable Start to Life on Philippine Coast," Phys.org, 29 February 2016, https://phys.org/news/2016-02-turtles-vulnerable-life-philippine-coast.html, accessed 29 March 2019.

[77] MTT_IOSEA, 12, para. 31.

[78] Romeo Trono, quoted in Morella.

reef-biota.[79] Coral reefs were formally defined at the international level in 2002 by the Contracting Parties to the 1971 Convention on Wetlands of International Importance especially as Waterfowl Habitats ("Ramsar Convention") as "massive carbonate structures built by biological activity of stony corals (true corals) and the associated complex of marine organisms that make up a coral reef ecosystem."[80] They are found throughout the oceans on mud-free coastlines, between the latitudes 30°N and 30°S. Their estimated total area in the world is 617,000 square kilometers, forming about 15% of marine shallow shelves.[81] The Ramsar definition identified three general types of coral reefs. Fringing reefs are found close against the coast. Barrier reefs are separated from land by a lagoon. Atolls are ring-shaped coral reefs that enclose a lagoon and have been formed where an island (often volcanic in origin) has progressively sunk below the sea surface. Coral reefs that developed on continental coastlines are often complex and contain features that are difficult to categorize.[82] An alternative classification, based on the relationship to landmass and the depth of the surrounding water, identifies five types of coral reefs. A fringing reef is a linear reef with a reef flat, some tens of meters across, growing along shelving coastlines and across embayments.[83] A bank barrier reef is another form of linear coastal reef, a little further from shore than a fringing reef, and sometimes coalescing with it. A barrier reef is also a linear structure, but fronting deep oceanic waters and broader—usually hundreds of meters across—and separated

[79] Terry Done, "Coral Reef, Definition," in David Hopley (ed.), *Encyclopedia of Modern Coral Reefs. Structure, Form and Process* (Dordrecht: Springer, 2011), 261.

[80] Ramsar Convention, 8th Meeting of the Conference of the Contracting Parties to the Convention on Wetlands (Ramsar, Iran, 1971), "Wetlands: Water, Life, and Culture," Valencia, Spain, 18–26 November 2002, Resolution VIII.11, Guidance for Identifying and Designating Peatlands, Wet Grasslands, Mangroves and Coral Reefs as Wetlands of International Importance (2002), 10, para. 52, http://archive.ramsar.org/cda/en/ramsar-documents-resol-resolution-viii-11/main/ramsar/1-31-107%5E21521_4000_0__, accessed 31 March 2019. *Ramsar Convention. Convention on Wetlands of International Importance Especially as Waterfowl Habitat,* signed at Ramsar, Iran, on 2 February 1971, as amended by the Protocol of 3 December 1982 and the Amendments of 28 May 1987, https://www.ramsar.org/sites/default/files/documents/library/current_convention_text_e.pdf, accessed 24 March 2019.

[81] Ramsar Convention, Resolution VIII-11 (2002), 11, para. 53.

[82] Ibid.

[83] An embayment is a shape resembling a bay. *Collins English Dictionary* (Glasgow: HarperCollins Publishers, 2019), https://www.collinsdictionary.com/dictionary/english/embayment, accessed 10 May 2019.

from the coastline by navigable waters. An atoll is a broadly circular reef enclosing a wide lagoon. A bank reef or a platform reef is a substantial reef fitting none of the above categories, occurring in oceanic and coastal settings.[84]

Reefs are both geological structures and living ecosystems.[85] The geological structure is formed from the skeletons of dead corals, which are made of calcium carbonate and rest in place after their death. The bodies of scleractinian (reef-building) corals are sac-like polyps that usually grow together to form colonies. The calcium carbonate derived from coral skeletons is cemented into wave-resistant structures by coralline algae, which flourish in shallow, turbulent, well-lit environments. The accumulation of calcium carbonate can form limestone reefs if corals grow in sufficient quantity, and the rate of skeleton production (calcification) and algal cementation exceed that of erosion. Coral reefs are the biggest structures ever made by living organisms.[86] The rates of growth of coral colonies and of reefs are different. The coral *Porites* grow at a rate of about 1 centimeter per year, while the staghorn and plate-forming *Acropora* can grow up to 30 centimeters per year. In the Pacific, corals grow only 3 to 165 millimeters per year.[87] Reefs grow at the rate of

---

[84] Done, 261.

[85] Unless otherwise indicated, the following paragraphs are summarized from John E. N. Veron, "Corals, Biology, Skeletal Disposition and Reef-Building," in David Hopley (ed.), *Encyclopedia of Modern Coral Reefs. Structure, Form and Process* (Dordrecht: Springer, 2011), 275–81. Reef formation is also described in Done, 261–66; *South China Sea Arbitration*, Independent Expert Report. Assessment of the Potential Environmental Consequences of Construction Activities on Seven Reefs in the Spratly Islands in the South China Sea, by Sebastian C. A. Ferse, Peter Mumby, and Selina Ward, 26 April 2016, 11–12 ("Ferse Report"), https://pcacases.com/web/sendAttach/1809, accessed 24 March 2019; John W. McManus, "Offshore Coral Reef Damage, Overfishing and Paths to Peace in the South China Sea," *The International Journal of Marine and Coastal Law* 32 (2017): 199–237; also accessible as *South China Sea Arbitration, The Philippines Annexes Cited During the Merits Hearing (Annexes 820-59)* (30 November 2015), Annex 820, "Offshore Coral Reef Damage, Overfishing and Paths to Peace in the South China Sea," by John W. McManus, 578–608, https://pcacases.com/web/view/7, accessed 26 March 2019.

[86] Done, 261.

[87] *MP*, Annex 240, "Eastern South China Sea Environmental Disturbances and Irresponsible Fishing Practices and Their Effects on Coral Reefs and Fisheries," by Kent E. Carpenter, vol. VII (30 March 2014), 410 ("First Carpenter Report"), https://files.pca-cpa.org/pcadocs/The%20Philippines%27%20Memorial%20-%20Volume%20VII%20%28Annexes%20222-255%29.pdf, accessed 31 March 2019.

0.6 meters per century. In optimal conditions, the rate might be three times faster.[88]

Coral reefs are among the most biodiverse ecosystems on earth. They cover only 0.1–0.5% of the ocean floor, yet they harbor nearly one-third of marine fishes and represent 10% of fish consumption.[89] Coral reefs form an ecosystem, in the sense that they are groups of species living cooperatively for joint survival. Corals produce the building blocks. Coralline algae cement the blocks together. Herbivores prevent macroalgae from taking over the reef. Photosynthetic algae (zooxanthellae) live inside the coral tissues and produce food.[90] The coral provides the algae with a protected environment and compounds that they need for photosynthesis. In return, the algae produce oxygen and help the coral to remove wastes. Most importantly, zooxanthellae supply the coral with glucose, glycerol, and amino acids, which are the products of photosynthesis.[91] The coral uses these products to make proteins, fats, and carbohydrates, as well as to produce calcium carbonate. The relationship between the algae and coral polyp facilitates a tight recycling of nutrients in nutrient-poor tropical waters. In fact, as much as 90% of the organic material photosynthetically produced by the zooxanthellae is transferred to the host coral tissue, this is the driving force behind the growth and productivity of coral reefs.[92] Light availability is critical for reef-building because of the dependence of corals on algal symbiosis.[93]

---

[88] Veron, 280.

[89] Fredrik Moberg and Carl Folke, "Ecological Goods and Services of Coral Reef Ecosystems," *Ecological Economic* 29 (1999): 216. The Philippines relied heavily on this document for its arguments. See *South China Sea Arbitration*, Hearing on the Merits and Remaining Issues of Jurisdiction and Admissibility, Transcript, Day 3 (26 November 2015), 15–17, https://pcacases.com/web/sendAttach/1401, accessed 31 March 2019. It is reproduced in *Supplemental Documents of the Philippines* (19 November 2015), Annex 262(bis), vol. III, 3–21, https://files.pca-cpa.org/pcadocs/The%20Philippines%27%20 Supplemental%20Documents%20-%20Volume%20III%20%28Annexes%20710-756%29.pdf, accessed 31 March 2019.

[90] Veron, 277.

[91] As corals do not make their own food, they are in fact animals. "Corals Are Animals," *Florida Keys National Marine Sanctuary*, 2011, http://floridakeys.noaa.gov/corals/coral-animals.html, accessed 31 March 2019. See also First Carpenter Report, 264.

[92] "Corals. Zooxanthellae… What's That," *NOAA [National Oceanic and Atmospheric Administration] Ocean Service Education*, 2008, https://oceanservice.noaa.gov/education/tutorial_corals/coral02_zooxanthellae.html, accessed 31 March 2019.

[93] Veron, 281.

Giant clams on coral reefs strain small plants and animals from surrounding water and turn them into biomass that contributes to food on the reef.[94]

As geological structures and as living ecosystems, healthy coral reefs are essential to the marine environment.

2. Coral Reefs and the Marine Environment

The ecological roles of coral reefs are multiple and diverse.[95] They provide physical structure services, biotic services (services that are essential for the functioning of ecosystems) within ecosystems and between ecosystems, biogeochemical services, and information services.

The physical structure services involve the protection of shoreline from currents, waves and storms, which would otherwise erode land[96]; the building up of land, on which many tropical countries in the Indian and Pacific Oceans are situated; the promotion of growth of mangroves and seagrass beds; and the generation of fine coral sand, which supplies shores with white sand.

Among the biotic services that they perform within ecosystems, coral reefs maintain habitats by functioning as important spawning, breeding, and feeding areas for a multitude of organisms. Coral reefs are species-rich habitats: up to 60,000 reef-living animals and plants have been described. As habitats, coral reefs are important for maintaining biodiversity and niche diversification. Between ecosystems, coral reefs provide support through mobile links, as some coral reef organisms migrate back and forth between adjacent ecosystems. For example, herbivorous fishes and sea urchins from coral reefs that move to seagrass beds for grazing may serve as food for predators and human beings; may indirectly control the productivity of benthic algae and seagrass assemblages by reducing self-shading, weeding out large algae with low productivity, and enhancing nutrient exchange with the water; and may influence the nutrient cycles of seagrass beds and mangroves through their excretion and defecation. Reef organisms either drift in their pelagic juvenile stages into adjacent systems, where they are sources of food for commercially important fishes; or settle and mature until they are harvested.

---

[94]First Carpenter Report, 398.

[95]Unless otherwise indicated, the following paragraphs are summarized from Moberg and Folke.

[96]Surface-breaking coral reefs reduce wave energy by 97%. McManus, 215.

Coral reefs provide four crucial biogeochemical services. First, they function as nitrogen fixers in nutrient-poor environments, as the microbial and cyanobacterial associations in reef-bottom biotopes as well as cyanobacteria in the water column assimilate atmospheric nitrogen.[97] Nitrogen is a major component of chlorophyll, the most important pigment needed for photosynthesis, and of amino acids, the key building blocks of proteins. Conversion of atmospheric nitrogen into compounds by natural processes is one of the ways through which plants acquire reduced or "combined" forms of nitrogen that they can utilize.[98] Second, coral reefs appear to act as sinks for carbon dioxide over geological (though not human) time-scales.[99] Third, biochemical processes on coral reefs play a significant role in the world's calcium balance. Reefs precipitate nearly half of the 1.2×10^{13} mol of calcium delivered to the ocean every year.[100] Fourth, coral reefs can transform, detoxify and sequester wastes, such as petroleum products and persistent pollutants that are released by human beings.

Coral reefs perform information services because they are highly sensitive systems used for monitoring of recent changes in the marine environment and as pollution records. For instance, the skeletons of reef-building corals act as long-term records of levels of metals in seawater, sea surface temperatures in the tropics, and variations in salinity. The layers of skeleton deposited by long-lined massive corals can be counted and used to trace periods of monsoon flooding.

The economic importance of coral reefs is commensurate with their ecological roles.

---

[97] A biotope is a small habitat in a large community. *Larousse Dictionary of Science and Technology*, 110.

[98] Stephen C. Wagner, "Biological Nitrogen Fixation," *Nature Education Knowledge* 3 (2011): 15, http://www.nature.com/scitable/knowledge/library/biological-nitrogen-fixation-23570419, accessed 31 March 2019.

[99] Carbon sinks are reservoirs that retain carbon and keep it from entering Earth's atmosphere. Noelle Eckley Selin, "Carbon Sequestration," *Encyclopædia Britannica*, 2018, https://www.britannica.com/technology/carbon-sequestration, accessed 31 March 2019.

[100] Mole, also spelled mol, in chemistry, a standard scientific unit for measuring large quantities of very small entities such as atoms, molecules, or other specified particles. The mole designates an extremely large number of units, 6.02214179×1023, which is the number of atoms determined experimentally to be found in 12 grams of carbon-12. The Editors of Encyclopædia Britannica, "Mole: Chemistry," *Encyclopædia Britannica*, 2018, https://www.britannica.com/science/mole-chemistry, accessed 31 March 2019.

B. The Economic Importance of Coral Reefs

In the Philippines, coral reefs are sources of livelihood for impoverished coastal communities. Overexploitation and the use of destructive fishing methods are the major threats to coral reefs.

1. Coral Reefs and Coastal Communities in the Philippines

The overall economic benefits of coral reefs are widely recognized. In the Philippines, they provide livelihoods to fishing communities that seem to combine all of the problems of developing countries: overpopulation, low technological levels, and extreme poverty.[101]

Coral reefs are not only sources of living resources and mining materials; they are also major tourist attractions.[102] Coral reefs generate a variety of seafood products such as fish, mussels, crustaceans, sea cucumbers, and seaweed. Mother-of-pearl shells (*TrochIus spp.*) and giant clams (*Tridacna spp.*) found on coral reefs are collected not only as food but also to sell as jewelry and as souvenirs. Live fish and corals are collected for the aquarium trade. Potentially useful substances with anticancer, AIDS-inhibiting, antimicrobial, anti-inflammatory, and anticoagulating properties are found among the seaweeds, sponges, mollusks, corals, and anemones of coral reefs. Many seaweed species are collected from reefs, to be used in the production of agar and carrageenan and as manure. Coral skeletons are promising in bone graft operations.

Coral blocks, sand, rubble, and sand are used for building. Lime is used as a pH regulator in agriculture. In some regions of the world, coral debris is collected and crushed to be used as fertilizer. Coral reefs are also used to produce lime and cement. Mineral oil and gas, resulting from the conversion of biomass of reef organisms, are thought to exist in large quantities below living reefs. Last but not least, coral reefs support tourism and recreation services.

---

[101] Porfirio M. Aliño et al., "Philippine Coral Reef Fisheries: Diversity in Adversity," in Geronimo Silvestre et al. (eds.), *In Turbulent Seas: The Status of Philippine Marine Fisheries* (Cebu City: Department of Agriculture—Bureau of Fisheries and Aquatic Resources, 2004), 67, http://oneocean.org/download/db_files/fshprofl.pdf, accessed 25 March 2019.

[102] Unless otherwise indicated, the following paragraphs are summarized from Moberg and Folke.

The Philippines has the third largest reef cover in the world (25,000 square kilometers), equivalent to 9% of the total reef area.[103] The country has 13 sites of nearshore fringing reefs. The more developed and extensive reef areas, which are also the least explored, are in the Kalayaan Island Group (KIG), comprising maritime features claimed by the Philippines in the Spratly Islands. These offshore reefs, which comprise approximately one-fourth of the total area of coral reefs in the Philippines, are characterized by high species richness.[104] To date, 468 species of scleractinian (reef-building) corals, 1755 species of reef-associated fishes, 648 species of mollusk, 19 species of seagrass, and 820 species of algae have been counted.[105] Coral reefs comprise 12% of the Philippine Exclusive Economic Zone ("EEZ") of 2.2 million square kilometers (including areas in the South China Sea).[106]

It was estimated in 2005 that the total quantifiable economic value of coral reefs to the country was PhP2901 million (US$53 million). Fisheries, tourism, and research values accounted for about 27% of the total net economic value. One square kilometer of coral reef can generate US$11,366 direct and indirect values. Philippine coral reefs had an estimated value of PhP1064 million per year (US$19.3 million) and the value of coral reefs for the South China Sea biogeographic region was estimated at PhP52.7 million per year.[107] Calculated over 20 years, with a discount rate of 10%, the net present value of benefits of Philippine coral reefs in the South China Sea basin was estimated at PhP24,700 million

---

[103] Republic of the Philippines, *The Fifth National Report to the Convention on Biological Diversity 2014* [Quezon City: Department of Environment and Natural Resources—Biodiversity Management Bureau, 2014], 22 (*"Philippines Fifth National Report to the CBD"*), https://www.cbd.int/reports/search/, accessed 25 March 2019; Wilfredo Y. Licuanan et al., "The Philippines," in Charles Sheppard (ed.), *World Seas: An Environmental Evaluation*, vol. II. *The Indian Ocean to the* Pacific (London: Academic Press, 2019), 517 ("Licuanan et al., 'The Philippines'").

[104] Porfirio Aliño, "National Report on Coral Reefs in the Coastal Waters of the South China Sea: Philippines," in United Nations Environment Programme (UNEP), *Reversing Environmental Degradation Trends in the South China Sea and the Gulf of Thailand. National Reports on Coastal Reefs in the Coastal Waters of the South China Sea* (Bangkok: UNEP, 2007), 55–57, http://unepscs.org/remository/startdown/1961.html, accessed 31 March 2019; Licuanan et al., "The Philippines," 517.

[105] *Philippines Fourth National Report to the CBD*, 45.

[106] Licuanan et al., "The Philippines," 522.

[107] Aliño, 62.

(US$449 million), which was equivalent to approximately PhP5.3 million per square kilometer net present value, or PhP266,112 per square kilometer per year on an annualized basis. These figures were based on an estimated Philippine coral reef area within the South China Sea basin of 4640.94 square kilometers.

The extent of coral cover in the Philippines means that coral reefs sustain the livelihood of many coastal communities. Coral reefs account for between 8 and 20% of total fishery production and up to 70% for some islands. One million small fishers and 62% of coastal populations depend on coral reefs.[108] Municipal coral reef fisheries represent 10–20% of total municipal fisheries production.[109] Approximately 55% of fish consumed by Filipinos depends on coral reefs.[110]

Reef fisheries are dominated by small- to medium-bodied families, such as damselfishes, fusiliers, parrotfishes, and wrasses. Many large-bodied species important to commerce and food security are rarely found in many reefs in the Philippines.[111] Historically most of the live reef fish trade was for ornamental fish (for aquariums) and marine organisms exported to the United States and Europe. This activity targeted 400 species of coral reef fish and some invertebrates, mostly for sale to hobbyists in the United States and Europe.[112] Coral fishes are still collected for this trade in San Salvador Island, Laguna province, on the main island of Luzon.[113] In 2005, the latest year for which data is available, the Philippines exported 5.7 million aquarium fishes to the United States, making the Philippines the world's biggest supplier of aquarium fishes.[114]

---

[108] *Philippines Fourth National Report to the CBD*, 46. See also Herman S. J. Cesar, "Coral Reefs: Their Threats, Functions and Economic Value," in Herman S. J. Cesar (ed.), *Collected Essays on the Economics of Coral Reefs* (Kalmar: Linnaeus University, 2002), 14–39, http://www.reefbase.org/resource_center/publication/pub_12370.aspx, accessed 31 March 2019.

[109] Aliño, 55.

[110] Henrylito D. Tacio, "Assessing the Status of Endangered Coral Reefs," *EdgeDavao*, 18 October 2018, http://edgedavao.net/science/2018/10/18/assessing-the-status-of-endangered-coral-reefs/, accessed 15 March 2019.

[111] *Philippines Fifth National Report to the CBD*, 22.

[112] Licuanan et al., "The Philippines," 526.

[113] Aliño, 60.

[114] Kristine L. Alave, "PH Center of 'Illegal' Live Reef, Aquarium Fish Trade," *Philippine Daily Inquirer*, 12 July 2012, https://globalnation.inquirer.net/43917/ph-center-of-%E2%80%98illegal%E2%80%99-live-reef-aquarium-fish-trade, accessed 31 March 2019.

In the early 1990s, trade shifted to live reef food fish, primarily consisting of groupers (*Serranidae*), wrasses (*Labridae*) and snappers (*Lutjanidae*).[115] The live reef food fish trade, in which reef fish are caught and kept alive until the moment they are cooked in a restaurant or hotel, is a lucrative activity for Filipino fishermen.[116] They can receive between PhP2400 ($48.42) and PhP2800 per kilogram ($56.49) for "good size" or preferred "plate-sized" fish (around 500 grams) to around PhP1200 ($24.21) to PhP1400 ($28.24) per fish for larger specimens. In contrast, for yellowfin tuna they receive only PhP220 ($4.44) per kilogram.[117] Exports of live reef food fish increased rapidly during the 1990s peaked in 1995 and have been declining since that time. Between 1991 and 1998, export value was estimated to be $7.2 million per year on average, representing 1.74% of the country's total fishing export value. From 2001 to 2003, export earnings increased to a yearly average value of $11.1 million. Almost all of the live reef food fish exported from the Philippines goes to Hong Kong and Taiwan, with small quantities going to Malaysia and Singapore. The distinctive reddish color of the leopard coral grouper is especially appealing to Chinese consumers, particularly around the time of Chinese New Year.[118]

There are many threats to coral reefs, but the Philippine government has singled out overexploitation and the use of destructive fishing methods as the most important.

2. Overexploitation and Dynamite and Cyanide Fishing as Threats to Coral Reefs

A study of species richness in 51 reef sites in 17 municipalities and four biogeographic regions (including the South China Sea) in the Philippines showed that most reefs were still species-rich, but exhibited

---

[115] Robert S. Pomeroy et al., "Evaluation of Policy Options for the Live Reef Food Fish Trade in the Province of Palawan, Western Philippines," *Marine Policy* 32 (2008): 56.

[116] WWF-Philippines, *Palawan's Live Reef Food Fish Trade* (Quezon City: WWF-Philippines, 2019), https://wwf.org.ph/what-we-do/food/lrfft/, accessed 31 March 2019.

[117] Yvonne Sadovy de Mitcheson et al., *The Trade in Live Reef Food Fish—Going, Going, Gone*, vol. I, *Main Report* (Hong Kong: ADM Capital Foundation and The University of Hong Kong, 2017), 110, https://www.chooserighttoday.org/wp-content/uploads/2018/02/LRFFTVol1_Final_12022018.pdf, accessed 31 March 2019.

[118] Pomeroy et al., 56.

signs of depletion, such as low abundance and biomass per unit area.[119] Overall, the percentage of Philippine reefs that were in excellent condition declined from 5% in the 1970s to 0.2% in the second decade of the twenty-first century.[120]

In its *Fifth National Report* on the implementation of the CBD (2014), the Philippines identified overfishing, destructive fishing practices, sedimentation and pollution due to inappropriate land use, irresponsible mining, and deforestation as the major threats to coral reefs. Among these, the first two were unambiguously declared to be the largest threats.[121]

There is no doubt that overfishing in certain coral reef sites is a problem in and of itself for the coastal communities concerned. To give an idea of the magnitude of the problem, in 2007, around Lingayen Gulf in the north, there were said to be 23,000 fishermen, for an average of 7 fishermen per square meter of coastline or 23 fishermen per square kilometer of fishing ground. Overfishing remained a problem in Pangasinan, Zambales, and Palawan provinces on the main island of Luzon.[122] Depletion of stocks is fast becoming a reality looming for the species that are the object of the live reef food fish trade. In the province of Palawan in the western part of the country, the earlier fishing grounds have already been "extirpated," as one recent scientific study described the situation. Fishing activity is thus moving to the south of the province and to other provinces.[123] In parts of the country, stocks of the Leopard Coralgrouper are depleted to the extent that the trade is no longer considered economically viable. In response, the trade is coming to rely on the capture of juveniles that are then grown out in captivity until they reach market size.[124] Over the long-term, the effects of overfishing are

---

[119] *Philippines Fourth National Report to the CBD*, 47; *Philippines Fifth National Report to the CBD*, 22.

[120] Licuanan et al., "The Philippines," 517; see also Ardea M. Licuanan et al., "Initial Findings of the Nationwide Assessment of Philippine Coral Reefs," *Philippine Journal of Science* 146 (2017): 177–85 ("Licuanan et al., 'Initial Findings'"), http://philjournalsci.dost.gov.ph/images/pdf/pjs_pdf/vol146no2/initial_findings_of_the_nationwide_assessment_of_philippine_coral_reefs.pdf, accessed 1 March 2019.

[121] *Philippines Fifth National Report to the CBD*, 22.

[122] Ibid., 60.

[123] Sadovy de Mitcheson et al., 26.

[124] Ibid., 47.

well-nigh catastrophic: Between the 1950s and 2014, at least 42 coral reef fish species disappeared from local catches.[125]

To make matters worse, overfishing is causing the degradation of the habitats of the fish, the coral reefs. Overfishing can alter trophic interactions, causing indirect environmental effects.[126] For example, if top predators are largely removed, the populations of their prey may increase and thus destabilize the ecosystem. Overexploitation can also reduce genetic variation in a population, making it harder for species to adapt to environmental change.[127] Once the larger species are depleted, fishermen turn to smaller, often herbivorous fish. The reefs may end up being left with numbers of mostly small fish. Such reefs are then prone to algal overgrowth, without herbivores to graze the algae as they grow. They become less resilient to stressors, more vulnerable to disease, and slower to recover from other human impacts.[128]

In the competition for diminishing stocks, fishermen resort to destructive and illegal fishing practices. The use of cyanide to stun and capture the fish was reported to be widespread in Palawan Province between 1994 and 2001.[129] In the first decade of the 21st century, sporadic cyanide fishing continued to occur in the country.[130] Cyanide in solution form is sprayed directly into the habitat of certain high-value live fish such as the Humphead Wrasse and Leopard Coralgrouper, which

---

[125] Licuanan et al., "The Philippines," 525.

[126] A trophic level refers to a broad class of organisms within an ecosystem characterized by mode of supply. The first trophic level comprises the green plants, the second is the herbivores, and the third is the carnivores that eat the herbivores. *Larousse Dictionary of Science and Technology*, 1134.

[127] Jon C. Day, "Conservation and Marine Protected Areas," in David Hopley (ed.), *Encyclopedia of Modern Coral Reefs. Structure, Form and Process* (Dordrecht: Springer, 2011), 231.

[128] Lauretta Burke et al., *Reefs at Risk Revisited* (Washington, DC: World Resources Institute, 2011), 26, https://wriorg.s3.amazonaws.com/s3fs-public/pdf/reefs_at_risk_revisited_hi-res.pdf?_ga=2.243625165.954175598.1554184854-1064952833.1552820007, accessed 17 March 2019.

[129] Samuel Mamauag, "The Live Reef Food Fish Trade in the Philippines," in Geronimo Silvestre et al. (eds.), *In Turbulent Seas: The Status of Philippine Marine Fisheries* (Cebu City: Department of Agriculture—Bureau of Fisheries and Aquatic Resources, 2004), 57, http://oneocean.org/download/db_files/fshprofl.pdf, accessed 25 March 2019; Michael Fabinyi and Dante Dalabajan, "Policy and Practice in the Live Reef Fish for Food Trade: A Case Study from Palawan, Philippines," *Marine Policy* 35 (2011): 373.

[130] Aliño, 60.

may be difficult to catch efficiently using other gears (hook and line).[131] Cyanide may destroy coral colonies due to the expulsion of symbiotic zooxanthellae.[132]

Blast fishing, which aims to stun or kill fish, destroys coral and flattens the reef structure.[133] In 2007, the use of dynamite and fine-meshed nets was reported as rampant in Pangasinan province on the main island of Luzon. Blast fishing was also carried out in San Salvador Island, Laguna Province in Luzon. Fortunately, blast fishing stopped in Nagabugan Bay, Davila, Ilocos Norte province, following the implementation of alternative livelihood projects in aquaculture, livestock raising, and farming.[134]

The creation of Marine Protected Areas (MPAs) and fishery management in some coral reef sites led to declining use of destructive fishing methods by 2012. To complement the MPAs, the Philippines is implementing a series of initiatives at the national level and participating in international programs in order to promote the recovery of coral reefs. At the national level, a Marine Sanctuary Strategy adopted in 2004 envisioned that 10% of coral reefs would be in no-take MPAs by 2020. By 2014, only 5% of municipal waters were in MPAs, but the very small area of coral reefs that is in excellent condition may be found in four MPAs, two national MPAs (Tubbataha Reef Marine Park in Palawan Province and Apo Island in Negros Oriental Province), and two local MPAs (Apo Reef in Puerto Galera, Mindoro Province, and Verde Island Passage off Batangas Province).[135] In the Tubbataha Reefs Natural Park, declared a UNESCO World Heritage Site in December 1993, the reef ecosystems support over 350 species of coral and almost 500 species of fish.[136] A Sustainable Coral Reef Ecosystem Management Program (SCREMP), set to run from 2012 to 2020, aims to increase by 5% the hard coral cover and fish density as well as to increase by 10% the income of fishing communities and project sites.[137] The Philippines has also been experimenting since 2012 with the rehabilitation of degraded coral reefs through

---

[131] Sadovy de Mitcheson et al., 51.

[132] Mamauag, 57; Fabinyi and Dalabajan, 373.

[133] Day, 231.

[134] Aliño, 60.

[135] Tacio; *Philippines Fifth National Report to the CBD*, 22.

[136] *The Tubbataha Reefs Natural Park* (Paris: UNESCO World Heritage Centre, 2019), https://whc.unesco.org/en/list/653, accessed 31 March 2019.

[137] *Philippines Fifth National Report to the CBD*, 22.

asexual reproduction of corals, involving the collection of dislodged live coral fragments and attaching them to coral nursery units.[138] The experiment has been criticized by coral reef experts as ineffective, unsustainable, and likely infeasible due to the scale needed.[139]

At the international level, the Philippines cooperates with five other states (Indonesia, Malaysia, Papua New Guinea, Solomon Islands, and Timor-Leste) in the Coral Triangle Initiative on Coral Reefs Fisheries and Food Security (CTI-CFF), launched in 2009. According to a recent evaluation, the Philippines has made considerable progress in achieving three of the five goals of the initiative, consisting of the establishment of seascapes, the promotion of an Ecosystems Approach to Fisheries Management (EAFM), and the improvement of MPA management. Progress has been much more limited in the two other objectives, the initiation of local climate adaptation planning and improvement in the status of threatened species.[140]

Whatever may be our assessment of the Philippine government's efforts to preserve marine biodiversity in the country, these can in no conceivable way be assisted by China's toleration of harvesting of sea turtles and corals and the use of destructive fishing methods by its nationals, nor from the Chinese government's construction activities on Spratly Island reefs. Indeed, one could point to instances in which the activities of Chinese fishermen undermined the government's conservation efforts. In February 2002, 95 Chinese fishermen were apprehended in the act of fishing in the vicinity of Tubbataha Reef National Marine Park and World Heritage Site in Palawan Province. The fishermen threw overboard an unknown number of sea turtles as they were about to be apprehended. On board the Chinese vessel, the Philippine authorities

---

[138] Melvin B. Carlos et al., *Restoring the Country's Coral Reefs Through Filipinnovation*, DOST-PCAARRD [Department of Science and Technology—Philippine Council for Aquatic and Natural Resources Research and Development], 12 February 2016, http://www.pcaarrd.dost.gov.ph/home/portal/index.php/quick-information-dispatch/2664-restoring-the-country-s-coral-reefs-through-filipinnovation, accessed 1 April 2019.

[139] Licuanan et al., "The Philippines," 518. See also Jonathan L. Mayuga, "Mission: Reduce Threats to Coral Reefs," *Business Mirror*, 25 February 2018 ("Mayuga, 'Mission'"), https://businessmirror.com.ph/2018/02/25/mission-reduce-threats-to-coral-reefs/, accessed 15 March 2019; Pia Ranada, "Is Coral Transplantation the Way to Save PH Corals?" *Rappler*, 16 April 2015, https://www.rappler.com/science-nature/environment/90155-coral-transplantation-philippines, accessed 15 March 2019.

[140] Licuanan et al., "The Philippines," 530.

found endangered species of marine flora and fauna, such as fan coral and giant clams (*Tridacna* sp.), 30 sacks of dried meat of giant clams, 4 sacks of dried sea cucumber and 13 pieces of live wrasses. They also found sodium cyanide, blasting caps, and detonating cords, which in the Philippine authorities' view constituted prima facie evidence of the use of explosives. Three months later, the local authorities apprehended 20 Chinese fishermen fishing within the municipal waters of the town of Cagayancillo in Palawan Province. In the course of the operation, the fishing vessel ran aground, damaging some 40 square meters of coral reef areas. The authorities confiscated various species of live wrasses and grouper estimated to be more than 600 kilograms, together with more than 30 bottles and 3 containers of solution and 34 tablets of sodium cyanide.[141]

In spite of their gravity, it was not incidents of this type that were encompassed by the Philippine Submission No. 11 to the Arbitral Tribunal. Given that they took place within Philippine territorial waters, there could be no doubt that the Philippines had jurisdiction over them. It is probably for this reason that no mention was made in the Philippine Submission of the incidents referred to above. The incidents that the Philippines brought to the Tribunal's attention occurred mostly in the waters around Scarborough Shoal, which is the object of a sovereignty dispute between the Philippines and China. As we have seen, the exercise of Philippine jurisdiction with respect to such incidents inevitably triggered Chinese reaffirmation of its sovereignty over Scarborough Shoal and created friction in bilateral relations. The diplomatic communications show that the Philippines consistently cited the CBD, in addition to or in lieu of the Convention, as a legal basis for its protests against China's toleration of fishing of endangered species by Chinese nationals and China's construction activities in the Spratly Islands. The question that must now be examined is whether the CBD offers an appropriate framework for the resolution of disputes between States concerning the conservation of biodiversity, particularly endangered species and fragile ecosystems

---

[141] Adelina B. Benavente-Villena and Michael N. Pido, "Poaching in Philippine Marine Waters: Intrusion of Chinese Fishing Vessels in Palawan Waters," in Geronimo Silvestre et al. (eds.), *In Turbulent Seas: The Status of Philippine Marine Fisheries* (Cebu City: Department of Agriculture—Bureau of Fisheries and Aquatic Resources, 2004), 267, http://oneocean.org/download/db_files/fshprofl.pdf, accessed 25 March 2019.

ANNEX 2.1: IUCN CRITERIA FOR ENDANGERED SPECIES

A taxon is Endangered when the best available evidence indicates that it meets any of the following criteria (A–E), and it is therefore considered to be facing a very high risk of extinction in the wild:

A. Reduction in population size based on any of the following:
  1. An observed, estimated, inferred, or suspected population size reduction of ≥70% over the last 10 years or three generations, whichever is the longer, where the causes of the reduction are clearly reversible AND understood AND ceased, based on (and specifying) any of the following:
  (a) direct observation;
  (b) an index of abundance appropriate to the taxon;
  (c) a decline in area of occupancy, extent of occurrence, and/or quality of habitat;
  (d) actual or potential levels of exploitation;
  (e) the effects of introduced taxa, hybridization, pathogens, pollutants, competitors, or parasites.
  2. An observed, estimated, inferred, or suspected population size reduction of ≥50% over the last 10 years or three generations, whichever is the longer, where the reduction or its causes may not have ceased OR may not be understood OR may not be reversible, based on (and specifying) any of (a) to (e) under A1.
  3. A population size reduction of ³50%, projected or suspected to be met within the next 10 years or three generations, whichever is the longer (up to a maximum of 100 years), based on (and specifying) any of (b) to (e) under A1.
  4. An observed, estimated, inferred, projected, or suspected population size reduction of ≥50% over any 10-year or three-generation period, whichever is longer (up to a maximum of 100 years in the future), where the time period must include both the past and the future, and where the reduction or its causes may not have ceased OR may not be understood OR may not be reversible, based on (and specifying) any of (a) to (e) under A1.
B. Geographic range in the form of either B1 (extent of occurrence) OR B2 (area of occupancy) OR both:
  1. Extent of occurrence estimated to be less than 5000 km², and estimates indicating at least two of a–c:

    a. Severely fragmented or known to exist at no more than five locations.

    b. Continuing decline, observed, inferred, or projected, in any of the following:

        (i) extent of occurrence;

        (ii) area of occupancy;

        (iii) area, extent and/or quality of habitat;

        (iv) number of locations or subpopulations;

        (v) number of mature individuals.

    c. Extreme fluctuations in any of the following:

        (i) extent of occurrence;

        (ii) area of occupancy;

        (iii) number of locations or subpopulations;

        (iv) number of mature individuals.

2. Area of occupancy estimated to be less than 500 km^2, and estimates indicating at least two of a–c:

    a. Severely fragmented or known to exist at no more than five locations.

    b. Continuing decline, observed, inferred, or projected, in any of the following:

        (i) extent of occurrence;

        (ii) area of occupancy;

        (iii) area, extent, and/or quality of habitat;

        (iv) number of locations or subpopulations;

        (v) number of mature individuals.

    c. Extreme fluctuations in any of the following:

        (i) extent of occurrence;

        (ii) area of occupancy;

        (iii) number of locations or subpopulations;

        (iv) number of mature individuals.

C. Population size estimated to number fewer than 2500 mature individuals and either:

1. An estimated continuing decline of at least 20% within five years or two generations, whichever is longer (up to a maximum of 100 years in the future), OR

2. A continuing decline, observed, projected, or inferred, in numbers of mature individuals AND at least one of the following (a–b):

    a. Population structure in the form of one of the following:
      (i) no subpopulation estimated to contain more than 250 mature individuals, OR
      (ii) at least 95% of mature individuals in one subpopulation.
    b. Extreme fluctuations in number of mature individuals.
D. Population size estimated to number fewer than 250 mature individuals.
E. Quantitative analysis showing the probability of extinction in the wild is at least 20% within 20 years or five generations, whichever is the longer (up to a maximum of 100 years).

Source IUCN Species Survival Commission (SSC), *IUCN Red List Categories and Criteria, Version 3.1* (Gland: IUCN, 2001), 18–20, https://portals.iucn.org/library/sites/library/files/documents/RL-2001-001.pdf, accessed 27 March 2019.

Annex 2.2: IUCN Criteria for Critically Endangered Species

A taxon is Critically Endangered when the best available evidence indicates that it meets any of the following criteria (A–E), and it is therefore considered to be facing an extremely high risk of extinction in the wild:

A. Reduction in population size based on any of the following:
    1. An observed, estimated, inferred, or suspected population size reduction of 90% over the last 10 years or three generations, whichever is the longer, where the causes of the reduction are clearly reversible AND understood AND ceased, based on (and specifying) any of the following:
      (a) direct observation;
      (b) an index of abundance appropriate to the taxon;
      (c) a decline in area of occupancy, extent of occurrence, and/or quality of habitat;
      (d) actual or potential levels of exploitation;
      (e) the effects of introduced taxa, hybridization, pathogens, pollutants, competitors, or parasites.
    2. An observed, estimated, inferred, or suspected population size reduction of $\geq 80\%$ over the last 10 years or three generations,

whichever is the longer, where the reduction or its causes may not have ceased OR may not be understood OR may not be reversible, based on (and specifying) any of (a) to (e) under A1.

3. A population size reduction of [3]80%, projected or suspected to be met within the next 10 years or three generations, whichever is the longer (up to a maximum of 100 years), based on (and specifying) any of (b) to (e) under A1.

4. An observed, estimated, inferred, projected, or suspected population size reduction of $\geq$80% over any 10-year or three-generation period, whichever is longer (up to a maximum of 100 years in the future), where the time period must include both the past and the future, and where the reduction or its causes may not have ceased OR may not be understood OR may not be reversible, based on (and specifying) any of (a) to (e) under A1.

B. Geographic range in the form of either B1 (extent of occurrence) OR B2 (area of occupancy) OR both:

1. Extent of occurrence estimated to be less than 100 km^2, and estimates indicating at least two of a–c:
   a. Severely fragmented or known to exist at only a single location.
   b. Continuing decline, observed, inferred or projected, in any of the following:
      (i)   extent of occurrence;
      (ii) area of occupancy;
      (iii) area, extent, and/or quality of habitat;
      (iv) number of locations or subpopulations;
      (v)   number of mature individuals.
   c. Extreme fluctuations in any of the following:
      (i)   extent of occurrence;
      (ii) area of occupancy;
      (iii) number of locations or subpopulations;
      (iv) number of mature individuals.

2. Area of occupancy estimated to be less than 10 km^2, and estimates indicating at least two of a–c:
   a. Severely fragmented or known to exist at only a single location.
   b. Continuing decline, observed, inferred, or projected, in any of the following:
      (i)   extent of occurrence;

(ii) area of occupancy;
(iii) area, extent, and/or quality of habitat;
(iv) number of locations or subpopulations;
(v) number of mature individuals.
   c. Extreme fluctuations in any of the following:
     (i) extent of occurrence;
     (ii) area of occupancy;
     (iii) number of locations or subpopulations;
     (iv) number of mature individuals.
C. Population size estimated to number fewer than 250 mature individuals and either:
   1. An estimated continuing decline of at least 25% within three years or one generation, whichever is longer (up to a maximum of 100 years in the future), OR
   2. A continuing decline, observed, projected, or inferred, in numbers of mature individuals AND at least one of the following (a–b):
     a. Population structure in the form of one of the following
       (i) no subpopulation estimated to contain more than 50 mature individuals, OR
       (ii) at least 90% of mature individuals in one subpopulation.
     b. Extreme fluctuations in number of mature individuals.
D. Population size estimated to number fewer than 50 mature individuals.
E. Quantitative analysis showing the probability of extinction in the wild is at least 50% within 10 years or three generations, whichever is the longer (up to a maximum of 100 years).

Source IUCN Species Survival Commission (SSC), *IUCN Red List Categories and Criteria, Version 3.1* (Gland: IUCN, 2001), 16–18, https://portals.iucn.org/library/sites/library/files/documents/RL-2001-001.pdf, accessed 27 March 2019.

REFERENCES

Alava, Monnyeen Nida R., and Jose Alfred B. Cantos. "Marine Protected Species in the Philippines." *In Turbulent Seas: The Status of Philippine Marine Fisheries*, 109–17. Eds. Geronimo Silvestre et al. Cebu City: Department of Agriculture—Bureau of Fisheries and Aquatic Resources, 2004, http://oneocean.org/download/db_files/fshprofl.pdf, accessed 25 March 2019.

Alave, Kristine L. "PH Center of 'Illegal' Live Reef, Aquarium Fish Trade." *Philippine Daily Inquirer*, 12 July 2012, https://globalnation.inquirer. net/43917/ph-center-of-%E2%80%98illegal%E2%80%99-live-reef-aquarium-fish-trade, accessed 31 March 2019.

Aliño, Porfirio M. "National Report on Coral Reefs in the Coastal Waters of the South China Sea: Philippines." *Reversing Environmental Degradation Trends in the South China Sea and the Gulf of Thailand. National Reports on Coastal Reefs in the Coastal Waters of the South China Sea*, 55–68. Bangkok: United Nations Environment Programme, 2007, http://unepscs.org/remository/startdown/1961.html, accessed 31 March 2019.

Aliño, Porfirio M. et al. "Philippine Coral Reef Fisheries: Diversity in Adversity." *In Turbulent Seas: The Status of Philippine Marine Fisheries*, 65–69. Eds. Geronimo Silvestre et al. Cebu City: Department of Agriculture—Bureau of Fisheries and Aquatic Resources, 2004, http://oneocean.org/download/db_files/fshprofl.pdf, accessed 25 March 2019.

Benavente-Villena, Adelina B., and Michael N. Pido. "Poaching in Philippine Marine Waters: Intrusion of Chinese Fishing Vessels in Palawan Waters." *In Turbulent Seas: The Status of Philippine Marine Fisheries*, 265–68. Eds. Geronimo Silvestre et al. Cebu City: Department of Agriculture—Bureau of Fisheries and Aquatic Resources, 2004, http://oneocean.org/download/db_files/fshprofl.pdf, accessed 25 March 2019.

Burke, Lauretta et al. *Reefs at Risk Revisited*. Washington, DC: World Resources Institute, 2011, https://wriorg.s3.amazonaws.com/s3fs-public/pdf/reefs_at_risk_revisited_hi-res.pdf?_ga=2.243625165.954175598.1554184854-1064952833.1552820007, accessed 17 March 2019.

Carlos, Melvin B. et al. *Restoring the Country's Coral Reefs Through Filipinnovation*, DOST-PCAARRD [Department of Science and Technology—Philippine Council for Aquatic and Natural Resources Research and Development], 12 February 2016, http://www.pcaarrd.dost.gov. ph/home/portal/index.php/quick-information-dispatch/2664-restoring-the-country-s-coral-reefs-through-filipinnovation, accessed 1 April 2019.

Cesar, Herman S. J. "Coral Reefs: Their Threats, Functions and Economic Value." *Collected Essays on the Economics of Coral Reefs*, 14–39. Ed. Herman S. J. Cesar. Kalmar: Linnaeus University, 2002, http://www.reefbase.org/resource_center/publication/pub_12370.aspx, accessed 31 March 2019.

Collins English Dictionary. Glasgow: HarperCollins Publishers, 2019, https://www.collinsdictionary.com/dictionary/english/embayment, accessed 10 May 2019.

Convention on Biological Diversity (CBD), signed at Rio de Janeiro on 5 June 1992, entered into force on 29 December 1993, https://www.cbd.int/doc/legal/cbd-en.pdf, accessed 26 March 2019.

Convention on International Trade in Endangered Species of Wild Fauna and Flora, signed at Washington, DC, on 3 March 1973, amended at Bonn, on

22 June 1979, amended at Gaborone, on 30 April 1983, https://www.cites. org/eng/disc/text.php, accessed 26 March 2019.

"Corals Are Animals." *Florida Keys National Marine Sanctuary*, 2011, http://floridakeys.noaa.gov/corals/coralanimals.html, accessed 31 March 2019.

"Corals. Zooxanthellae... What's That." *NOAA [National Oceanic and Atmospheric Administration] Ocean Service Education*, 2008, https://oceanservice.noaa.gov/education/tutorial_corals/coral02_zooxanthellae.html, accessed 31 March 2019.

Cruz, Renato D. "Marine Turtle Distribution in the Philippines." *Proceedings of the Western Pacific Sea Turtle Cooperative Research and Management Workshop. February 5–8, 2002, Honolulu, Hawaii, USA*, 57–65. Ed. Irene Kinan. Honolulu, HI: Western Pacific Regional Fishery Management Council, 2002, http://www.wpcouncil.org/library/docs/protectedspecies/1st%20WS%20Proceedings%20in%20full.pdf, accessed 25 March 2019.

Day, Jon C. "Conservation and Marine Protection Areas." *Encyclopedia of Modern Coral Reefs. Structure, Form and Process*, 230–36. Ed. David Hopley. Dordrecht: Springer, 2011.

Domantay, Jose S. "The Turtle Fisheries of the Turtle Islands." *Bulletin of the Fisheries Society of the Philippines* 3–4 (1953): 3–27.

Done, Terry. "Coral Reef, Definition." *Encyclopedia of Modern Coral Reefs. Structure, Form and Process*, 594–601. Ed. David Hopley. Dordrecht: Springer, 2011.

Eckert, Karen L. *The Biology and Population Status of Marine Turtles in the North Pacific Ocean*. NOAA Technical Memorandum NMFS, NOAA-TM-NMFS-SWFSC-186. Washington, DC: U.S. Department of Commerce, National Oceanic and Atmospheric Administration, National Marine Fisheries Service, Southwest Fisheries Science Center, 1993, https://repository.library.noaa. gov/view/noaa/6133/noaa_6133_DS1.pdf?, accessed 19 March 2019.

Editors of Encyclopædia Britannica. "Mole: Chemistry." *Encyclopædia Britannica*, 2018, https://www.britannica.com/science/mole-chemistry, accessed 31 March 2019.

Emirates Wildlife Society (EWS)-World Wildlife Fund (WWF). *Marine Turtle Conservation Project. Final Scientific Report. Arabian Region*. Abu Dhabi: EWS-WWF, 2015, https://www.cbd.int/doc/meetings/mar/ebsaws-2015-02/other/ebsaws-2015-02-ews-wwf-submission1-en.pdf, accessed 25 March 2019.

Esteban, Rolando C. "The Turtle Island Heritage Protected Area: The Possibilities and Limits of Transborder Conservation." *Ostrom Workshop Indiana University Bloomington*, July 2008, http://dlc.dlib.indiana.edu/dlc/bitstream/handle/10535/993/Esteban_223701.pdf?sequence=1, accessed 31 March 2019.

Fabinyi, Michael, and Dante Dalabajan. "Policy and Practice in the Live Reef Fish for Food Trade: A Case Study from Palawan, Philippines." *Marine Policy* 35 (2011): 371–78.

Hitchcock, Frank Harris. *Trade of the Philippine Islands.* Washington, DC: Government Printing Office, 1898, https://ia800200.us.archive.org/28/items/tradephilippine00hitcgoog/tradephilippine00hitcgoog.pdf, accessed 20 March 2019.

Hui, Dafeng. "Food Web: Concept and Applications." *Nature Education Knowledge* 3 (2012): 6, https://www.nature.com/scitable/knowledge/library/food-web-concept-and-applications-84077181, accessed 28 March 2019.

IUCN Species Survival Commission (SSC). *IUCN Red List Categories and Criteria, Version 3.1.* Gland: IUCN, 2001, https://portals.iucn.org/library/sites/library/files/documents/RL-2001-001.pdf, accessed 27 March 2019.

Lacuna-Richman, Celeste. "Subsistence Strategies of an Indigenous Minority in the Philippines: Nonwood Forest Product Use by the Tagbanua of Narra, Palawan." *Economic Botany* 58 (2004): 266–85.

Larousse Dictionary of Science and Technology. Edinburgh: Larousse plc, 1995.

Lejano, Raul P., and Helen Ingram. "Place-Based Conservation: Lessons from the Turtle Islands." *Environment: Science and Policy for Sustainable Development* 49 (2007): 18–27.

Licuanan, Ardea M. et al. "Initial Findings of the Nationwide Assessment of Philippine Coral Reefs." *Philippine Journal of Science* 146 (2017): 177–85, http://philjournalsci.dost.gov.ph/images/pdf/pjs_pdf/vol146no2/initial_findings_of_the_nationwide_assessment_of_philippine_coral_reefs.pdf, accessed 1 March 2019.

Licuanan, Wilfredo Y. et al. "The Philippines." *World Seas: An Environmental Evaluation.* Vol. II. *The Indian Ocean to the Pacific,* 515–38. Ed. Charles Sheppard. London: Academic Press, 2019.

Maliwanag, Dennis Atienza. "DOH Team Accused of Turtle Eggs Poaching in Tawi-Tawi." *Philippine Daily Inquirer,* 2 March 2017, https://newsinfo.inquirer.net/876941/doh-team-accused-of-turtle-eggs-poaching?utm_expid=.XqNwTug2W6nwDVUSgFJXed.1, accessed 18 March 2019.

—————. "Turtle Islands: Hundreds of Nautical Miles away from Justice." *Philippine Daily Inquirer,* 22 August 2010, https://www.pressreader.com/philippines/philippine-daily-inquirer-1109/20100822/281775625465558, accessed 29 March 2019.

Mamauag, Samuel. "The Live Reef Food Fish Trade in the Philippines." *In Turbulent Seas: The Status of Philippine Marine Fisheries,* 53–59. Eds. Geronimo Silvestre et al. Cebu City: Department of Agriculture—Bureau of Fisheries and Aquatic Resources, 2004, http://oneocean.org/download/db_files/fshprofl.pdf, accessed 25 March 2019.

Marine Wildlife Watch of the Philippines. *Philippine Aquatic Wildlife Rescue and Response Manual to Marine Turtle Incidents.* Taguig City: Marine Wildlife Watch of the Philippines, 2014, http://www.mwwphilippines.org/downloads/rm-marineturtles.pdf, accessed 19 March 2019.

Matz, Nele. *Wege zur Koordinierung völkerrechtlicher Verträge. Völkervertragsrechtliche und institutionelle Ansätze* [Means to Co-ordinate International Treaties. International Treaty Law Approaches And Institutional Approaches]. Heidelberg: Springer Verlag, 2005.

Mayuga, Jonathan L. "From Predators to Protectors of 'Pawikan [Sea Turtles]'." 26 November 2018, https://businessmirror.com.ph/2018/11/26/from-predators-to-protectors-of-pawikan/, accessed 18 March 2019.

_____. "Mission: Reduce Threats to Coral Reefs." *Business Mirror*, 25 February 2018, https://businessmirror.com.ph/2018/02/25/mission-reduce-threats-to-coral-reefs/, accessed 15 March 2019.

_____. "Saving the Endangered 'Pawikan [Sea Turtles]'." *Business Mirror*, 17 January 2016, http://www.businessmirror.com.ph/saving-the-endangered-pawikan-2/, accessed 10 March 2019.

McManus, John W. "Offshore Coral Reef Damage, Overfishing and Paths to Peace in the South China Sea." *The International Journal of Marine and Coastal Law* 32 (2017): 199–237.

Miclat, Evangeline, and Enrique Nunez. "The Philippine-Sabah Turtle Island Heritage Protected Area (TIHPA)." *Marine Transboundary Conservation and Protected Areas*, 32–47. Ed. Peter Mackelworth. New York: Routledge, 2016.

Milliken, Tom, and Hideomi Tokunaga. *The Japanese Sea Turtle Trade 1970–1986.* Tokyo: TRAFFIC (Japan) [Wildlife Trade Monitoring Network], 1987, https://www.traffic.org/site/assets/files/9659/japanese-sea-turtle-trade-1970-1986.pdf, accessed 19 March 2019.

Moberg, Fredrik, and Carl Folke. "Ecological Goods and Services of Coral Reef Ecosystems." *Ecological Economics* 29 (1999): 215–33.

Morella, Cecil. "'Turtles' Vulnerable Start to Life on Philippine Coast." *Phys. org*, 29 February 2016, https://phys.org/news/2016-02-turtles-vulnerable-life-philippine-coast.html, accessed 29 March 2019.

Ocean Ambassadors. Track a Turtle. *The Philippine Turtle Islands* [2001], http://www.oneocean.org/ambassadors/track_a_turtle/tihpa/pti.html, accessed 31 March 2019.

Pejabat Perikanan Negeri and Siow Kuan Tow. "Observations on the Exploitation of Turtles in the Philippines." *Marine Turtle Newsletter*, 3 (1977), http://www.seaturtle.org/mtn/PDF/MTN3.pdf, accessed 19 March 2019.

Pomeroy Robert S. et al. "Evaluation of Policy Options for the Live Reef Food Fish Trade in the Province of Palawan, Western Philippines." *Marine Policy* 32 (2008): 55–65.

Poonian, Christopher et al. "Diversity, Habitat Distribution, and Indigenous Hunting of Marine Turtles in the Calamian Islands, Palawan, Republic of the Philippines." *Journal of Asia-Pacific Biodiversity* 9 (2016): 69–73.

Public Radio International. "High Demand for Sea Turtles in China Sends Poachers Toward Philippines." *PRI's* [Public Radio International] *The World*, 9 February 2012, https://www.pri.org/stories/2012-02-09/high-demand-sea-turtles-china-sends-poachers-toward-philippines, accessed 29 March 2019.

Ramsar Convention. *Ramsar Convention. Convention on Wetlands of International Importance Especially as Waterfowl Habitat*, signed at Ramsar, Iran, on 2 February 1971, as amended by the Protocol of 3 December 1982 and the Amendments of 28 May 1987, https://www.ramsar.org/sites/default/files/documents/library/current_convention_text_e.pdf, accessed 24 March 2019.

_____. 8th Meeting of the Conference of the Contracting Parties to the Convention on Wetlands (Ramsar, Iran, 1971). "Wetlands: Water, Life, and Culture," Valencia, Spain, 18–26 November 2002. Resolution VIII.11. Guidance for Identifying and Designating Peatlands, Wet Grasslands, Mangroves and Coral Reefs as Wetlands of International Importance (2002), http://archive.ramsar.org/cda/en/ramsar-documents-resol-resolution-viii-11/main/ramsar/1-31-107%5E21521_4000_0__, accessed 31 March 2019.

Ranada, Pia. "Is Coral Transplantation the Way to Save PH Corals?" *Rappler*, 16 April 2015, https://www.rappler.com/science-nature/environment/90155-coral-transplantation-philippines, accessed 15 March 2019.

Republic of the Philippines. *Assessing Progress Towards the 2010 Biodiversity Targets. The Fourth National Report to the Convention on Biological Diversity* [Quezon City: Department of Environment and Natural Resources—Protected Areas and Wildlife Bureau, 2009], https://www.cbd.int/reports/, accessed 25 March 2019.

_____. *The First Philippine National Report to the Convention on Biological Diversity May 1998.* [Quezon City: Department of Environment and Natural Resources—Protected Areas and Wildlife Bureau, 1998], https://www.cbd.int/reports/, accessed 25 March 2019.

_____. *The Fifth National Report to the Convention on Biological Diversity 2014* [Quezon City: Department of Environment and Natural Resources—Biodiversity Management Bureau, 2014], https://www.cbd.int/reports/search/, accessed 25 March 2019.

_____. *The Second Philippine National Report to the Convention on Biological Diversity* [Quezon City: Department of Environment and Natural Resources—Protected Areas and Wildlife Bureau, 2002], https://www.cbd.int/reports/search/, accessed 3 May 2019.

_____. Department of Environment and Natural Resources, Biodiversity Management Bureau. *Philippines-Malaysia Partnership in Marine Turtle Conservation*, 2016, http://www.bmb.gov.ph/index.php/

mainmenu-news-events/mainmenu-news/396-philippine-malaysia-partner-ship-in-marine-turtle-conservation, accessed 29 March 2010.

_____. National Statistics Office. *2010 Census of Population and Housing. Report No. 2A—Demographic and Housing Characteristics (Non-Sample Variables). Palawan.* Manila: National Statistics Office, 2013, https://psa.gov.ph/sites/default/files/PALAWAN_FINAL%20PDF.pdf, accessed 18 March 2019.

_____. *Statistical Tables on Sample Variables from the Results of 2010 Census of Population and Housing—Tawi-Tawi.* Manila: National Statistics Office, 31 July 2014, https://psa.gov.ph/content/statistical-tables-sample-variables-results-2010-census-population-and-housing-tawi-tawi, accessed 18 March 2019.

Reyes, Jonas. "215 'Pawikan [Sea Turtle]' Hatchlings Released in Subic." *Manila Bulletin*, 21 December 2017, https://news.mb.com.ph/2017/12/20/215-pawikan-hatchlings-released-in-subic/, accessed 29 March 2019.

Sadovy de Mitcheson, Yvonne et al. *The Trade in Live Reef Food Fish—Going, Going, Gone.* Vol. I. *Main Report.* Hong Kong: ADM Capital Foundation and The University of Hong Kong, 2017, https://www.chooserighttoday.org/wp-content/uploads/2018/02/LRFFTVol1_Final_12022018.pdf, accessed 31 March 2019.

Selin, Noelle Eckley. "Carbon Sequestration." *Encyclopædia Britannica*, 2018, https://www.britannica.com/technology/carbon-sequestration, accessed 31 March 2019.

South China Sea Arbitration. Hearing on the Merits and Remaining Issues of Jurisdiction and Admissibility. Transcript, Day 3 (26 November 2015), https://pcacases.com/web/sendAttach/1401, accessed 31 March 2019.

_____. Transcript, Day 4 (30 November 2015), https://pcacases.com/web/sendAttach/1550, accessed 27 March 2019.

_____. Independent Expert Report. Assessment of the Potential Environmental Consequence of Construction Activities on Seven Reefs in the Spratly Islands in the South China Sea, by Sebastian C. A. Ferse, Peter Mumby, and Selina Ward, 26 April 2016, https://pcacases.com/web/sendAttach/1809, accessed 24 March 2019.

_____. *Memorial of the Philippines.* Vol. I (30 March 2014), https://files.pca-cpa.org/pcadocs/Memorial%20of%20the%20Philippines%20Volume%20I.pdf, accessed 27 March 2019.

_____. Annex 19. Memorandum from Erlinda F. Basilio, Acting Assistant Secretary, Office of Asian and Pacific Affairs, Department of Foreign Affairs, Republic of the Philippines, to the Secretary of Foreign Affairs of the Republic of the Philippines (29 March 1995). Vol. III, 213–17, https://files.pca-cpa.org/pcadocs/The%20Philippines%27%20Memorial%20-%20Volume%20III%20%28Annexes%201-60%29.pdf, accessed 26 March 2019.

_____. Annex 20. Memorandum from Lauro L. Baja, Jr., Assistant Secretary, Office of Asian and Pacific Affairs, Department of Foreign Affairs, Republic of the Philippines, to the Secretary of Foreign Affairs of the Republic of the Philippines (7 April 1995). Vol. III, 219–24, https://files.pca-cpa.org/pcadocs/The%20Philippines%27%20Memorial%20-%20Volume%20III%20%28Annexes%201-60%29.pdf, accessed 26 March 2019.

_____. Annex 45. Memorandum from Willy C. Gaa, Assistant Secretary of Foreign Affairs, Republic of the Philippines to Secretary of Foreign Affairs, Republic of the Philippines (14 February 2001). Vol. III, 363–67. https://files.pca-cpa.org/pcadocs/The%20Philippines%27%20Memorial%20-%20Volume%20III%20%28Annexes%201-60%29.pdf, accessed 26 March 2019.

_____. Annex 48. Memorandum from Josue L. Villa, Embassy of the Republic of the Philippines in Beijing, to the Secretary of Foreign Affairs of the Republic of the Philippines (21 May 2001). Vol. III, 379–94, https://files.pca-cpa.org/pcadocs/The%20Philippines%27%20Memorial%20-%20Volume%20III%20%28Annexes%201-60%29.pdf, accessed 26 March 2019.

_____. Annex 181. Government of the Republic of the Philippines, Transcript of Proceedings Republic of the Philippines-People's Republic of China Bilateral Talks (10 August 1995). Vol. VI, 193–215, https://files.pca-cpa.org/pcadocs/The%20Philippines%27%20Memorial%20-%20Volume%20VI%20%28Annexes%20158-221%29.pdf, accessed 26 March 2019.

_____. Annex 186. Note Verbale from the Department of Foreign Affairs of the Republic of the Philippines to the Embassy of the People's Republic of China in Manila No. 2000100 (14 January 2000). Vol. VI, 259–61, https://files.pca-cpa.org/pcadocs/The%20Philippines%27%20Memorial%20-%20Volume%20VI%20%28Annexes%20158-221%29.pdf, accessed 26 March 2019.

_____. Annex 205. Note Verbale from the Department of Foreign Affairs of the Republic of Philippines to the Embassy of the People's Republic of China in Manila, No. 12-0894 (11 April 2012). Vol. VI, 377–80, https://files.pca-cpa.org/pcadocs/The%20Philippines%27%20Memorial%20-%20Volume%20VI%20%28Annexes%20158-221%29.pdf, accessed 26 March 2019.

_____. Annex 210. Note Verbale from the Department of Foreign Affairs of the Philippines to the Embassies of ASEAN Member States in Manila, No. 12-1372 (21 May 2012). Vol. VI, 397–400, https://files.pca-cpa.org/pcadocs/The%20Philippines%27%20Memorial%20-%20Volume%20VI%20%28Annexes%20158-221%29.pdf, accessed 26 March 2019.

_____. Annex 240. "Eastern South China Sea Environmental Disturbances and Irresponsible Fishing Practices and Their Effects on Coral Reefs and Fisheries," by Kent E. Carpenter. Vol. VII, 389–437, https://files.pca-cpa.org/pcadocs/The%20Philippines%27%20Memorial%20-%20Volume%20VII%20%28Annexes%20222-255%29.pdf, accessed 26 March 2019.

_____. *The Philippines Annexes Cited During the Merits Hearing (Annexes 820-59)* (30 November 2015). Annex 820. "Offshore Coral Reef Damage, Overfishing and Paths to Peace in the South China Sea," by John W. McManus, 578–608, https://pcacases.com/web/view/7, accessed 26 March 2019.

_____. *Supplemental Documents of the Philippines* (19 November 2015). Annex 262(bis). "Ecological Goods and Services of Coral Reef Ecosystems," by Fredrik Moberg and Carl Folke, *Ecological Economics*, 29 (1999). Vol. III, 3–21, https://files.pca-cpa.org/pcadocs/The%20Philippines%27%20Supplemental%20Documents%20-%20Volume%20III%20%28Annexes%20710-756%29.pdf, accessed 31 March 2019.

Tacio, Henrylito D. "Assessing the Status of Endangered Coral Reefs." *EdgeDavao*, 18 October 2018, http://edgedavao.net/science/2018/10/18/assessing-the-status-of-endangered-coral-reefs/, accessed 15 March 2019.

Trono, Romeo B. "Management and Conservation Program of a Protected Wildlife. Species." *Philippine Journal of Public Administration* 26 (1990): 388–413, http://lynchlibrary.pssc.org.ph:8081/bitstream/handle/0/4078/10_Management%20and%20Conservation%20Program.pdf?sequence=1, accessed 25 March 2019.

The Tubbataha Reefs Natural Park. Paris: UNESCO World Heritage Centre, 2019, https://whc.unesco.org/en/list/653, accessed 31 March 2019.

United Nations Convention on the Law of the Sea, concluded at Montego Bay on 10 December, entered into force on 16 November 1994, http://www.un.org/Depts/los/convention_agreements/texts/unclos/closindx.htm, accessed 21 March 2019.

Veron, John E. N. "Corals, Biology, Skeletal Disposition and Reef-Building." *Encyclopedia of Modern Coral Reefs. Structure, Form and Process*, 275–81. Ed. David Hopley. Dordrecht: Springer, 2011.

Virata, John B. "Philippine Police Arrest Sea Turtle Poachers from China." *Reptiles Magazine*, 11 May 2014, http://www.reptilesmagazine.com/Turtles-Tortoises/Information-News/Philippine-Police-Arrest-Chinese-Sea-Turtle-Poachers/, accessed 1 April 2019.

Wagner, Stephen C. "Biological Nitrogen Fixation." *Nature Education Knowledge* 3 (2011): 15, http://www.nature.com/scitable/knowledge/library/biological-nitrogen-fixation-23570419, accessed 31 March 2019.

Wilson, E. D. et al. *Why Healthy Oceans Need Sea Turtles: The Importance of Sea Turtles to Marine Ecosystems*. Washington, DC: Oceana, 2010, http://www.oceana.org/sites/default/files/reports/Why_Healthy_Oceans_Need_Sea_Turtles.pdf, accessed 25 March 2019.

Wolfrum, Rüdiger, and Nele Matz. "The Interplay of the United Nations Convention on the Law of the Sea and the Convention on Biological Diversity." *Max Planck Yearbook of United Nations Law* 4 (2000): 445–80.

World Wildlife Fund (WWF)-Philippines. *Turtle Islands: Resources and Livelihoods Under Threat. A Case Study on the Philippines.* Quezon City: WWF-Philippines, 2005, https://wwf.org.ph/wp-content/uploads/2017/11/Turtle-Islands-2005.pdf, accessed 28 March 2019.

WWF-Philippines. *Palawan's Live Reef Food Fish Trade.* Quezon City: WWF-Philippines, 2019, https://wwf.org.ph/what-we-do/food/lrfft/, accessed 31 March 2019.

"Zambales Turtle Conservation Program 2004–2005 Technical Turtle Seminar." *Environmental Protection of Asia,* 2006, http://www.environmentalprotectionofasia.com/ztcp/reports/2004_2005.htm, accessed 13 March 2017.

CHAPTER 3

The International Frameworks for the Conservation of Biodiversity and of the Marine Environment

In the second half of the 1990s, Philippine diplomatic communications with China following the arrest of Chinese fishermen nearly always stressed the violation of Philippine laws together with the violation of international agreements.[1] The logical consequence of the stress on violation of Philippine laws was the detention and trial of Chinese fishermen.[2] The disadvantage of this approach was that it inevitably triggered

---

[1] *South China Sea Arbitration, Memorial of the Philippines* (30 March 2014), Annex 19, Memorandum from Erlinda F. Basilio, Acting Assistant Secretary, Office of Asian and Pacific Affairs, Department of Foreign Affairs, Republic of the Philippines, to the Secretary of Foreign Affairs of the Republic of the Philippines (29 March 1995), vol. III, 216 ("*MP*"), https://files.pca-cpa.org/pcadocs/The%20Philippines%27%20Memorial%20-%20 Volume%20III%20%28Annexes%201-60%29.pdf, accessed 26 March 2019; *MP*, Annex 20, Memorandum from Lauro L. Baja, Jr., Assistant Secretary, Office of Asian and Pacific Affairs, Department of Foreign Affairs, Republic of the Philippines, to the Secretary of Foreign Affairs of the Republic of the Philippines (7 April 1995), vol. III, 223, https:// files.pca-cpa.org/pcadocs/The%20Philippines%27%20Memorial%20-%20Volume%20 III%20%28Annexes%201-60%29.pdf, accessed 26 March 2019.

[2] *MP*, Annex 28, Memorandum from Fact Finding Committee, National Police Commission, Republic of the Philippines, to Chairman and Members of the Regional Committee on Illegal Entrants for Region 1, Republic of the Philippines (28 January 1998), vol. III, 271–76, https://files.pca-cpa.org/pcadocs/The%20Philippines%27%20 Memorial%20-%20Volume%20III%20%28Annexes%201-60%29.pdf, accessed 26 March 2019; and *MP*, Annex 30, People of the Philippines v. Shin Ye Fen et al., Criminal Case No. RTC 2357-I, Decision, Regional Trial Court, Third Judicial Region, Branch 69, Iba, Zambales, Philippines (29 April 1998), vol. III, 283–86, https://files.pca-cpa.org/

© The Author(s) 2020
A. C. Robles Jr., *Endangered Species and Fragile Ecosystems in the South China Sea*,
https://doi.org/10.1007/978-981-13-9813-1_3

assertions of Chinese sovereignty over the maritime areas in which the Chinese fishermen were apprehended.[3] Invoking international agreements enabled the Philippines to circumvent Chinese objections based on assertions of sovereignty and to focus on the damage to the marine environment independently of the question of sovereignty, without the Philippines abandoning its own claims of sovereignty.[4] Even at this stage,

pcadocs/The%20Philippines%27%20Memorial%20-%20Volume%20III%20%28Annexes%20 1-60%29.pdf, accessed 26 March 2019; *MP*, Annex 31, People of the Philippines v. Wuh Tsu Kai, et al., Criminal Case No. RTC 2362-I, Decision, Regional Trial Court, Third Judicial Region, Branch 69, Iba, Zambales, Philippines (29 April 1998), vol. III, 287– 90, https://files.pca-cpa.org/pcadocs/The%20Philippines%27%20Memorial%20-%20 Volume%20III%20%28Annexes%201-60%29.pdf, accessed 26 March 2019; *MP*, Annex 32, People of the Philippines v. Zin Dao Guo, et al., Criminal Case No. RTC 2363-I, Decision, Regional Trial Court, Third Judicial Region, Branch 69, Iba, Zambales, Philippines (29 April 1998), vol. III, 291–94, https://files.pca-cpa.org/pcadocs/The%20 Philippines%27%20Memorial%20-%20Volume%20III%20%28Annexes%201-60%29.pdf, accessed 26 March 2019.

[3] *MP*, Annex 29, Memorandum from Assistant Secretary of the Department of Foreign Affairs, Republic of the Philippines to the Secretary of Foreign Affairs of the Republic of the Philippines (23 March 1998), vol. III, 277–82, https://files.pca-cpa.org/pcadocs/ The%20Philippines%27%20Memorial%20-%20Volume%20III%20%28Annexes%201-60%29. pdf, accessed 26 March 2019.

[4] *MP*, Annex 37, Memorandum from the Embassy of the Republic of the Philippines in Beijing to the Secretary of Foreign Affairs of the Republic of the Philippines, No. ZPE-85-98-S (4 December 1998), vol. III, 315–19, https://files.pca-pa.org/pcadocs/The%20 Philippines%27%20Memorial%20-%20Volume%20III%20%28Annexes%201-60%29. pdf, accessed 26 March 2019; *MP*, Annex 43, *Memorandum* from the Embassy of the Republic of the Philippines in Beijing to the Secretary of Foreign Affairs of the Republic of the Philippines, No. ZPE-06-2001-S (13 February 2001), vol. III, 351–55, https:// files.pca-pa.org/pcadocs/The%20Philippines%27%20Memorial%20-%20Volume%20III%20 %28Annexes%201-60%29.pdf, accessed 26 March 2019; *MP*, Annex 45, Memorandum from Willy C. Gaa, Assistant Secretary of Foreign Affairs, Republic of the Philippines to Secretary of Foreign Affairs, Republic of the Philippines (14 February 2001), vol. III, 365–66, https://files.pca-pa.org/pcadocs/The%20Philippines%27%20Memorial%20 -%20Volume%20III%20%28Annexes%201-60%29.pdf, accessed 26 March 2019; *MP*, Annex 47, Memorandum from the Embassy of the Republic of the Philippines in Beijing to the Secretary of Foreign Affairs of the Republic of the Philippines, No. ZPE-09-2001-S (17 March 2001) vol. III, 378, https://files.pca-pa.org/pcadocs/The%20 Philippines%27%20Memorial%20-%20Volume%20III%20%28Annexes%201-60%29.pdf, accessed 26 March 2019; *MP*, Annex 48, Memorandum from Josue L. Villa, Embassy of the Republic of the Philippines in Beijing, to the Secretary of Foreign Affairs of the Republic of the Philippines (21 May 2001), vol. III, 390, https://files.pca-pa.org/pca-docs/The%20Philippines%27%20Memorial%20-%20Volume%20III%20%28Annexes%20

the Philippine arguments suggested its belief that the obligation to protect and preserve the marine environment was independent of the question of sovereignty. Other than (or in addition to) the United Nations Convention on the Law of the Sea ("the Convention"),[5] the Philippines identified three treaties to which both the Philippines and China were Parties. The Philippines referred to the 1973 Convention on International Trade in Endangered Species of Wildlife and Fauna ("CITES") in 1999, 2000, 2012, and 2015,[6] the 1971 Convention on Wetlands of International Importance especially as Waterfowl Habitat ("Ramsar Convention") in 2001,[7] and the Convention on Biological Diversity ("CBD") in 1999

1-60%29.pdf, accessed 26 March 2019; *MP*, Annex 51, Memorandum from Josue L. Villa, Embassy of the Republic of the Philippines in Beijing, to the Secretary of Foreign Affairs of the Republic of the Philippines (19 August 2002), vol. III, 408, https://files. pca-pa.org/pcadocs/The%20Philippines%27%20Memorial%20-%20Volume%20III%20 %28Annexes%201-60%29.pdf, accessed 26 March 2019; and *MP*, Annex 58 Memorandum from the Secretary of Foreign Affairs of the Republic of the Philippines to the President of the Republic of the Philippines (11 January 2006), vol. III, 483, https://files.pca-pa.org/ pcadocs/The%20Philippines%27%20Memorial%20-%20Volume%20III%20%28Annexes%20 1-60%29.pdf, accessed 26 March 2019.

[5] *United Nations Convention on the Law of the Sea*, concluded at Montego Bay on 10 December, entered into force on 16 November 1994, http://www.un.org/Depts/los/ convention_agreements/texts/unclos/closindx.htm, accessed 21 March 2019.

[6] *MP*, Annex 45, vol. III, 365–66; *MP*, Annex 205, Memorandum from Erlinda F. Basilio, Assistant Secretary of Foreign Affairs to the Secretary of Foreign Affairs, 11 January 2006, annexed to Memorandum from the Secretary of Foreign Affairs to the President of the Republic of the Philippines (11 January 2006), vol. VI, 377–81, https://files. pca-cpa.org/pcadocs/The%20Philippines%27%20Memorial%20-%20Volume%20VI%20 %28Annexes%20158-221%29.pdf, accessed 26 March 2019; *Supplemental Documents of the Philippines* (19 November 2015), Annex 608, Department of Foreign Affairs—Republic of the Philippines. Statement on China's Reclamation Activities and their Impact on the Region's Marine Environment (13 April 2015), vol. I, 7–9 ("*SDP*"), https://files.pca-cpa.org/pcadocs/The%20Philippines%27%20Supplemental%20Documents%20-%20 Volume%20I%20%28Annexes%20607-667%29.pdf, accessed 2 April 2019. *Convention on International Trade in Endangered Species of Wild Fauna and Flora*, signed at Washington, DC, on 3 March 1973, amended at Bonn, on 22 June 1979, amended at Gaborone, on 30 April 1983, https://www.cites.org/eng/disc/text.php, accessed 26 March 2019.

[7] *MP*, Annex 186, Note Verbale from the Department of Foreign Affairs of the Republic of the Philippines to the Embassy of the People's Republic of China in Manila No. 2000100 (14 January 2000), vol. VI, 259–61. https://files.pca-cpa.org/pcadocs/The%20 Philippines%27%20Memorial%20-%20Volume%20VI%20%28Annexes%20158-221%29.pdf,

and 2015.[8] The search for a legal basis is understandable, for as the Arbitral Tribunal pointed out, it is "not uncommon in international law for treaties to often mirror each other in content and consequently for more than one treaty to have a bearing on a dispute."[9] The identification of legal basis is also crucial, in part because each treaty, in the absence of a unified and hierarchical system of courts at the international level, establishes its own mechanism for the settlement of disputes relating to compliance with the treaty by States Parties to the treaty, and these mechanisms differ widely in their nature.

One can imagine the reason why the Philippines invoked the Ramsar Convention only once as the legal basis for its protests against Chinese fishermen's activities. The Ramsar Convention aims "to stem the progressive encroachment on and loss of wetlands now and in the future."[10] Wetlands are defined as "areas of marsh, fen, peatland or water, whether natural or artificial, permanent or temporary, with water that is static or flowing, fresh, brackish or salt, including areas of marine water the depth of which at low tide does not exceed six metres."[11] This extremely broad definition includes coral reefs.[12] A Contracting Party's main obligations are (1) to designate within its territory wetlands for inclusion in a List of Wetlands of International Importance, one of which must be designated at the time it becomes a Party to the Ramsar Convention and (2) "to formulate and implement their planning so as to promote the conservation of the wetlands included in the List, and as far as possible the wise

accessed 26 March 2019. *Convention on Wetlands of International Importance Especially as Waterfowl Habitat*, signed at Ramsar, Iran, on 2 February 1971, as amended by the Protocol of 3 December 1982 and the Amendments of 28 May 1987, Preamble, https://www.ramsar.org/sites/default/files/documents/library/current_convention_text_e.pdf, accessed 24 March 2019.

[8] *MP*, Annex 186, vol. VI, 259–61; *SDP*, Annex 608, vol. I, 7–9. *Convention on Biological Diversity* (CBD), signed at Rio de Janeiro on 5 June 1992, entered into force on 29 December 1993, https://www.cbd.int/doc/legal/cbd-en.pdf, accessed 26 March 2019.

[9] *South China Sea Arbitration*, Award on Jurisdiction and Admissibility (29 October 2015), 69–70, paras. 176–178 ("Award on Jurisdiction"), https://pcacases.com/web/sendAttach/1506, accessed 26 March 2019.

[10] Ramsar Convention, Preamble.

[11] Ibid., Article 1(1).

[12] Michael Bowman et al., *Lyster's International Wildlife Law* (2nd ed.; Cambridge: Cambridge University Press, 2011), 405.

use of wetlands in their territory."[13] In case a wetland extends "over the territories of more than one Contracting Party or where a water system is shared by Contracting Parties," their obligation is to "consult with each other and endeavour to coordinate and support present and future policies and regulations concerning the conservation of wetlands and their flora and fauna."[14]

China, which has been a Party to the Ramsar Convention since 31 July 1992, designated 49 sites, with a surface area of 411,224 hectares.[15] For its part, the Philippines has been bound by the Ramsar Convention since 8 November 1994 and has designated 7 sites, with a surface area of 244,017 hectares.[16] Currently, there are currently 277 Wetlands of International Importance that host coral formations, but none is found in waters of the South China Sea that are claimed by either country.[17] As a result, even assuming that the two countries could agree on the question of jurisdiction and control over land territory and maritime spaces in the South China Sea, they would not be bound by the obligations to consult each other and to support present and future policies regarding the conservation of coral reefs in the South China Sea.

It may have made more sense to invoke CITES to protest against the illegal harvesting of sea turtles and corals. Both the Philippines and China are Parties to CITES (since 18 August 1981 and 8 January 1981, respectively). CITES seeks to protect certain species of wild fauna and flora against overexploitation through international trade.[18] Trade in all species "threatened with extinction which are or may be affected by trade" is subject to "particularly strict regulation in order not to endanger further their survival" and is authorized only in exceptional circumstances.[19] Trade in these species, which are listed in CITES Appendix I,

---

[13] Ramsar Convention, Articles 2(1), 2(4) and 3(1).

[14] Ibid., Article 5.

[15] "China," *Ramsar Sites Information Service*, http://www.ramsar.org/wetland/china, accessed 2 April 2019.

[16] "Philippines," *Ramsar Sites Information Service*, http://www.ramsar.org/wetland/philippines, accessed 2 April 2019.

[17] The Ramsar Convention Secretariat, *Coral Reefs: Critical Wetlands in Severe Danger*, RAMSAR CoP12 Doc. 25 (1 April 2015), https://www.ramsar.org/sites/default/files/documents/library/factsheet_5_coral_reefs_en.pdf, accessed 2 April 2019.

[18] CITES, Preamble.

[19] Ibid., Article II(1).

requires both an export permit and an import permit.[20] Species which "although not necessarily now threatened with extinction may become so unless trade in specimens of such species" are listed in CITES Appendix II. Trade in these species is subject to strict regulation "in order to avoid utilization incompatible with their survival," but requires only an export permit.[21] CITES Appendix III includes species that are subject to regulation in a Contracting Party, which requires the cooperation of other Parties in the control of trade. Contracting Parties are required to penalize trade in and/or the possession of specimens of the species in violation of CITES, to provide for confiscation or return to the State of export of such specimens, and to maintain records of trade in specimens of species included in Appendixes I, II, and III.[22] The ultimate sanction for failure to comply with obligations under CITES is suspension of commercial or all trade with the non-compliant Party.[23]

CITES appears to be relevant because certain species of sea turtles (*Cheloniidae spp.*) and corals (Blue corals, *Heliopora coerulea*; Stony corals, *Scleractinia spp.*; and Organ-pipe corals, *Tubiporidae spp.*) are listed in Appendixes I and II, respectively, of CITES.[24] But the Philippines would have had to prove that illegal trade had taken place. The question would then have arisen which State party to CITES, the Philippines or China, had the right to issue the export permits, and ultimately, which State had sovereignty or sovereign rights over the waters in which they were harvested, an issue excluded from the jurisdiction of the Arbitral Tribunal.

Assuming these preliminary issues could have been satisfactorily resolved, CITES dispute settlement mechanisms are rather simple. The Parties to the dispute are required to first enter into negotiations. Should negotiations prove unsuccessful, they may agree to submit the dispute to arbitration.[25] In the absence of such an agreement, CITES provides no further guidance as to the means of settling the dispute. These provisions add very little to the mechanisms available to States under general

---

[20] Ibid., Article III.

[21] Ibid., Articles II(2) and IV.

[22] Ibid., Article XVIII(1) and (6).

[23] Bowman et al., 518.

[24] Appendixes I, II and III valid from 2 January 2017, https://www.cites.org/eng/app/appendices.php, accessed 2 April 2019.

[25] CITES, Article XVIII(3).

international law. Negotiation is by far the preferred mode of dispute settlement of States. They do not need authorization from an international instrument to engage in negotiations, although a particular treaty may require them to do so as a preliminary to other modes of dispute settlement. In case negotiations fail, States are free to choose other modes of dispute settlement, provided no international instrument requires that they adopt a specific mode for that particular dispute. If no other agreement obliges the parties to the dispute to settle that dispute by a specific mode, a dispute will remain unresolved in the face of intransigence shown by one or both parties. This means that even if CITES had been the appropriate basis for a Philippine case against China, China's refusal to enter into negotiations would have blocked the dispute settlement process.

By the time the Philippines filed its *Memorial* in March 2014, the CBD was the only one of the three instruments, in addition to the Convention, that it still believed to be relevant for its case. In the mid-1990s, the Philippine protests to China were based on the concept of "conservation of natural resources," an expression widely used in the Convention. As the incidents kept recurring, the Philippines increasingly referred to the protection of endangered species, habitat, and ecosystems, concepts that are associated with the conservation of biodiversity, without necessarily invoking explicitly the CBD.[26] When the Chinese government undertook construction on Mischief Reef, the Philippines repeatedly protested on the ground that the construction damaged biodiversity. The use of concepts associated with biodiversity reflected the influence of the CBD on the environmental awareness of the States Parties to the CBD. By the turn of the century, the Philippines was already familiar with the concepts of ecosystem and species biodiversity as well as threats to diversity, having submitted its first report on the implementation of the CBD in 1998.[27] Certainly, these concepts seemed more appropriate for describing the damage to the marine environment caused by Chinese nationals.

Be that as it may, when alleging that China had violated Philippine rights under the "the Convention," the Philippines argued that China

---

[26] *MP*, Annex 37, vol. III, 315–19; *MP*, Annex 43, vol. III, 351–55; *MP*, Annex 45, vol. III, 363–67; *MP*, Annex 48, vol. III, 379–94.

[27] Republic of the Philippines, Department of Environment and Natural Resources, Protected Areas and Wildlife Bureau, *The First Philippine National Report to the Convention on Biological Diversity* (Quezon City: Department of Environment and Natural Resources, 1998), https://www.cbd.int/reports/, accessed 25 April 2010.

had violated the CBD. Both the Philippines and China are Parties to the CBD (since 12 June 1992 and 11 June 1992, respectively). In subsequent submissions, the Philippines was at pains to point out that it was not seeking to allege that China was in breach of the CBD or that the dispute over China's conduct was a dispute regarding the interpretation and application of the CBD.[28] In seeking to characterize China's conduct, the Philippines found the CBD more useful than the Convention, to the extent that the nature and jurisdictional scope of the obligations to conserve biodiversity seemed to be broader than those of the obligation to protect and preserve marine living resources under the Convention. Yet it was the dispute settlement mechanism of the Convention that offered the possibility of resort to compulsory mechanisms entailing binding decisions and rendered the Convention more attractive as a framework for settling the Philippines' environmental claims against China.

I. The Content and Jurisdictional Scope of the Obligation to Conserve Biodiversity and of the Obligation to Protect and Preserve Marine Living Resources

It seems that whether we consider the content of the obligation imposed on the respective States Parties or the jurisdictional scope of the obligations undertaken by the respective States Parties, the CBD is a more attractive instrument than the Convention for States seeking a legal basis for a case involving the protection and preservation of endangered species and fragile marine ecosystems.

A. *The Content of the Obligation to Conserve Biodiversity and of the Obligation to Conserve Marine Living Resources*

The CBD requires that States conserve all components of biodiversity in all areas under national jurisdiction. In contrast, the Convention appears to mandate the conservation of a limited number of species of marine living resources only in the exclusive economic zone ("EEZ").

---

[28] *South China Sea Arbitration*, Hearing on Jurisdiction and Admissibility, Transcript, Day 2 (8 July 2015), 101, 110 ("Hearing on Jurisdiction"), https://pcacases.com/web/sendAttach/1400, accessed 3 April 2019.

1. The Obligation to Conserve All Components of Biodiversity Under the CBD

The CBD defines "biological diversity" as

> the variability among living organisms from all sources including, *inter alia*, terrestrial, marine and other aquatic ecosystems and the ecological complexes of which they are part: This includes diversity within species, between species, and of ecosystems.[29]

Diversity within species (genetic diversity), diversity between species, and diversity of ecosystems are the three components of biodiversity.[30] All three components of biodiversity are threatened by habitat loss and fragmentation resulting from urbanization, agricultural, and development projects; overexploitation of plants and animals from fishing and hunting; trade in animals and plants; air, water, and soil pollution; the introduction of alien species; and global atmospheric changes.[31]

Genetic diversity results from the contribution of genes, the principal units of heredity, to the different attributes of an organism. Every individual of a species that originates from sexual reproduction has a slightly different combination of genes. Genetic diversity is the genetic variation within living organisms, that is, the genetic differences among populations of a single species and those among individuals within a population. Genetic diversity allows species to adapt over time to the environmental stresses that they face.[32] "Species diversity" describes the variety of species—whether wild or domesticated—within a geographical area. A species is a population of organisms which are able to interbreed freely under natural conditions. It represents a group of organisms which has evolved distinct inheritable features and occupies a unique geographical area.[33] Ecosystem diversity refers to the variety and frequency of different ecosystems. The CBD defines an "ecosystem" as "a dynamic complex of plant, animal and micro-organism communities and their non-living

---

[29] CBD, Article 2.

[30] Lyle Glowka et al., *A Guide to the Convention on Biological Diversity* (Gland: IUCN-The World Conservation Union, 1994), 16, https://www.iucn.org/content/a-guide-convention-biological-diversity, accessed 3 April 2019.

[31] Ibid., 21.

[32] Ibid.

[33] Ibid., 17.

environment interacting as a functional unit."[34] Plants, animals and micro-organisms are the living (biotic) components of an ecosystem, while the non-living (abiotic) components are sunlight, air, water, minerals, and nutrients. The living components interact with each other and with the non-living components. Their interactions are the basis of the ecosystem's "functioning," which taken together with the functions of other ecosystems, provides "services" on which life on earth depends. These functions include creating the soil, maintaining hydrographic cycles, and recycling nutrients. An ecosystem can be small and ephemeral (a water-filled tree hole), or long-lived (forests and lakes). Among the most threatened ecosystems and habitat types are freshwaters (rivers and lakes), coastal areas, wetlands, coral reefs, oceanic islands, temperate moist forests, temperate grasslands, tropical dry forests, and tropical moist forests.[35]

The CBD's premise is that biological diversity as such is crucial "for evolution and for maintaining life-sustaining systems of the biosphere."[36] Biodiversity is now widely believed to be the best guarantee of life on earth. Only diversity within and among plant and animal species will make it possible for them to adapt to changes in the conditions for survival caused by drought, overpopulation, temperatures, or disease; otherwise, they will die out. If biodiversity is destroyed, the pool of genes in reserve—recessive genes, genes for traits not currently in demand or genes found in wild flora or fauna—will be diminished. In future crises, nature may not be able to regenerate or respond to changing conditions.[37] Ecosystem diversity must also be preserved, since one major cause of the rapid rate of extinction of many animal and plant species is the destruction of their habitats. In the marine environment, habitats may be destroyed by bottom trawling, fishing by using explosives, the laying of cables and pipelines, and the exploitation of seabed mineral resources.[38]

---

[34] CBD, Article 2.

[35] Glowka et al., 21, 41.

[36] CBD, Preamble.

[37] Catherine Tinker, "A 'New Breed' of Treaty: The United Nations Convention on Biological Diversity," *Pace Environmental Law Review* 13 (Fall 1995): 197, http://digital-commons.pace.edu/cgi/viewcontent.cgi?article=1397&context=pelr, accessed 3 April 2019.

[38] Robin Churchill, "The LOS Regime for Protection of the Marine Environment—Fit for the Twenty-First Century?" in Rosemary Rayfuse (ed.), *Research Handbook for International Marine Environmental Law* (Cheltenham, Gloucestershire: Edward Elgar, 2015), 20.

Within the CBD framework, the Philippines could in theory have presented claims against China for the failure to conserve species and genetic diversity (sea turtles, giant clams, and corals) and ecosystem diversity (coral reefs as ecosystems). Whether the Convention provisions on the protection and preservation of the marine environment, and more specifically, on the conservation of marine living resources, would have been applicable to these species and ecosystems is at first sight much less clear.

2. The Obligation to Conserve Marine Living Resources Under the Convention

Part XII of the Convention, entitled "Protection and Preservation of the Marine Environment," lays down the general principle that "States have the obligation to protect and preserve the marine environment."[39] Article 194(5) provides that "the measures taken in accordance with this Part shall include those necessary to protect and preserve rare or fragile ecosystems as well as the habitat of depleted, threatened or endangered species and other forms of marine life." The reference to "ecosystems" in Article 194(5), which was described by the UN Secretary- General as "a contribution to the development of further concepts for the effective protection of the environment,"[40] is not accompanied by a definition of the concept. This paragraph was proposed by the United States in 1978, at the 7th session of the Third United Nations Conference on the Law of the Sea ("UNCLOS III"), which drafted the Convention between 1973 and 1982.[41] Since all negotiations on Article 194 took place in informal meetings, there was no record, making it difficult to explain the rationale for the United States proposal at the time.[42] This is the only

---

[39] Convention, Article 192.

[40] *Law of the Sea. Protection and Preservation of the Marine Environment. Report of the Secretary General*, UN Doc. A/44/461 (18 September 1989), 5, para. 6, https://digitallibrary.un.org/record/76086?ln=en, accessed 3 April 2019.

[41] United Nations Office of Legal Affairs, Codification Division, *United Nations Diplomatic Conferences. Third United Nations Conference on the Law of the Sea, 1973–1982*, 2019, http://legal.un.org/diplomaticconferences/1973_los/, accessed 4 April 2019.

[42] Myron H. Norquist et al. (eds.), *United Nations Convention on the Law of the Sea 1982 Commentary*, vol. IV, *Third Committee: Protection and Preservation of the Marine Environment, Marine Scientific Research, and Development and Transfer of Marine Technology* (Leiden: Martinus Nijhoff, 1991), 64. States may be willing to accept the US proposal because in other informal meetings it had been agreed that the concept of "marine environment" included "marine life." Detlef Czybulka, "Article 194. Measures to

reference to "ecosystems" in the Convention. The other 46 articles in Part XII are concerned with marine pollution.

The observer seeking to identify provisions that would be relevant to sea turtles, giant clams, and corals would have to turn to provisions in other parts of the Convention, particularly Part V on the EEZ. The Convention requires that the coastal State, which has sovereign rights over the marine living resources in its EEZ up to 200 nautical miles from the baselines of the territorial sea, cooperate with other States to conserve five groups of species.[43]

1. Shared and straddling fish stocks. Shared stocks occur within the EEZ of two or more coastal States, which must cooperate to conserve them. Straddling stocks occur both within the EEZ and in an area beyond and adjacent to the EEZ. The coastal State and the States fishing in the adjacent area must cooperate to conserve the stocks in the adjacent area.[44]
2. Highly migratory species (e.g., tuna, marlin, sail-fishes, swordfish, dolphin, shark, and cetaceans). States must cooperate to conserve these species in the EEZ and the high seas.[45]
3. Marine mammals. States may regulate their exploitation more strictly than other living resources in the EEZ.[46]
4. Anadromous stocks (e.g., salmon and sturgeon), which spend most of their life in the sea but spawn in freshwater. Conservation is the primary responsibility of the State of origin, defined as the State in whose waters the fish spawn.[47]
5. Catadromous stocks (e.g., eels), which spend most of their lives in freshwater but spawn in the ocean. Their conservation is under the overall management responsibility of the host State, the State in which they spend the greater part of their life cycle.[48]

Prevent, Reduce and Control Pollution in the Marine Environment," in Alexander Proelss (ed.), *The United Nations Convention on the Law of the Sea: A Commentary* (München: Verlag C.H. Beck oHG, 2017), 1301.

[43] This paragraph is summarized from Yoshifumi Tanaka, *The International Law of the Sea* (2nd ed.; Cambridge: Cambridge University Press, 2015), 239–46.

[44] Convention, Article 63.

[45] Ibid., Article 64. These species are listed in Annex I of the Convention.

[46] Ibid., Article 65. Marine mammals are not listed in the Convention; they include whales, small cetaceans, dolphins, porpoises, seals, dugong, and marine otters.

[47] Ibid., Article 66.

[48] Ibid., Article 67.

The Convention provisions might be applicable to the sharks (highly migratory species) and eels (catadromous species) that Philippine authorities occasionally found on Chinese fishing vessels. But if one has to identify the host State that has sovereign rights over the living resources of the EEZ, it will be necessary to determine whether it is China or the Philippines that has sovereign rights over the EEZ, an issue over which the Arbitral Tribunal did not have jurisdiction. To complicate matters further, it is unclear to what extent these provisions would encompass sea turtles, giant clams, and corals. Coral reefs as geological structures would probably be excluded from the conservation obligations under these provisions.

The broad jurisdictional scope of obligations under the CBD also contrasts with the approach differentiated by maritime areas of the Convention.

B. The Jurisdictional Scope of the Obligation to Conserve Biodiversity and of the Obligation to Conserve Marine Living Resources

A State party to the CBD is under an obligation to conserve all components of biodiversity within the limits of its national jurisdiction and to regulate the activities and processes carried out by their nationals in areas outside its national jurisdiction. The obligations to conserve and manage marine living resources undertaken by a State party to the Convention are differentiated by maritime zones.

1. The Jurisdictional Scope of the Obligation to Conserve Biodiversity

Article 4 of the CBD, entitled "Jurisdictional Scope," clarifies in what instances and in what geographical areas a CBD Contracting Party is obliged to act. It indicates where or how each type of obligation applies.[49]

The jurisdictional scope of the obligations undertaken by the CBD Contracting Parties to conserve the components of biodiversity is quite broad. If the components fall within the limits of its national jurisdiction, a State party to the CBD is under an obligation to conserve all three components of biodiversity (species, genetic, and ecosystem). The areas within the limits of a State's national jurisdiction are (1) the land

---

[49] Glowka et al., 27.

territory within its internationally recognized borders and (2) for any coastal State, its territorial waters, as well as the various maritime zones adjacent to them (e.g., the EEZ and the continental shelf).[50] No distinction is made between land territory and maritime areas. In the areas within national jurisdiction, States may exploit their natural resources, as Article 3 of the CBD recognizes, but they must do so "pursuant to their own environmental policies." This formulation has been interpreted to mean that there exist obligations for all States to protect their environment, and not just to prevent utilization of their territory in ways that harm the environment of other States.[51] In areas beyond the limits of national jurisdiction, on the high seas and the upper atmosphere, States may not exercise territorial jurisdiction.[52] States, by definition, do not have the right to regulate the utilization of the components of biodiversity. Nevertheless, they must regulate the activities and processes carried out by their nationals within these areas that may have an impact on biodiversity, and they must cooperate with other States for this purpose.[53]

The far-reaching implications of such a broad jurisdictional scope become clearer if one examines the obligations relating to the conservation of ecosystems. The main obligations of a State party under the CBD with respect to ecosystems are listed under the heading of "in situ conservation" defined as

> the conservation of ecosystems and natural habitats and the maintenance and recovery of viable populations of species in their natural surroundings and, in the case of domesticated or cultivated species, in the surroundings where they have developed their distinctive properties.[54]

These obligations are to

> (d) Promote the protection of ecosystems, natural habitats and the maintenance of viable populations of species in natural surroundings;

---

[50] Ibid., 28.
[51] Ibid., 26.
[52] Ibid., 27.
[53] CBD, Article 5.
[54] Ibid., Article 2.

(f) Rehabilitate and restore degraded ecosystems and promote the recovery of threatened species, *inter alia*, through the development and implementation of plans or other management strategies;
(h) Prevent the introduction of, control or eradicate those alien species which threaten ecosystems, habitats or species...[55]

Marine ecosystems were not singled out for attention during the negotiations for the CBD, nor in the CBD itself, but it seems that within the CBD framework coastal States have been willing to assume stronger obligations concerning marine and coastal biodiversity. A work program on the subject was adopted in 1995, which included sustainable use of marine and coastal living resources among its five thematic areas.[56] In 2004, the rehabilitation of damaged coral reefs was made one of the components of an Elaborated Programme of Work.[57] In 2010, the CBD Conference of the Parties ("COP") set as one of the 20 Aichi Biodiversity Targets, the goal of minimizing by 2015 "the multiple anthropogenic pressures on coral reefs and other vulnerable ecosystems impacted by climate change or ocean acidification...so as to

---

[55] Ibid., Article 8(d), (f) and (h). Many commentators argue (1) that the CBD contains no substantive obligation to protect biodiversity, e.g., Lakshman D. Guruswamy, "The Convention on Biological Diversity: Exposing the Flawed Foundations," *Environmental Conservation* 26 (1999): 79–82, http://globalseminarhealth.wdfiles.com/local-files/pharmaceutical-harvesting/Guruswamy.pdf, accessed 3 April 2019, and Marie-Ange Hermitte, "La Convention sur la diversité biologique [The Convention on Biological Diversity]," *Annuaire Français de Droit International* [French Yearbook of International Law]" 38 (1992): 844–70 ("*AFDI*"), http://www.persee.fr/docAsPDF/afdi_0066-3085_1992_num_38_1_3098.pdf, accessed 3 April 2019; or (2) that the Convention imposes very few obligations, e.g., Lee A. Kimball, "The Biodiversity Convention: How to Make It Work," *Vanderbilt Journal of Transnational Law* 28 (1995): 763–75, http://69.90.183.227/doc/articles/2002-/A-00497.pdf, accessed 9 December 2016; or (3) that the obligations are merely procedural, e.g., Catherine Tinker, "Responsibility for Biological Diversity under International Law," *Vanderbilt Journal of Transnational Law* 28 (1995): 777–821. The first and second views seem to be unfounded, while the third is only partially true.

[56] Rüdiger Wolfrum and Nele Matz, "The Interplay of UNCLOS and CBD," *Max Planck Yearbook of United Nations Law* 4 (2000): 459–60. Wolfrum was one of the judges in the *South China Sea Arbitration*.

[57] CBD, Conference of the Parties ("COP") 7, Decision VII/5, Marine and Coastal Biological Diversity, Doc. UNEP/CBD/COP/DEC/VII/5 (13 April 2004), Annex I, I.A, 10; III, 14, https://www.cbd.int/decision/cop/default.shtml?id=7742, accessed 3 April 2019.

maintain their integrity and functioning."[58] In 2014, the COP called for actions to reduce the impacts of multiple stressors on coral reefs and enhance the resilience of coral reefs and closely associated ecosystems; to reduce unsustainable fishing practices for multispecies reef fisheries; to sustainably manage populations of key reef fish and invertebrate species targeted by export-driven fisheries or by the aquarium and curio trades; and prioritize the recovery and sustainable management of reef species with key ecological functions, in particular herbivorous reef fish population.[59]

In contrast to the CBD, the Convention's approach to the conservation and management of marine living resources is differentiated by maritime zone.

2. *The Jurisdictional Scope of the Obligation to Conserve Marine Living Resources*

The scholarly consensus is that in internal waters, the territorial sea, and archipelagic waters, the Convention imposes no explicit obligation to conserve marine living resources.[60] Internal waters consist of a State's harbors, ports, and roadsteads and of its internal gulfs and bays, straits, lakes, and rivers.[61] The territorial sea is a belt of water adjacent to the coast, extending up to 12 nautical miles from the baselines drawn by the coastal State and over which the latter enjoys sovereignty, subject to the right of innocent passage of foreign vessels.[62] Archipelagic waters are the waters enclosed by baselines drawn by archipelagic States in conformity with the rules promulgated by the Convention. Their status, not being identical to that of the internal waters, territorial sea, or international straits, is best described as *sui generis*.[63] In these areas, the coastal State can exercise exclusive jurisdiction over living resources. The only limitations on the coastal State's jurisdiction over the conservation of

---

[58]CBD, COP 10, Decision X/2, The Strategic Plan for Biodiversity 2011–2020 and the Aichi Biodiversity Targets, Doc. UNEP/CBD/COP/DEC/X/2 (29 October 2010), 9, https://www.cbd.int/decisions/cop/?m=cop-10, accessed 3 April 2019.

[59]Ibid., 7.

[60]Unless otherwise indicated, this paragraph is summarized from Tanaka, 234–38.

[61]John P. Grant and J. Craig Barker, *Parry & Grant Encyclopedic Dictionary of International Law* (3rd ed.; Oxford: Oxford University Press, 2009), 284.

[62]Ibid., 598.

[63]Ibid., 24.

marine living resources in the territorial sea would be those that flow from general principles of international environmental law and treaties in force for the coastal State.[64]

Two other maritime zones need to be considered. The continental shelf is a maritime zone comprising

> the seabed and subsoil of the submarine areas that extend beyond its territorial sea throughout the natural prolongation of its land territory to the outer edge of the continental margin, or to a distance of 200 nautical miles from the baselines [of the territorial sea] where the outer edge of the continental margin does not extend up to that distance.[65]

The only living resources on the continental shelf are sedentary species, defined as "organisms which, at the harvestable stage, either are immobile on or under the sea-bed or are unable to move except in constant physical contact with the sea-bed or subsoil."[66] The Convention imposes no obligation to conserve sedentary species. On the high seas, States must take conservation measures applicable to their nationals or cooperate with other States for conservation purposes.[67]

Within the EEZ, extending up to 200 nautical miles from the baseline of the territorial sea, a coastal State has conservation obligations relating to living resources, harvested species, species associated with harvested species, and species dependent upon harvested species.[68] The expression "living resources" appears to refer to marine species that either are commercially exploited (in which case they are "harvested species") or have the potential to be commercially exploited. The coastal State's obligation is to ensure that the maintenance of the living resources is not endangered by overexploitation as well as to maintain or restore populations of harvested species at levels that can produce the maximum sustainable yield.[69] Species "associated with harvested species" include

---

[64] Carl-August Fleischer, "La pêche [Fisheries]," in René-Jean Dupuy and Daniel Vignes (eds.), *Traité du Nouveau Droit de la mer* [Treatise on the New Law of the Sea] (Paris: Éditions Économica, 1985), 876–77.

[65] Convention, Article 76(1).

[66] Ibid., Article 77(4).

[67] Ibid., Article 117.

[68] This paragraph is summarized from Churchill, "The LOS Regime," 3–9.

[69] Convention, Article 61(2) and (3).

those that may be caught incidentally when fishing, such as certain fish species, smaller marine mammals (particularly dolphins and porpoises) and amphibians (such as turtles) as well as some seabirds (such as albatrosses). Species "dependent upon harvested species" are the predators of the latter. The aim is to maintain or restore populations of the associated or dependent species above levels at which their reproduction may become seriously threatened. In relation to these last two categories, the coastal State's obligation is merely to take into consideration the effects on them of the conservation measures adopted for harvested species.[70]

Coral reefs as geological structures would, by definition, be excluded from these categories. Corals and giant clams could presumably be classified as "harvested species, and sea turtles as "associated species." Other species that live on the reef that are or may not be commercially exploited would seem to be excluded from these categories. Yet these species might be more numerous than the associated and dependent species. They might also be adversely affected by damage to the coral reef, even if the coastal State's policies for the three categories comply with the Convention.[71] In any case, to assess a State's conformity with these provisions of the Convention, one would have to identify the coastal State that is responsible for conservation. Doing so would require that an international court or tribunal examine questions of sovereignty. This preliminary question would be an insuperable obstacle for settlement of the dispute between the Philippines and China.

The content and scope of the obligation to conserve biodiversity under the CBD may justify the use of concepts associated with the CBD in Philippine diplomatic communications in the years preceding the arbitration and in its submissions to the Tribunal. An examination of the mechanisms for compliance and dispute settlement available under the CBD and the Convention may help us to understand the reasons why despite this circumstance, it may be preferable to seek a resolution of a dispute concerning marine biodiversity, formulated as a dispute relating to the protection and preservation of the marine environment, within the Convention framework.

---

[70] Ibid., Article 61(4).
[71] Wolfrum and Matz, 450.

II. Compliance and Dispute Settlement Under the CBD and Under the Convention

Compliance, rather than dispute settlement, may be described as the main concern of multilateral environmental agreements ("MEAs"), as compliance is the only means to prevent harm to the environment. If the harm to the environment is irreversible, for example, extinction of a species, restitution would by definition be impossible and compensation would be of little help.[72] Non-compliance mechanisms ("NCMs") set up within the framework of MEAs facilitate and secure compliance because unlike dispute settlement mechanisms, they are non-confrontational, transparent, cost-effective, simple, flexible, non-binding and aimed at helping Parties to implement treaty provisions.[73] The desire of MEAs to promote compliance would explain the existence of non-compliance mechanisms and the relative weakness of dispute settlement mechanisms. These mechanisms may be compulsory, but they do not entail binding decisions. The CBD seems to be unique in that only one component of NCMs, a reporting system, exists within its CBD framework, while at the same time its dispute settlement mechanism is as weak as that of other MEAs. The Convention lacks even a reporting system, but alongside this lacuna there exists a system of compulsory dispute settlement entailing binding decisions.

A. Compliance and Dispute Settlement Within the CBD Framework

The compliance and dispute settlement mechanisms within the CBD framework are quite simple. In the first place, the CBD reporting system does not lead to a mechanism for monitoring compliance of States Parties with the CBD. In the second place, the compulsory mechanism for settlement of disputes that may arise between States Parties

---

[72] Malgosia Fitzmaurice, "International Protection of the Environment," *Recueil des Cours de l'Académie de Droit International de La Haye* [Collected Courses of the Hague Academy of International Law], vol. 293 (2001-VI), 337–39.

[73] Susana Borràs, "Comparative Analysis of Selected Compliance Procedures Under Multilateral Environmental Agreements," in Sandrine Maljean-Dubois and Lavanya Rajamani (eds.), *La mise en oeuvre du droit international de l'environnement* [The Implementation of International Law of the Environment] (Leiden: Martinus Nijhoff Publishers, 2011), 328.

concerning the interpretation and implementation of the CBD is limited to conciliation, which does not entail binding decisions.

1. Compliance Within the CBD Framework

Reporting by States Parties is usually the first step in monitoring and ensuring compliance of treaties in the fields of arms control, human rights, and the environment, but in most such treaties, reporting is the basis for subsequent procedures designed to ensure compliance. The CBD differs from such treaties in that it requires only reporting and does not provide for an NCM based on reporting.

Each Contracting Party to the CBD is required under Article 26 to present reports to the COP "on measures which it has taken for the implementation of the provisions of this Convention and their effectiveness in meeting the objectives of this Convention." The success of any reporting system depends on the good faith of the reporting State.[74] The first national reports, submitted in 1997, varied in length and scope.[75] Delay in submission is a recurrent concern.[76] The COP has sought to improve the quality of reports by setting guidelines that indicate the format and specify the information that the State party must provide.[77]

---

[74]David M. Ong, "International Environmental Law Governing Threats to Biological Diversity," in Malgosia Fitzmaurice, David M. Ong, and Panos Merkouris (eds.), *Research Handbook on International Environmental Law* (Cheltenham, Gloucestershire: Edward Elgar, 2010), 536.

[75]CBD, COP 4, Decision IV/14, National Reports by Parties, Doc. UNEP/CBD/COP/4/27 (15 June 1998), 127–28, https://www.cbd.int/doc/meetings/cop/cop-04/official/cop-04-27-en.pdf, accessed 3 April 2019. The quality of the reports ranged from a 12-page type-written report to a 200-page report on glossy paper. Gudrun Henne and Saleem Fakir, "The Regime-Building of the Convention on Biological Diversity on the Road to Nairobi," *Max Planck Yearbook on United Nations Law* 3 (1999): 358.

[76]CBD, COP 7, Decision VII/25, National Reporting, Doc. UNEP/CBD/COP/DEC/VII/25 (13 April 2004), https://www.cbd.int/decisions/cop/?m=cop-07, accessed 3 April 2019.

[77]See, for example, CBD, COP 2, Decision II/17, Form and Intervals of National Reports by Parties, Doc. UNEP/CBD/COP/2/19 (30 November 1995), 72–74, https://www.cbd.int/doc/meetings/cop/cop-02/official/cop-02-19-en.pdf, accessed 3 April 2019; CBD, COP 5, Decision V/19, National Reporting, UNEP/CBD/COP/5/23 (22 June 2000), 151–52, https://www.cbd.int/doc/meetings/cop/cop-05/official/cop-05-23-en.pdf, accessed 3 April 2019; CBD, COP 8, Decision VIII/14, National Reporting and the Next Global Biodiversity Outlook, Doc. UNEP/CBD/COP/DEC/VIII/14 (15 June 2006), https://www.cbd.int/decisions/cop/?m=cop-08,

The Contracting Parties have so far been asked to submit six national reports (1998, 2002, 2006, 2009, 2014, and 2018), and it seems that both the Philippines and China have complied with their reporting obligation in timely fashion.[78]

The Philippines would have found interesting information in China's National Reports. In 2001, China reported that since 1979, it carried out Environmental Impact Assessments ("EIA") that allegedly took full consideration of the possible negative impact of construction projects on biodiversity and formulated measures to minimize the negative impacts of construction.[79] The Philippines could have asked China whether it carried out an EIA before it initiated construction of typhoon shelters in 1995 and construction of a multi-story structure, wharves, and a helipad in 1999 at Mischief Reef. In 2005, China reported that in the previous year, it had established 15 ecological monitoring offshore ecologically sensitive areas, including coral reefs.[80] The Philippines would surely have been interested in learning of China's assessment of the role of dynamite fishing by Chinese fishermen in the degradation of coral reefs in the South China Sea. Perhaps the *Fifth National Report* (2014) would have been the most interesting. China reported that China and the United States had streamlined the management of trade in turtles and developed policies for zero trade in wild species, promoting the monitoring

accessed 3 April 2019; and CBD, COP 10, Decision X/10, National Reporting: Review of Experience and Proposals for the Fifth National Report, Doc. UNEP/CBD/COP/ DEC/X/10 (29 October 2010), https://www.cbd.int/decisions/cop/?m=cop-10, accessed 3 April 2019. Some of the earlier criticisms of the reporting system no longer appear to be valid. See Adelheidur Jóhansdóttir et al., "The Current Framework for International Governance of Biodiversity: Is It Doing More Harm Than Good?" *Review of European Community and International Environmental Law* 19 (2010): 139–49; Philippe Le Prestre, "The Convention on Biological Diversity: Negotiating the Turn to Effective Implementation," *ISUMA: Canadian Journal of Policy Research* 3 (2002): 92–98.

[78] A database of national reports is maintained in https://www.cbd.int/reports/, accessed 3 April 2019.

[79] People's Republic of China, State Environmental Protection Administration of China, *China's Second National Report on Implementing the Convention on Biological Diversity* (Beijing: China National Environmental Science Press, 2001), 55, https://www.cbd.int/ reports, accessed 3 April 2019.

[80] Ibid., *China's Third National Report on Implementation of the Convention on Biological Diversity* (Beijing: State Environmental Protection Administration of China, 2005), 192, https://www.cbd.int/reports, accessed 3 April 2019.

and replenishment of wild turtles and promoting farming to nurture wild populations. According to China, it had strengthened cooperation with Russia, India, Mongolia, Vietnam, Laos, Indonesia, and Thailand in law enforcement and the implementation of CITES.[81] The Philippines would surely have wanted to know the reasons for China's willingness to cooperate with these countries to stamp out the trade in wild turtles and to implement CITES, and yet was unwilling to cooperate with the Philippines to prevent harvesting of sea turtles in the South China Sea.

The problem that the Philippines would have faced is that there is no forum within the CBD that makes it possible for States Parties to exert pressure for compliance on the basis of national reports. Unlike other MEAs, the CBD does not provide for the submission of cases of non-compliance, either by a State party or by the CBD Secretariat, to a body, whether a compliance committee or the COP, that would facilitate compliance by providing assistance to the non-compliant State or imposing sanctions on it, in case of persistent non-compliance.[82] The national reports are merely synthesized by the CBD Secretariat, with the syntheses focusing on the mere submission of reports and on the quantitative analysis of legislative developments (such as the percentage of Parties with biodiversity-related legislation and action plans). The syntheses do not undertake a qualitative analysis of the content of national reports, by inquiring whether State measures have actually conserved biodiversity and achieved CBD goals.[83]

At COP 6 (2002), the EU proposed the establishment of a monitoring and evaluation mechanism, but the proposal was rejected by the

---

[81] Ibid., Ministry of Environmental Protection, *China's Fifth National Report on Implementation of the Convention on Biological Diversity* (Beijing: Ministry of Environmental Protection, 2014), 98, https://www.cbd.int/reports, accessed 3 April 2019.

[82] The non-compliance mechanism established within the framework of the Montreal Protocol on Substances That Deplete the Ozone Layer is the model often cited in this respect. "Annex IV, Non-compliance Procedure," *Report of the Fourth Meeting of the Parties to the Montreal Protocol on Substances That Deplete the Ozone Layer*, Copenhagen, 23–25 November 1992, Doc. UNEP/OzL.Pro.4/15 (25 November 1992), 44–45, https://unep.ch/ozone/Meeting_Documents/mop/04mop/MOP_4.shtml, accessed 3 April 2019. *Montreal Protocol on Substances That Deplete the Ozone Layer*, concluded at Montreal, 16 September 1987, entered into force 26 August 1989, https://ozone.unep.org/sites/default/files/Montreal-Protocol-English_0.pdf, accessed 3 April 2019.

[83] Elisa Morgera and Elsa Tsioumani, "Yesterday, Today and Tomorrow: Looking Afresh at the Convention on Biological Diversity," *Yearbook of International Environmental Law* 21 (2010): 9–10.

developing countries.[84] COP 9 (2008) and COP 11 (2012) initiated a process of review of National Biodiversity Strategies and Action Plans ("NBSAP"), which would make recommendations to States with the aim of improving their capacities to effectively implement the CBD. The outcome of the review would be recommendations for developing, updating and implementing the NBSAP of the State under review.[85] This review process, as presently envisaged, would be hardly comparable to an NCM, in that it would not be compulsory. Another limitation is that it would only indirectly determine whether biodiversity (genetic, species and/or ecosystem) had been lost or conserved in the State Party under review. Even this very limited process runs the risk of resistance from the States Parties to the CBD, if we are to judge by the violent reaction of India, the first country in which the methodology for review was tested.[86]

Could a Contracting State like the Philippines that is dissatisfied with the record of compliance of another Contracting State resort to the CBD's dispute settlement mechanism?

2. Dispute Settlement Within the CBD Framework

Article 27 of the CBD provides for dispute settlement involving both diplomatic means and compulsory mechanisms. Diplomatic means may involve only the two parties—negotiation—or may involve a third party—good offices or mediation.[87] If diplomatic means fail, compulsory mechanisms intervene in the second phase. Two scenarios are

[84] Jean-Pierre Le Danff, "La Convention sur la diversité biologique – tentative de bilan depuis le sommet de Rio de Janeiro [The Convention on Biological Diversity. A Preliminary Balance Sheet Since the Rio de Janeiro Summit]," *Vertigo. La revue juridique en sciences de l'environnement* [Vertigo. The Legal Journal on Environmental Sciences] 3 (2002), https://journals.openedition.org/vertigo/4168, accessed 3 April 2019.

[85] CBD, Subsidiary Body on Implementation, *Voluntary Peer-Review Mechanism for National Biodiversity Strategies and Action Plan. Note by the Executive Secretary*, Doc. UNEP/CBD/SBI/1/10/ADD1 (8 March 2016), https://www.cbd.int/doc/?meeting=sbi-01, accessed 3 April 2019.

[86] COP 13, Decision XIII/25, Modus Operandi of the Subsidiary Body on Implementation and Mechanisms to Support the Review of Implementation, Doc. UNEP/CBD/COP/DEC/XIII/25 (9 December 2016), 1, paras. 2, 3, https://www.cbd.int/decisions/cop/?m=cop-12, accessed 3 April 2019.

[87] Ibid., Article 27(1) and (2). Good offices consist in various kinds of action tending to call negotiations between parties to a dispute into existence. Mediation consists in direct conduct of negotiations between the parties at issue on the basis of proposals made by the mediator. Grant and Barker, 247–48.

possible. In the first, the dispute may be submitted to arbitration or to the International Court of Justice ("ICJ") provided the two parties to the dispute have declared in advance (at the time of ratification or at any time thereafter), that they accept arbitration or adjudication by the ICJ as compulsory.[88] Since the decision of an arbitral tribunal and the judgment of the ICJ is binding on the parties to the dispute, the mechanism could be described as "dispute settlement entailing a binding decision," although this expression does not appear in the CBD itself. In the second scenario, the two States Parties have not accepted the same procedure, or they have not accepted any procedure at all. In either circumstance, the applicable compulsory procedure is conciliation. Either party to the dispute may submit the dispute to conciliation, without the consent of the other party.[89] Since it is the recourse to the procedure that is compulsory, but the outcome remains non-binding, it is compulsory dispute settlement, not entailing binding decisions. It has sometimes been called "compulsory conciliation."[90] The system is a compromise between States who do not wish a dispute to remain unresolved and States that refuse to accept the compulsory jurisdiction of international courts or tribunals.

When a dispute is submitted to conciliation, the work of settlement is undertaken by a commission, permanent or ad hoc, which examines the

---

[88] CBD, Article 27(3). Austria, Georgia, Latvia, and the Netherlands have accepted both arbitration and submission of a dispute to the ICJ as compulsory. Cuba has accepted arbitration as compulsory. If diplomatic methods fail to resolve a dispute, only disputes between any two of the first four States Parties to the CBD may be submitted to the ICJ. Only disputes between any two of the five may be submitted to arbitration. United Nations Treaty Collection, Convention on Biological Diversity (16 April 2017), https://treaties.un.org/Pages/ViewDetails.aspx?src=TREATY&mtdsg_no=XXVII-8&chapter=27&clang=_en, accessed 3 April 2019.

[89] CBD, Article 27(4).

[90] Adolfo Maresca, *Il diritto dei trattati. La Convenzione codificatrice di Viena del 23 maggio 1969* [The Law of Treaties. The Codifying Convention of Vienna, 23 May 1969] (Milano: Dott. A. Giuffrè Editore, 1971), 752. Compulsory arbitration (i.e., at the request of either party to the dispute) had been proposed in an earlier draft of the CBD as one of two alternatives, but the option was ruled out in favor of compulsory conciliation in the CBD's final draft. See United Nations Environment Programme (UNEP), Intergovernmental Negotiating Committee for a Convention on Biological Diversity, *Fifth Revised Draft Convention on Biological Diversity. Explanatory Note*, Doc. UNEP/Bio. Div/N7-INC.5/2 (20 February 1992), Article 30, 26–27, https://www.cbd.int/doc/meetings/iccbd/bdn-07-inc-05/official/bdn-07-inc-05-02-en.pdf, accessed 3 April 2019.

dispute and makes non-binding proposals for settlement.[91] Throughout its history, less than 20 bilateral disputes have been resolved through conciliation, despite the fact that over 200 conciliation treaties were signed in the inter-war period and presumably remain in force.[92] Apart from compulsory conciliation provided for in many multilateral treaties, such as the CBD, the only other recent attempt to revive conciliation was made through the establishment of the Court of Conciliation and Arbitration within the Organisation for Security and Co-operation in Europe ("OSCE") by the Stockholm Convention in 1992. This Court has remained a "sleeping beauty," as no dispute has ever been submitted to it.[93]

One reputed advantage of conciliation is its flexibility, which may be illustrated by the resolution of the conflict between Iceland and

---

[91] Institut de Droit International [Institute of International Law], Session of Salzburg, *International Conciliation*, 1961, http://www.idi-iil.org/app/uploads/2017/06/1961_salz_02_en.pdf, accessed 3 April 2019. The English text is a translation; the French text is authoritative.

[92] J. G. Merrills, *International Dispute Settlement* (4th ed.; Cambridge: Cambridge University Press, 2011), 59–60, 69–72.

[93] On the Court of Conciliation and Arbitration within the OSCE, see Robert Badinter, "La Cour de conciliation et d'arbitrage au sein de l'O.S.C.E. [The OSCE Court of Conciliation and Arbitration]," *International Law FORUM du droit inter-national* 1 (1999): 99–102; Lucius Caflisch and Laurence Cuny, "Der Vergleichs- und Schiedsgerichtshof der OSZE: Aktuelle Probleme [The OSCE Court of Conciliation and Arbitration: Current Problems]," *OSZE-Jahrbuch* [OSCE Yearbook] 3 (1997): 373–82, https://ifsh.de/core/publikationen/osze-jahrbuch/1997/, accessed 3 April 2019; Court of Conciliation and Arbitration, *Court of Conciliation and Arbitration Within the CSCE: Periodic Report 2013–2016 to the States Parties to the Convention on Conciliation and Arbitration Within the OSCE*, http://www.osce.org/cca, accessed 3 April 2019; Dieter S. Lutz, "Der OSZE-Gerichtshof [The OSCE Court of Conciliation and Arbitration]," *OSZE-Jahrbuch* [OSCE Yearbook] 1 (1995): 241–53, https://ifsh.de/core/publika-tionen/osze-jahrbuch/1995/, accessed 3 April 2019; Heinrich B. Reimann, "Le règle-ment pacifique des différends à l'OSCE [The Peaceful Settlement of Disputes at the OSCE]," in Marcelo G. Kohen (ed.), *La promotion de la justice, des droits de l'homme et du règlement des conflits par le droit international. Liber Amicorum Lucius Caflisch* [Promoting Justice, Human Rights and Conflict Resolution Through International Law. Liber Amicorum Lucius Caflisch] (Leiden: Martinus Nijhoff Publishers, 2007), 891–96; Christian Tomuschat, "Sleeping Beauty: The OSCE Court of Conciliation and Arbitration," *Security Community. The OSCE Magazine* 2 (2014): 36–37, http://www.osce.org/cca, accessed 3 April 2019; and Christian Tomuschat et al. (eds.), *Conciliation in International Law: The OSCE Court of Conciliation and Arbitration* (Leiden: Martinus Nijhoff Publishers 2016).

Norway over the delimitation of the continental shelf in the area of the Norwegian island of Jan Mayen.[94] The island is situated about 540 nautical miles from the Norwegian mainland and 293 nautical miles from Iceland. Following an agreement on the delimitation of the two countries' respective EEZs, Iceland claimed a continental shelf extending beyond a 200-nautical mile limit even in the areas where the distance between Iceland and Jan Mayen was only 280–290 nautical miles and up to 12 nautical miles from Jan Mayen. Norway refused the idea of two different delimitation lines, one for the EEZ and one for the continental shelf.[95]

The flexibility of conciliation made it possible for the two States to stipulate that the Commission take into account Iceland's strong economic interests in these sea areas, the existing geographic and geological factors, and other special circumstances; that the Commission's recommendations should be adopted unanimously; and that the Commission's work should be completed within five months of its appointment. The Commission dispensed with written and oral submissions from the two States, since the two national members had participated in all diplomatic negotiations prior to the conciliation proceedings. The Commission sought out information on the geology of the continental shelf areas in question by convening a meeting of scientific experts at Columbia University in New York.[96] Most importantly, the Commission proposed an imaginative solution that would not have been possible if its decision had had to be based on law. It decided not to propose a delimitation line for the continental shelf different from that of the EEZ, because Iceland had already been granted a considerable area beyond the median line and that the resource potential of the area was uncertain. Instead, the Commission recommended the adoption of a joint development

---

[94]For a recent summary, see Ulf Linderfalk, "The Jan Mayen Case (Iceland/Norway): An Example of Successful Conciliation," paper presented at the Symposium "Conciliation in the Globalized World of Today," organized by the Court of Conciliation and Arbitration within the O.S.C.E. (10–11 June 2015), https://papers.ssrn.com/sol3/papers.cfm?abstract_id=2783622, accessed 3 April 2019.

[95]Jens Evensen, "La délimitation entre la Norvège et l'Islande du plateau continental dans le secteur de Jan Mayen [The Delimitation between Norway and Iceland of the Continental Shelf in the Jan Mayen Sector]," *AFDI* 27 (1981): 721, http://www.persee.fr/doc/afdi_0066-3085_1981_num_27_1_2469, accessed 3 April 2019.

[96]Elliott L. Richardson, "Jan Mayen in Perspective," *American Journal of International Law* 82 (1988): 445.

agreement covering substantially all of the area covering any significant prospect of hydrocarbon production.[97] The proposal was accepted by the two States.[98] Conciliation proved to be an inexpensive way of resolving a complex dispute within a relatively short period of time.[99]

Conciliation is likely to succeed if favorable political conditions exist. First, the parties to the dispute have generally good relations.[100] In the past States who chose conciliation were near neighbors with an interest in maintaining friendly relations.[101] Compulsory conciliation is less likely to succeed in cases when the relations between the parties to the dispute are already tense.[102] The other condition of a political nature is the salience of the issue for the two parties to the dispute. In the Iceland/Norway conciliation, the proposed solution was accepted

[97] "Conciliation Commission on the Continental Shelf Area Between Iceland and Jan Mayen: Report and Recommendations to the Governments of Iceland and Norway. Decision of June 1981," *Reports of International Arbitral Awards*, vol. 27 (New York: United Nations, 2008), 7–9, 22–25, http://legal.un.org/riaa/cases/vol_XXVII/1-34. pdf, accessed 3 April 2019.

[98] Evensen, 723–24.

[99] Ibid., 724.

[100] Ibid. For conciliation between neighbors with friendly relations, see Suzanne Bastid, "La Commission de conciliation franco-suisse [The Franco-Swiss Conciliation Commission]," *AFDI* 2 (1956): 436–40, http://www.persee.fr/doc/afdi_0066-3085_1956_num_2_1_1256; Marcelle Breton-Jokl, "La Commission de conciliation ita-lo-suisse [The Italo-Swiss Conciliation Commission]," *AFDI* 3 (1957): 210–21, http://www.persee.fr/doc/afdi_0066-3085_1957_num_3_1_1322; and Daniel Vignes, "La sentence de la Commission de conciliation franco-italienne dans l'affaire du différend sur les biens immeubles appartenant à l'ordre de Saint-Maurice et de Saint Lazare (Hospice du Petit Saint-Bernard) [The Award of the Franco-Italian Conciliation Commission in the Case relating to the Dispute concerning Immovable Property Belonging to the Order of Saint Maurice and Saint Lazarus (Little Saint Bernard Hospice)]," *AFDI* 11 (1965): 319–32, http://www.persee.fr/doc/afdi_0066-3085_1965_num_11_1_1822, accessed 3 April 2019.

[101] Merrills, 80.

[102] Hans Wehberg, "Die Vergleichskommissionen im modernen Völkerrecht [Conciliation Commissions in Modern International Law]," *Zeitschrift für ausländisches öffentliches Recht und Völkerrecht* [Journal for Foreign Public Law and International Law] 19 (1958): 586, http://www.zaoerv.de/19_1958/19_1958_1_3_a_551_593.pdf, accessed 3 April 2019. An example would be the conciliation between France and Morocco during the Algerian war of independence from France. "L'affaire du F. OABV (Maroc c. France) [The Case of F. OABV (Morocco v. France)]," *AFDI* 4 (1958): 282–95, http://www.persee.fr/doc/afdi_0066-3085_1958_num_4_1_1382, accessed 3 April 2019.

because it did not affect substantially any pressure group in either country, and the groundwork had already been laid by the earlier agreement to delimit the EEZ.[103] Where the issue involves a substantial interest of a political nature, a compromise is more difficult to achieve.[104] Even if conciliation is successful, the outcome cannot guarantee legal certainty.

The weakness of the CBD dispute settlement mechanism may be better appreciated when one examines the dispute settlement mechanism of the Convention.

B. Compliance and Dispute Settlement Within the Convention Framework

The Convention lacks a non-compliance mechanism, or even a reporting mechanism, but does provide for compulsory dispute settlement entailing binding decisions under certain conditions.

1. Compliance Within the Convention Framework

The Convention, like several conventions for the codification and progressive development of international law concluded under UN auspices, does not establish a reporting system for States Parties. Article 319(2) (a) indirectly provides for the creation of a Meeting of States Parties to the Convention by requiring that the UN Secretary-General, as depositary of the Convention, "report to all States Parties, the [International Seabed] Authority and competent international organizations on issues of a general nature that have arisen with respect to this Convention." This provision has given rise to divergent views relating to the power of the Meeting of the States Parties to act as a mechanism for the follow-up of the implementation of the Convention.[105]

---

[103] Linderfalk, 21–22.

[104] Merrills, 81. An example is the conciliation between France and Siam (Thailand) concerning the adjustment of the latter's borders with French Indochina. "Rapport de la Commission de conciliation Franco-Siamoise, Washington, 27 Juin 1947 [Report of the Franco-Siamese Conciliation Commission, Washington, 27 June 1947]," *Reports of International Arbitral Awards*, vol. 28 (New York: United Nations, 2007), 433–50, http://legal.un.org/riaa/cases/vol_XXVIII/433-450.pdf, accessed 3 April 2019.

[105] Aurélie Tardieu, "Les conferences des États parties [Conferences of States Parties]," *AFDI* 57 (2011): 116, http://www.persee.fr/docAsPDF/afdi_0066-3085_2011_num_57_1_4178.pdf, accessed 3 April 2019.

Some States Parties believe that Article 319(2)(a) is the legal basis for the Meeting to consider issues relating to the implementation of the Convention. The Meeting seems to them to be the logical forum for discussion on all issues relating to the Convention's implementation. In their view, the Meeting could provide a forum for an exchange of information on State practice, promote cooperation and further debates on issues of interest to the States Parties, and contribute to finding consensus on emerging issues.[106] These States point out that the Meeting has already adopted substantive decisions relating to the Convention's implementation.[107] Some States Parties go further and argue that the Meeting has the full mandate to consider all issues pertaining to the application and implementation of the Convention.[108]

Other States believe that the Meeting should limit itself to a consideration of financial and administrative matters related to the bodies established by the Convention—the ITLOS, the International Seabed Authority, and the Commission on the Limits of the Continental Shelf (CLCS). They warn that the meeting should not be regarded as a forum for the discussion and resolution of particular disputes relating to the interpretation and application of the Convention.[109] They point out that the parts of the Convention that referred to the Meeting were Annexes II and VI, requiring the meeting to elect the members of the CLCS and of the ITLOS as well as to approve the ITLOS budget. Article 319 should therefore be interpreted as giving the meeting only an

---

[106] United Nations Convention on the Law of the Sea, Meeting of the States Parties, Fifteenth Meeting, New York, 16–24 June 2005, *Report of the Fifteenth Meeting of States Parties*, Doc. SPLOS/135 (25 July 2005), 16, para. 81, https://documents-dds-ny.un.org/doc/UNDOC/GEN/N05/439/16/PDF/N0543916.pdf?OpenElement, accessed 3 April 2019.

[107] Ibid., Twenty-First Meeting, New York, 13–17 June 2011, *Report of the Twenty-First Meeting of States Parties*, Doc. SPLOS/231 (29 June 2011), 18, para. 119, https://documents-dds-ny.un.org/doc/UNDOC/GEN/N11/393/68/PDF/N1139368.pdf?OpenElement, accessed 3 April 2019.

[108] Ibid., Twenty-Fifth Meeting, New York, 8–12 June 2015, *Report of the Twenty-Fifth Meeting of States Parties*, 16, para. 81, Doc. SPLOS/287 (13 July 2015), https://documents-dds-ny.un.org/doc/UNDOC/GEN/N15/217/42/PDF/N1521742.pdf?OpenElement, accessed 3 April 2019.

[109] Ibid., Twenty-Sixth Meeting, New York, 20–24 June 2016, *Report of the Twenty-Sixth Meeting of States Parties*, Doc. SPLOS/30 (2 August 2016), 16, para. 92, https://documents-dds-ny.un.org/doc/UNDOC/GEN/N16/245/62/PDF/N1624562.pdf?OpenElement, accessed 3 April 2019.

administrative and budgetary role. Treaties that contained a mechanism to oversee their implementation explicitly provided for it, which was not the case for the Convention. In fact, the Convention's negotiating history showed that a proposal to establish a mechanism to review common problems and address new uses of the sea had been rejected. The UN General Assembly, in the view of these States, is the only inclusive mechanism in which to discuss the Convention's implementation.[110]

The issue remains unresolved in principle and is discussed at every Meeting of the States Parties. In practice, it is the second, restrictive, interpretation that has prevailed.[111] States do bring to the Meeting's attention their disputes regarding the implementation of the Convention, but the Meeting does not review them, much less make recommendations to the States Parties concerned on the measures that they need to take to ensure conformity of their acts with the Convention.

Meanwhile, observers have noted the proliferation of cases of non-compliance with the Convention. Among the most serious of these cases of non-compliance are the overexploitation of fish stocks and the failure to protect and preserve rare or fragile ecosystems. The utilization of the Convention's compulsory dispute mechanism is one of the solutions proposed to address the problem of non-compliance.[112]

2. Dispute Settlement Within the Convention Framework

It is not possible to do justice here to the voluminous literature on dispute settlement within the Convention framework.[113] Here, one can

[110]Ibid., *Report of the Fifteenth Meeting of States Parties*, 16, para. 82.

[111]Tardieu, 115.

[112]Robin Churchill, "The Persisting Problem of Non-compliance with the Law of the Sea Convention: Disorder in the Oceans," in David Freestone (ed.), *The 1982 Law of the Sea Convention at 30: Successes, Challenges and New Agendas* (Leiden: Martinus Nijhoff Publishers, 2013), 141–42.

[113]For brief introductions see Claude-Albert Colliard, "Problèmes et solutions en matière de règlement des différends [Problems and Solutions in Dispute Settlement]," in Société Française pour le Droit International [French Society for International Law], Colloque de Rouen [Rouen Colloquium], *Perspectives du droit de la mer à l'issue de la 3ᵉ Conférence des Nations Unies* [Perspectives on the Law of the Sea at the Conclusion of the Third United Nations Conference] (Paris: Éditions Pedone, 1984), 174–89; Myron H. Nordquist et al. (eds.), *United Nations Convention on the Law of the Sea 1982 Commentary*, vol. V, *Settlement of Disputes, General and Final Provisions and related Annexes and Resolutions* (Leiden: Martinus Nijhoff Publishers, 1989), 1–146; Raymond Ranjeva, "Le

only stress the differences between the dispute settlement systems of the CBD and of the Convention.

Like the CBD, the Convention provides for a first phase in dispute settlement through negotiation and other peaceful means chosen by the parties to a dispute (Section 1 of Part XV of the Convention). States are allowed to resolve their disputes relating to the interpretation and application of the Convention through means of settlement previously agreed upon outside the Convention framework and to exclude the application of the Convention framework even if they have failed to reach settlement through these alternative procedures.[114] States are also allowed to resolve their disputes through procedures entailing binding decisions provided for in general, regional, or bilateral agreements in lieu of the procedures in Part XV.[115] Within the Convention framework, conciliation is an option available to States Parties at this stage but it is not compulsory: If the State invited to submit the dispute to conciliation refuses to participate in the procedure, conciliation cannot proceed.[116]

Like the CBD, the Convention contains provisions that are applicable in a second phase, when diplomatic means of dispute settlement have failed. Two main scenarios are envisaged, with compulsory conciliation being limited to specific categories of disputes. In the first scenario, one party to the dispute may submit it unilaterally to a tribunal that the two parties to the dispute have previously chosen by means of a declaration. The CBD offered a choice between the ICJ and an ad hoc

règlement des différends [Dispute Settlement]," in René-Jean Dupuy and Daniel Vignes (eds.), *Traité du Nouveau Droit de la mer* [Treatise on the New Law of the Sea] (Paris: Éditions Economica, 1985), 1105–68; Louis B. Sohn, "Settlement of Disputes Relating to the Interpretation and Application of Treaties," *Recueil des Cours de l'Académie de Droit International de La Haye* [Collected Courses of the Hague Academy of International Law], vol. (1976-II), 280–92; and Ugo Villani, "Osservazioni sulla soluzione delle controversie nelle convenzioni di codificazione del diritto internazionale [Observations on Dispute Settlement in Conventions Codifying International Law]," in *Le droit international à l'heure de sa codification. Études en l'honneur de Roberto Ago* [International Law at the Time of Its Codification. Essays in Honour of Roberto Ago], vol. III, *Les différends entre États et la responsabilité* [Inter-State Disputes and Responsibility] (Milano: Dott. A. Giuffrè Editore, 1987), 508–21.

[114] Convention, Article 281.

[115] Ibid., Article 282.

[116] Ibid., Article 284.

arbitral tribunal.[117] States Parties to the Convention have a wider choice: the ITLOS, the ICJ, arbitration under Annex VII of the Convention, and special arbitration under Annex VIII of the Convention.[118] By definition, all these procedures entail decisions that are binding on both parties to the dispute—they are binding even on the State that did not give its consent for the particular dispute to be submitted to the court or tribunal that it had previously chosen by means of a declaration. In the second scenario, the two parties to a dispute have not chosen the same procedure or have not chosen any procedure at all. Under the CBD, the dispute may be submitted by either party to the dispute (without the consent of the other) only to conciliation, the outcome of which consists only of recommendations to the parties to the dispute.[119] In this scenario, dispute settlement is compulsory, but it does not entail binding decisions. Under the Convention, if the two parties to a dispute have not chosen the same procedure or have not chosen any procedure at all, the dispute may be submitted unilaterally by one party to arbitration. The outcome will be legally binding on both parties to the dispute—it will be binding even on the State that did not give its consent to the submission of the particular dispute to arbitration.[120] Dispute settlement in the Convention framework is not only compulsory, it also entails binding decisions.

Conciliation is compulsory for disputes arising from an allegation by a State carrying out a marine scientific research project that the coastal State, with respect to a specific request for permission to carry out the research, is not exercising its rights under Articles 246 and 253 in a manner compatible with the Convention.[121] Conciliation is also compulsory for disputes concerning the sovereign right of the coastal State over the living resources in its EEZ, including its discretionary powers to determine the allowable catch, its harvesting capacity, the allocation of surpluses to other States, and the terms and conditions established in its conservation and management laws and regulations. Such disputes they may not be submitted to procedures entailing binding decisions.[122] Finally, conciliation

---

[117] CBD, Article 27(3).

[118] Convention, Article 287(1).

[119] CBD, Article 27(4).

[120] Convention, Article 287(3) and (5).

[121] Ibid., Article 297(2)(b).

[122] Ibid., Article 297(3).

is compulsory in the event of disputes concerning the interpretation or application of the Convention provisions relating to the delimitation of the territorial sea, the EEZ, and the continental shelf or disputes involving historic bays or titles, between States that have excluded these categories of disputes from compulsory procedures entailing binding decisions.[123]

Subject to a number of preconditions as well as to a limited number of limitations and optional exceptions, the Convention's dispute settlement procedures entailing binding decisions extend to all questions relating to the protection and preservation of the marine environment. With this in mind, a State that believes that another State is violating the obligation to conserve marine biodiversity will find it advantageous to formulate its claims pertaining to marine biodiversity in terms of the more general obligation to protect and preserve the marine environment. In this manner, it can hope to submit the dispute unilaterally to a procedure entailing binding decisions within the Convention framework.

It is not clear from the documents that the Philippines submitted to the Arbitral Tribunal whether the Philippines ever seriously considered dispute settlement within the CBD framework. The Philippines, responding to a question from the Tribunal on this point during the Hearing on Jurisdiction and Admissibility, declared that the Philippines made no attempt to invoke Article 27(4) of the CBD because it had no dispute with China relating to compliance with the CBD. The Philippines insisted that the only dispute that its Submissions No. 11 and 12(b) required the Tribunal to decide was a dispute concerning Part XII of the Convention and related provisions.[124]

Assuming that conciliation had been seriously considered, the absence of favorable political conditions would have led to the conclusion that the chances of success were doubtful. The generally friendly relations between China and the Philippines had gradually been adversely affected over the years by the persistence of this and other disputed issues in the South China Sea. Clearly, the conservation of biodiversity was of great salience for the Philippines; apparently, this was not the case for China.

---

[123] Ibid., Article 298(1)(a). The parties to the dispute shall enter into negotiations on the basis of the report of the Conciliation Commission. If the negotiations fail, the consent of both parties to the dispute is required for submission of the dispute to a procedure entailing binding decisions.

[124] *South China Sea Arbitration*, Hearing on Jurisdiction and Admissibility, Transcript, Day 3 (13 July 2015), 42, https://pcacases.com/web/sendAttach/1401, accessed 3 April 2019.

This divergence of views would have made it difficult for the Philippines to envisage a compromise. At any rate, the risk would have been great that a compromise would imperil marine biodiversity.

The existence of conditions favoring the success of conciliation would not necessarily compensate for the major shortcoming of conciliation, its inability to offer legal certainty.[125] It has been plausibly argued that in the Iceland/Norway conciliation, Norway, which has always insisted on the equidistance line in its negotiations with other States concerning the continental shelf, accepted conciliation precisely because it would not constitute a precedent for the other negotiations. The non-binding outcome of the Iceland/Norway conciliation would not have compelled Norway to change this position.[126]

Judicial means of settling international environmental disputes are often portrayed as confrontational, time-consuming, expensive, and inadequate. A flexible and not very costly conciliation procedure that results in a compromise may be a tempting alternative to protracted legal proceedings that produce an outcome unsatisfactory for one or even both parties to the dispute. Unfortunately, transactional solutions that apply extralegal principles, even if they lead to the solution of a dispute, would not contribute to the affirmation of a rule that has been codified in a treaty or to the certainty of law in the subject matter governed by the treaty.[127]

More broadly, judicial settlement of a dispute, in certain circumstances, offers advantages that cannot be underestimated. It can determine the meaning and scope of principles and norms contained in a treaty, particularly the extent to which they confer rights or impose obligations on a State vis-à-vis another State. It can elucidate the implications of a particular principle and its application to a situation or activity that may not have been contemplated at the time the treaty was drafted. It can resolve a conflict between two (contradictory) principles that are applicable to a particular situation. It can determine the application of an agreement to parties to a dispute or to the facts of a particular dispute.[128]

[125] Maresca, 752.

[126] Linderfalk, 18.

[127] Villani, 498.

[128] Thomas A. Mensah, "Using Judicial Bodies for the Implementation and Enforcement of International Environmental Law," in Isabelle Buffard et al. (eds.), *International Law Between Universalism and Fragmentation. Festschrift in Honour of Gerhard Hafner* (Leiden: Martinus Nijhoff Publishers, 2008), 801–803. Mensah was the president of the Arbitral Tribunal in the *South China Sea Arbitration*.

A State that wishes to unilaterally submit its dispute to a procedure entailing binding decisions under the Convention and formulates its claims of damage to biodiversity must grapple with two challenges. First, to convince the tribunal or court that it has jurisdiction over the dispute, it must demonstrate that the preconditions to compulsory dispute settlement have been fulfilled and that the limited number of limitations and optional exceptions provided for in the Convention do not apply to its marine environmental claims. Second, it must also demonstrate that the obligation to conserve marine biodiversity may be subsumed under the obligation to protect and preserve the marine environment, or to put it differently, that the obligation to protect and preserve the marine environment encompasses the obligation to conserve marine biodiversity. These two challenges will be the object of the next two chapters.

REFERENCES

Badinter, Robert. "La Cour de conciliation et d'arbitrage au sein de l'O.S.C.E [The OSCE Court of Conciliation and Arbitration]." *International Law FORUM du droit international* 1 (1999): 99–102.

Bastid, Suzanne. "La Commission de conciliation franco-suisse [The Franco-Swiss Conciliation Commission]." *Annuaire Français de Droit International* [French Yearbook of International Law] 2 (1956): 436–40, http://www.persee.fr/doc/afdi_0066-3085_1956_num_2_1_1256, accessed 3 April 2019.

Borràs, Susana. "Comparative Analysis of Selected Compliance Procedures Under Multilateral Environmental Agreements." *La mise en oeuvre du droit international de l'environnement* [The Implementation of International Law of the Environment], 319–71. Eds. Sandrine Maljean-Dubois and Lavanya Rajamani. Leiden: Martinus Nijhoff Publishers, 2011.

Bowman, Michael, et al. *Lyster's International Wildlife Law*, 2nd ed. Cambridge: Cambridge University Press, 2011.

Breton-Jokl, Marcelle. "La Commission de conciliation italo-suisse [The Italo-Swiss Conciliation Commission]." *Annuaire Français de Droit International* [French Yearbook of International Law] 3 (1957): 210–21, http://www.persee.fr/doc/afdi_0066-3085_1957_num_3_1_1322, accessed 3 April 2019.

Caflisch, Lucius, and Laurence Cuny. "Der Vergleichs- und Schiedsgerichtshof der OSZE: Aktuelle Probleme [The OSCE Court of Conciliation and Arbitration: Current Problems]." *OSZE-Jahrbuch* [OSCE Yearbook] 3 (1997): 373–82, https://ifsh.de/core/publikationen/osze-jahrbuch/1997/, accessed 3 April 2019.

Churchill, Robin. "The Persisting Problem of Non-compliance with the Law of the Sea Convention: Disorder in the Oceans." *The 1982 Law of the Sea Convention at 30: Successes, Challenges and New Agendas*, 139–46. Ed. David Freestone. Leiden: Martinus Nijhoff Publishers, 2013.

———. "The LOS Regime for Protection of the Marine Environment—Fit for the Twenty-First Century?" *Research Handbook for International Marine Environmental Law*, 3–30. Ed. Rosemary Rayfuse. Cheltenham, Gloucestershire: Edward Elgar, 2015.

Colliard, Claude-Albert. "Problèmes et solutions en matière de règlement des différends [Problems and Solutions in Dispute Settlement]." *Perspectives du droit de la mer à l'issue de la 3ᵉ Conférence des Nations Unies* [Perspectives on the Law of the Sea at the Conclusion of the Third United Nations Conference], 174–89. Société Française pour le Droit International [French Society for International Law]. Colloque de Rouen [Rouen Colloquium]. Paris: Éditions Pedone, 1984.

"Conciliation Commission on the Continental Shelf Area Between Iceland and Jan Mayen: Report and Recommendations to the Governments of Iceland and Norway. Decision of June 1981." *Reports of International Arbitral Awards*. Vol. 28, 1–34. New York: United Nations, 2008, http://legal.un.org/riaa/cases/vol_XXVII/1-34.pdf, accessed 3 April 2019.

Convention on Biological Diversity (CBD), signed at Rio de Janeiro on 5 June 1992, entered into force on 29 December 1993, https://www.cbd.int/doc/legal/cbd-en.pdf, accessed 26 March 2019.

———. Conference of the Parties (COP) 2. Decision II/17. Form and Intervals of National Reports by Parties. Doc. UNEP/CBD/COP/2/19 (30 November 1995), https://www.cbd.int/doc/meetings/cop/cop-02/official/cop-02-19-en.pdf, accessed 3 April 2019.

———. COP 4. Decision IV/14. National Reports by Parties. Doc. UNEP/CBD/COP/4/27 (15 June 1998), https://www.cbd.int/doc/meetings/cop/cop-04/official/cop-04-27-en.pdf, accessed 3 April 2019.

———. COP 5. Decision V/19. National Reporting. UNEP/CBD/COP/5/23 (22 June 2000), https://www.cbd.int/doc/meetings/cop/cop-05/official/cop-05-23-en.pdf, accessed 3 April 2019.

———. COP 7. Decision VII/5. Marine and Coastal Biological Diversity. Doc. UNEP/CBD/COP/DEC/VII/5 (13 April 2004), https://www.cbd.int/doc/decisions/cop-07/cop-07-dec-05-en.pdf, accessed 3 April 2019.

———. COP 7. Decision VII/25. National Reporting, Doc. UNEP/CBD/COP/DEC/VII/25 (13 April 2004), https://www.cbd.int/decisions/cop/?m=cop-07, accessed 3 April 2019.

———. COP 8. Decision VIII/14, National Reporting and the Next Global Biodiversity Outlook, Doc. UNEP/CBD/COP/DEC/VIII/14 (15 June 2006), https://www.cbd.int/decisions/cop/?m=cop-08, accessed 3 April 2019.

———. COP 10. Decision X/2. The Strategic Plan for Biodiversity 2011–2020 and the Aichi Biodiversity Targets. Doc. UNEP/CBD/COP/DEC/X/2 (29 October 2010), https://www.cbd.int/decisions/cop/?m=cop-10, accessed 3 April 2019.

———. COP 10. Decision X/10, National Reporting: Review of Experience and Proposals for the Fifth National Report. Doc. UNEP/CBD/COP/DEC/X/10 (29 October 2010), https://www.cbd.int/decisions/cop/?m=cop-10, accessed 3 April 2019.

———. COP 13. Decision XIII/25. Modus Operandi of the Subsidiary Body on Implementation and Mechanisms to Support the Review of Implementation. Doc. UNEP/CBD/COP/DEC/XIII/25 (9 December 2016), https://www.cbd.int/decisions/cop/?m=cop-12, accessed 3 April 2019.

———. Subsidiary Body on Implementation. *Voluntary Peer-Review Mechanism for National Biodiversity Strategies and Action Plan. Note by the Executive Secretary.* Doc. UNEP/CBD/SBI/1/10/ADD1 (8 March 2016), https://www.cbd.int/doc/?meeting=sbi-01, accessed 3 April 2019.

Convention on International Trade in Endangered Species of Wild Fauna and Flora, signed at Washington, DC, on 3 March 1973, amended at Bonn, on 22 June 1979, amended at Gaborone, on 30 April 1983, https://www.cites.org/eng/disc/text.php, accessed 26 March 2019.

Court of Conciliation and Arbitration. *Court of Conciliation and Arbitration Within the OSCE: Periodic Report 2013–2016 to the States Parties to the Convention on Conciliation and Arbitration Within the OSCE*, http://www.osce.org/cca, accessed 3 April 2019.

Czybulka, Detlef. "Article 194: Measures to Prevent, Reduce and Control Pollution in the Marine Environment." *The United Nations Convention on the Law of the Sea: A Commentary*, 1295–1315. Ed. Alexander Proelss. München: Verlag C.H. Beck oHG, 2017.

Evensen, Jens. "La délimitation entre la Norvège et l'Islande du plateau continental dans le secteur de Jan Mayen [The Delimitation Between Norway and Iceland of the Continental Shelf in the Jan Mayen Sector]." *Annuaire Français de Droit International* [French Yearbook of International Law] 27 (1981): 711–38, http://www.persee.fr/doc/afdi_0066-3085_1981_num_27_1_2469, accessed 3 April 2019.

Fitzmaurice, Malgosia. "International Protection of the Environment." *Recueil des Cours de l'Académie de Droit International de La Haye* [Collected Courses of the Hague Academy of International Law]. Vol. 293 (2001-VI), 9–488.

Fleischer, Carl-August. "La pêche [Fisheries]." *Traité du Nouveau Droit de la mer* [Treatise on the New Law of the Sea], 819–956. Eds. René-Jean Dupuy and Daniel Vignes. Paris: Éditions Economica, 1985.

Glowka, Lyle, et al. *A Guide to the Convention on Biological Diversity*. Gland: IUCN-The World Conservation Union, 1994, https://www.iucn.org/content/a-guide-convention-biological-diversity, accessed 3 April 2019.

Grant, John P., and J. Craig Barker. *Parry & Grant Encyclopedic Dictionary of International Law*. Oxford: Oxford University Press, 2009.

Guruswamy, Lakshman D. "The Convention on Biological Diversity: Exposing the Flawed Foundations." *Environmental Conservation* 26 (1999): 79–82, http://globalseminarhealth.wdfiles.com/local--files/pharmaceutical-harvesting/Guruswamy.pdf, accessed 3 April 2019.

Henne, Gudrun, and Saleem Fakir. "The Regime-Building of the Convention on Biological Diversity on the Road to Nairobi." *Max Planck Yearbook on United Nations Law* 3 (1999): 315–61.

Hermitte, Marie-Angèle. "La Convention sur la diversité biologique [The Convention on Biological Diversity]." *Annuaire Français de Droit International* [French Yearbook of International Law] 38 (1992): 844–70, http://www.persee.fr/docAsPDF/afdi_0066-3085_1992_num_38_1_3098.pdf, accessed 3 April 2019.

Institut de Droit International [Institute of International Law]. Session of Salzburg, *International Conciliation*, 1961, http://www.idi-iil.org/app/uploads/2017/06/1961_salz_02_en.pdf, accessed 3 April 2019.

Jóhansdóttir, Adelheidur, et al. "The Current Framework for International Governance of Biodiversity: Is It Doing More Harm Than Good?" *Review of European Community and International Environmental Law* 19 (2010): 139–49.

Kimball, Lee A. "The Biodiversity Convention: How to Make It Work." *Vanderbilt Journal of Transnational Law* 28 (1995): 763–75, http://69.90.183.227/doc/articles/2002-/A-00497.pdf, accessed 9 December 2016.

"L'affaire du F. OABV (Maroc c. France) [The Case of F. OABV (Morocco v. France)]." *Annuaire Français de Droit International* [French Yearbook of International Law] 4 (1958): 282–95, http://www.persee.fr/doc/afdi_0066-3085_1958_num_4_1_1382, accessed 3 April 2019.

Le Danff, Jean-Pierre. "La Convention sur la diversité biologique – tentative de bilan depuis le sommet de Rio de Janeiro [The Convention on Biological Diversity. A Preliminary Balance Sheet Since the Rio de Janeiro Summit]." *Vertigo. La revue juridique en sciences de l'environnement* [Vertigo. The Legal Journal on Environmental Sciences] 3 (2002), https://journals.openedition.org/vertigo/4168, accessed 3 April 2019.

Le Prestre, Philippe. "The Convention on Biological Diversity: Negotiating the Turn to Effective Implementation." *ISUMA: Canadian Journal of Policy Research* 3 (2002): 92–98.

Linderfalk, Ulf. "The Jan Mayen Case (Iceland/Norway): An Example of Successful Conciliation." Paper presented at the Symposium "Conciliation in the Globalized World of Today," organized by the Court of Conciliation and Arbitration within the O.S.C.E. (10–11 June 2015), https://papers.ssrn.com/sol3/papers.cfm?abstract_id=2783622, accessed 3 April 2019.

Maresca, Adolfo. *Il diritto dei trattati. La Convenzione codificatrice di Viena del 23 maggio 1969* [The Law of Treaties. The Codifying Convention of Vienna, 23 May 1969]. Milano: Dott. A. Giuffrè Editore, 1971.

Mensah, Thomas A. "Using Judicial Bodies for the Implementation and Enforcement of International Environmental Law." *International Law Between Universalism and Fragmentation. Festschrift in Honour of Gerhard Hafner*, 797–815. Eds. Isabelle Buffard et al. Leiden: Martinus Nijhoff Publishers, 2008.

Merrills, J. G. *International Dispute Settlement*, 4th ed. Cambridge: Cambridge University Press, 2011.

Montreal Protocol on Substances that Deplete the Ozone Layer, concluded at Montreal, 16 September 1987, entered into force 26 August 1989, https://ozone.unep.org/sites/default/files/Montreal-Protocol-English_0.pdf, accessed 3 April 2019.

———. *Report of the Fourth Meeting of the Parties to the Montreal Protocol on Substances that Deplete the Ozone Layer*, Copenhagen, 23–25 November 1992. Doc. UNEP/OzL.Pro.4/15 (25 November 1992), https://unep.ch/ozone/Meeting_Documents/mop/04mop/MOP_4.shtml, accessed 3 April 2019.

Morgera, Elisa, and Elsa Tsioumani. "Yesterday, Today and Tomorrow: Looking Afresh at the Convention on Biological Diversity." *Yearbook of International Environmental Law* 21 (2010): 3–40.

Nordquist, Myron H., et al. (eds.). *United Nations Convention on the Law of the Sea 1982 Commentary*. Vol. V. *Settlement of Disputes, General and Final Provisions and related Annexes and resolutions*. Leiden: Martinus Nijhoff Publishers, 1989.

———. *United Nations Convention on the Law of the Sea 1982 Commentary*. Vol. IV. *Third Committee: Protection and Preservation of the Marine Environment, Marine Scientific Research, and Development and Transfer of Marine Technology*. Leiden: Martinus Nijhoff Publishers, 1991.

Ong, David M. "International Environmental Law Governing Threats to Biological Diversity." *Research Handbook on International Environmental Law*, 567–85. Eds. Malgosia Fitzmaurice, David M. Ong, and Panos Merkouris. Cheltenham, Gloucestershire: Edward Elgar, 2010.

People's Republic of China. Ministry of Environmental Protection. *China's Fifth National Report on Implementation of the Convention on Biological Diversity*. Beijing: Ministry of Environmental Protection, 2014, https://www.cbd.int/reports, accessed 3 April 2019.

————. State Environmental Protection Administration of China. *China's Second National Report on Implementing the Convention on Biological Diversity.* Beijing: China National Environmental Science Press, 2001, https://www. cbd.int/reports, accessed 3 April 2019.

————. *China's Third National Report on Implementation of the Convention on Biological Diversity.* Beijing: State Environmental Protection Administration of China, 2005, https://www.cbd.int/reports, accessed 3 April 2019.

Ramsar Convention. Convention on Wetlands of International Importance Especially as Waterfowl Habitat, signed at Ramsar, Iran, on 2 February 1971, as amended by the Protocol of 3 December 1982 and the Amendments of 28 May 1987, https://www.ramsar.org/sites/default/files/documents/library/current_convention_text_e.pdf, accessed 24 March 2019.

————. "China." *Ramsar Sites Information Service,* http://www.ramsar.org/wetland/china, accessed 2 April 2019.

————. "Philippines." *Ramsar Sites Information Service,* http://www.ramsar.org/wetland/philippines, accessed 2 April 2019.

————. The Ramsar Convention Secretariat. *Coral Reefs: Critical Wetlands in Severe Danger.* RAMSAR CoP12 Doc. 25 (1 April 2015), https://www.ramsar.org/sites/default/files/documents/library/factsheet_5_coral_reefs_en.pdf, accessed 2 April 2019.

Ranjeva, Raymond. "Le règlement des différends [Dispute Settlement]." *Traité du Nouveau Droit de la mer* [Treatise on the New Law of the Sea], 1105–68. Eds. René-Jean Dupuy and Daniel Vignes. Paris: Éditions Economica, 1985.

"Rapport de la Commission de conciliation Franco-Siamoise, Washington, 27 Juin 1947 [Report of the Franco-Siamese Conciliation Commission, Washington, 27 June 1947]." *Reports of International Arbitral Awards.* Vol. 28, 433–50. New York: United Nations, 2007, http://legal.un.org/riaa/cases/vol_XXVIII/433-450.pdf, accessed 3 April 2019.

Reimann, Heinrich B. "Le règlement pacifique des différends à l'OSCE [The Peaceful Settlement of Disputes at the OSCE]." *La promotion de la justice, des droits de l'homme et du règlement des conflits par le droit international. Liber Amicorum Lucius Caflisch* [Promoting Justice, Human Rights and Conflict Resolution through International Law. Liber Amicorum Lucius Caflisch], 891–96. Ed. Marcelo G. Kohen. Leiden: Martinus Nijhoff Publishers, 2007.

Republic of the Philippines. Department of Environment and Natural Resources. Protected Areas and Wildlife Bureau. *The First Philippine National Report to the Convention on Biological Diversity.* Quezon City: Department of Environment and Natural Resources, 1998, https://www.cbd.int/reports/, accessed 25 April 2010.

Richardson, Elliott L. "Jan Mayen in Perspective." *American Journal of International Law* 82 (1988): 443–58.

Sohn, Louis B. "Settlement of Disputes Relating to the Interpretation and Application of Treaties." *Recueil des Cours de l'Académie de Droit International de La Haye* [Collected Courses of the Hague Academy of International Law]. Vol. 150 (1976-II), 195–294.

South China Sea Arbitration. Award on Jurisdiction and Admissibility, 29 October 2015, https://pcacases.com/web/sendAttach/1506, accessed 26 March 2019.

———. Hearing on Jurisdiction and Admissibility. Transcript, Day 2 (8 July 2015), https://pcacases.com/web/sendAttach/1400, accessed 3 April 2019.

———. Transcript, Day 3 (13 July 2015), https://pcacases.com/web/sendAttach/1401, accessed 3 April 2019.

———. *Memorial of the Philippines* (30 March 2014). Annex 19. Memorandum from Erlinda F. Basilio, Acting Assistant Secretary, Office of Asian and Pacific Affairs, Department of Foreign Affairs, Republic of the Philippines, to the Secretary of Foreign Affairs of the Republic of the Philippines (29 March 1995). Vol. III, 213–17, https://files.pca-cpa.org/pcadocs/The%20 Philippines%27%20Memorial%20-%20Volume%20III%20%28Annexes%20 1-60%29.pdf, accessed 26 March 2019.

———. Annex 20. Memorandum from Lauro L. Baja, Jr., Assistant Secretary, Office of Asian and Pacific Affairs, Department of Foreign Affairs, Republic of the Philippines, to the Secretary of Foreign Affairs of the Republic of the Philippines (7 April 1995). Vol. III, 219–24, https://files.pca-cpa.org/ pcadocs/The%20Philippines%27%20Memorial%20-%20Volume%20III%20 %28Annexes%201-60%29.pdf, accessed 26 March 2019.

———. Annex 28. Memorandum from Fact Finding Committee, National Police Commission, Republic of the Philippines, to Chairman and Members of the Regional Committee on Illegal Entrants for Region 1, Republic of the Philippines (28 January 1998). Vol. III, 271–76, https://files.pca-cpa.org/ pcadocs/The%20Philippines%27%20Memorial%20-%20Volume%20III%20 %28Annexes%201-60%29.pdf, accessed 26 March 2019.

———. Annex 29. Memorandum from Assistant Secretary of the Department of Foreign Affairs, Republic of the Philippines to the Secretary of Foreign Affairs of the Republic of the Philippines (23 March 1998). Vol. III, 277–82, https://files.pca-cpa.org/pcadocs/The%20Philippines%27%20Memorial%20 -%20Volume%20III%20%28Annexes%201-60%29.pdf, accessed 26 March 2019.

———. Annex 30. People of the Philippines v. Shin Ye Fen, et al., Criminal Case No. RTC 2357-I, Decision, Regional Trial Court, Third Judicial Region, Branch 69, Iba, Zambales, Philippines (29 April 1998). Vol. III, 283–86, https://files.pca-cpa.org/pcadocs/The%20Philippines%27%20Memorial%20 -%20Volume%20III%20%28Annexes%201-60%29.pdf, accessed 26 March 2019.

————. Annex 31. People of the Philippines v. Wuh Tsu Kai, et al., Criminal Case No. RTC 2362-I, Decision, Regional Trial Court, Third Judicial Region, Branch 69, Iba, Zambales, Philippines (29 April 1998). Vol. III, 287–90, https://files.pca-cpa.org/pcadocs/The%20Philippines%27%20Memorial%20-%20Volume%20III%20%28Annexes%201-60%29.pdf, accessed 26 March 2019.

————. Annex 32. People of the Philippines v. Zin Dao Guo, et al., Criminal Case No. RTC 2363-I, Decision, Regional Trial Court, Third Judicial Region, Branch 69, Iba, Zambales, Philippines (29 April 1998). Vol. III, 291–94, https://files.pca-cpa.org/pcadocs/The%20Philippines%27%20Memorial%20-%20Volume%20III%20%28Annexes%201-60%29.pdf, accessed 26 March 2019.

————. Annex 37. Memorandum from the Embassy of the Republic of the Philippines in Beijing to the Secretary of Foreign Affairs of the Republic of the Philippines, No. ZPE-85-98-S (4 December 1998). Vol. III, 315–19, https://files.pca-pa.org/pcadocs/The%20Philippines%27%20Memorial%20-%20Volume%20III%20%28Annexes%201-60%29.pdf, accessed 26 March 2019.

————. Annex 43. Memorandum from the Embassy of the Republic of the Philippines in Beijing to the Secretary of Foreign Affairs of the Republic of the Philippines, No. ZPE-06-2001-S (13 February 2001). Vol. III, 351–55, https://files.pca-pa.org/pcadocs/The%20Philippines%27%20Memorial%20-%20Volume%20III%20%28Annexes%201-60%29.pdf, accessed 26 March 2019.

————. Annex 45. Memorandum from Willy C. Gaa, Assistant Secretary of Foreign Affairs, Republic of the Philippines to Secretary of Foreign Affairs, Republic of the Philippines (14 February 2001). Vol. III, 363–67, https://files.pca-pa.org/pcadocs/The%20Philippines%27%20Memorial%20-%20Volume%20III%20%28Annexes%201-60%29.pdf, accessed 26 March 2019.

————. Annex 47. Memorandum from the Embassy of the Republic of the Philippines in Beijing to the Secretary of Foreign Affairs of the Republic of the Philippines, No. ZPE-09-2001-S (17 March 2001). Vol. III, 375–78, https://files.pca-pa.org/pcadocs/The%20Philippines%27%20Memorial%20-%20Volume%20III%20%28Annexes%201-60%29.pdf, accessed 26 March 2019.

————. Annex 48. Memorandum from Josue L. Villa, Embassy of the Republic of the Philippines in Beijing, to the Secretary of Foreign Affairs of the Republic of the Philippines (21 May 2001). Vol. III, 379–94, https://files.pca-pa.org/pcadocs/The%20Philippines%27%20Memorial%20-%20Volume%20III%20%28Annexes%201-60%29.pdf, accessed 26 March 2019.

————. Annex 51. Memorandum from Josue L. Villa, Embassy of the Republic of the Philippines in Beijing, to the Secretary of Foreign Affairs of the

Republic of the Philippines (19 August 2002). Vol. III, 405–409, https://files.pca-pa.org/pcadocs/The%20Philippines%27%20Memorial%20-%20Volume%20III%20%28Annexes%201-60%29.pdf, accessed 26 March 2019.

———. Annex 58. Memorandum from the Secretary of Foreign Affairs of the Republic of the Philippines to the President of the Republic of the Philippines (11 January 2006). Vol. III, 481–84, https://files.pca-pa.org/pcadocs/The%20Philippines%27%20Memorial%20-%20Volume%20III%20%28Annexes%201-60%29.pdf, accessed 26 March 2019.

———. Annex 186. Note Verbale from the Department of Foreign Affairs of the Republic of the Philippines to the Embassy of the People's Republic of China in Manila No. 2000100 (14 January 2000). Vol. VI, 259–61, https://files.pca-cpa.org/pcadocs/The%20Philippines%27%20Memorial%20-%20Volume%20VI%20%28Annexes%20158-221%29.pdf, accessed 26 March 2019.

———. Annex 205. Memorandum from the Secretary of Foreign Affairs to the President of the Republic of the Philippines (11 January 2006). Vol. VI, 377–81, https://files.pca-cpa.org/pcadocs/The%20Philippines%27%20Memorial%20-%20Volume%20VI%20%28Annexes%20158-221%29.pdf, accessed 26 March 2019.

———. *Supplemental Documents of the Philippines* (19 November 2015). Annex 608. Department of Foreign Affairs—Republic of the Philippines. Statement on China's Reclamation Activities and their Impact on the Region's Marine Environment (13 April 2015). Vol. I, 7–9, https://files.pca-cpa.org/pcadocs/The%20Philippines%27%20Supplemental%20Documents%20-%20Volume%20I%20%28Annexes%20607-667%29.pdf, accessed 2 April 2019.

Tanaka, Yoshifumi. *The International Law of the Sea*, 2nd ed. Cambridge: Cambridge University Press, 2015.

Tardieu, Aurélie. "Les conferences des États parties [Conferences of States Parties]." *Annuaire Français de Droit International* [French Yearbook of International Law] 57 (2011): 111–43, http://www.persee.fr/docAsPDF/afdi_0066-3085_2011_num_57_1_4178.pdf, accessed 3 April 2019.

Tinker, Catherine. "A 'New Breed' of Treaty: The United Nations Convention on Biological Diversity." *Pace Environmental Law Review* 13 (1995): 191–218, http://digitalcommons.pace.edu/cgi/viewcontent.cgi?article397&context=pelr, accessed 2 April 2019.

———. "Responsibility for Biological Diversity Under International Law." *Vanderbilt Journal of Transnational Law* 28 (1995): 777–821.

Tomuschat, Christian. "Sleeping Beauty: The OSCE Court of Conciliation and Arbitration." *Security Community. The OSCE Magazine* 2 (2014): 36–37, http://www.osce.org/cca, accessed 3 April 2019.

Tomuschat, Christian, et al. (eds.). *Conciliation in International Law: The OSCE Court of Conciliation and Arbitration*. Leiden: Martinus Nijhoff Publishers, 2016.

United Nations. *Law of the Sea: Protection and Preservation of the Marine Environment. Report of the Secretary General.* UN Doc. A/44/461 (18 September 1989), https://digitallibrary.un.org/record/76086?ln=en, accessed 3 April 2019.

United Nations Convention on the Law of the Sea, concluded at Montego Bay on 10 December, entered into force on 16 November 1994, http://www.un.org/Depts/los/convention_agreements/texts/unclos/closindx.htm, accessed 21 March 2019.

———. Meeting of the States Parties. Fifteenth Meeting, New York, 16–24 June 2005. *Report of the Fifteenth Meeting of States Parties.* Doc. SPLOS/135 (25 July 2005), https://documents-dds-ny.un.org/doc/UNDOC/GEN/N05/439/16/PDF/N0543916.pdf?OpenElement, accessed 3 April 2019.

———. Twenty-First Meeting, New York, 13–17 June 2011. *Report of the Twenty-First Meeting of States Parties,* Doc. SPLOS/231 (29 June 2011), https://documents-dds-ny.un.org/doc/UNDOC/GEN/N11/393/68/PDF/N1139368.pdf?OpenElement, accessed 3 April 2019.

———. Twenty-Fifth Meeting, New York, 8–12 June 2015. *Report of the Twenty-Fifth Meeting of States Parties.* Doc. SPLOS/287 (13 July 2015), https://documents-dds-ny.un.org/doc/UNDOC/GEN/N15/217/42/PDF/N1521742.pdf?OpenElement, accessed 3 April 2019.

———. Twenty-Sixth Meeting, New York, 20–24 June 2016. *Report of the Twenty-Sixth Meeting of States Parties.* Doc. SPLOS/30 (2 August 2016), https://documents-dds-ny.un.org/doc/UNDOC/GEN/N16/245/62/PDF/N1624562.pdf?OpenElement, accessed 3 April 2019.

United Nations Environment Programme (UNEP). Intergovernmental Negotiating Committee for a Convention on Biological Diversity. *Fifth Revised Draft Convention on Biological Diversity. Explanatory Note.* Doc. UNEP/Bio.Div/N7-INC.5/2 (20 February 1992), https://www.cbd.int/doc/meetings/iccbd/bdn-07-inc-05/official/bdn-07-inc-05-02-en.pdf, accessed 3 April 2019.

United Nations Office of Legal Affairs. Codification Division. *United Nations Diplomatic Conferences. Third United Nations Conference on the Law of the Sea, 1973–1982,* 2019, http://legal.un.org/diplomaticconferences/1973_los/, accessed 4 April 2019.

Vignes, Daniel. "La sentence de la Commission de conciliation franco-italienne dans l'affaire du différend sur les biens immeubles appartenant à l'ordre de Saint-Maurice et de Saint Lazare (Hospice du Petit Saint-Bernard) [The Award of the Franco-Italian Conciliation Commission in the Case relating to the Dispute concerning Immovable Property belonging to the Order of Saint Maurice and Saint Lazarus (Little Saint Bernard Hospice)]." *Annuaire Français de Droit International* [French Yearbook of International Law] 11 (1965): 319–32, http://www.persee.fr/doc/afdi_0066-3085_1965_num_11_1_1822, accessed 3 April 2019.

Villani, Ugo. "Osservazioni sulla soluzione delle controversie nelle convenzioni di codificazione del diritto internazionale [Observations on the Settlement of Disputes in Conventions Codifying International Law]." *Le droit international à l'heure de sa codification. Études en l'honneur de Roberto Ago* [International Law at the Time of Its Codification. Studies in Honor of Roberto Ago]. Vol. III. *Les différends entre États et la responsabilité* [Inter-State Disputes and Responsibility], 497–521. Milano: Dott. A. Giuffrè Editore, 1987.

Wehberg, Hans. "Die Vergleichskommissionen im modernen Völkerrecht [Conciliation Commissions in Modern International Law]." *Zeitschrift für ausländisches öffentliches Recht und Völkerrecht* [Journal for Foreign Public Law and International Law] 19 (1958): 551–93, http://www.zaoerv.de/19_1958/19_1958_1_3_a_551_593.pdf, accessed 3 April 2019.

Wolfrum, Rüdiger, and Nele Matz. "The Interplay of the United Nations Convention on the Law of the Sea and the Convention on Biological Diversity." *Max Planck Yearbook of United Nations Law* 4 (2000): 445–80.

CHAPTER 4

The Protection and Preservation of Endangered Species and Fragile Ecosystems and the Jurisdiction of the Arbitral Tribunal

The recourse to compulsory dispute settlement procedures entailing binding decisions under Section 2 of Part XV of the United Nations Convention on the Law of the Sea, ("the Convention"), whether this be the International Tribunal for the Law of the Sea ("ITLOS"), the International Court of Justice ("ICJ"), arbitration under Annex VII of the Convention, or special arbitration under Annex VIII of the Convention, assumes that a dispute exists between two States Parties to the Convention and that the dispute concerns the interpretation and application of the Convention. A court or tribunal may exercise jurisdiction under Article 288 of the Convention only if these two assumptions are fulfilled.[1]

Anyone who reads the diplomatic exchanges can have no doubt that there existed a dispute between China and the Philippines regarding the activities of Chinese fishermen and of China in the South China Sea In its *Memorial*, the Philippines, when explaining its claim that China's toleration of the Chinese fishermen's practices that were particularly harmful to

---

[1] *South China Sea Arbitration*, Award on Jurisdiction and Admissibility, 29 October 2015, 45, para. 130, https://pcacases.com/web/sendAttach/1506, accessed 26 March 2019. The text of Article 288(1) is as follows: "1. A court or tribunal referred to in article 287 shall have jurisdiction over any dispute concerning the interpretation or application of this Convention which is submitted to it in accordance with this Part." *United Nations Convention on the Law of the Sea*, concluded at Montego Bay on 10 December, entered into force on 16 November 1994, http://www.un.org/Depts/los/convention_agreements/texts/unclos/closindx.htm, accessed 21 March 2019.

© The Author(s) 2020
A. C. Robles Jr., *Endangered Species
and Fragile Ecosystems in the South China Sea*,
https://doi.org/10.1007/978-981-13-9813-1_4

unique, fragile, and highly vulnerable ecosystems of the South China Sea violated China's international obligations to protect and not pollute the marine environment, argued that China had violated not just the Convention but also the Convention on Biological Diversity ("CBD").[2] When elaborating on its allegation that China's construction of artificial islands, installations, and structures at Mischief Reef breached China's obligation to protect and preserve the marine environment, the Philippines again alleged that China had also violated the CBD, for construction had harmed fragile ecosystems and resulted in significant damage to the habitats of vulnerable species.[3] The question that the Tribunal had to ask the Philippines was whether the Philippines was presenting a claim under the CBD.

In answer to questions posed by the Tribunal, the Philippines clarified that it was not presenting a claim arising under the CBD.[4] It explained that it was using the normative content of the CBD (e.g., the obligation to protect endangered species) to inform the scope and content of China's obligations under Articles 192 and 194 of the Convention.[5] The Philippines argued that this approach was entirely consistent with the scheme of the Convention, which at the time of its adoption had anticipated that its normative content would be further developed in other international instruments.[6]

In support of this approach, the Philippines cited the rule of treaty interpretation contained in Article 31(3)(c) of the Vienna Convention on the

[2] South China Sea Arbitration, Memorial of the Philippines (30 March 2014), vol. I, 174, 190–93, paras. 6.49, 6.82–6.89 ("MP"), https://files.pca-cpa.org/pcadocs/Memorial%20of%20the%20Philippines%20Volume%20I.pdf, accessed 27 March 2019. Convention on Biological Diversity (CBD), signed at Rio de Janeiro on 5 June 1992, entered into force on 29 December 1993, https://www.cbd.int/doc/legal/cbd-en.pdf, accessed 26 March 2019. The jurisdiction of international courts and tribunals is "the power...to hear the dispute, [the] legal aptitude to render a decision." Ibrahim F. I. Shihata, The Power of the International Court to Determine Its Own Jurisdiction: Compétence de la Compétence (Dordrecht: Springer Science+Business Media, 1965), 1.

[3] MP, vol. I, 200–201, paras. 6.108–6.110.

[4] South China Sea Arbitration, Supplemental Written Submission of the Philippines (16 March 2015), vol. I, 56, para. 11.3 ("SWSP"), http://www.pcacases.com/pcadocs/Supplemental%20Written%20Submission%20Volume%20I.pdf, accessed 5 April 2019. Question No. 11 was part of the Tribunal's "Request for Further Written Argument by the Philippines Pursuant to Article 25(2) of the Rules of Procedure," issued on 16 December 2014. The text of the Request itself is not available. The text of the questions may only be obtained by reading the SWSP.

[5] Ibid., 57, para. 11.8; 56, para. 11.3.

[6] Ibid., 56, para. 11.6.

Law of Treaties and the provision on applicable law in Article 293 of the Convention. The former provides that in interpreting a treaty, account shall be taken of "[a]ny relevant rules of international law applicable in the relations between the parties."[7] Now both the Philippines and China were States Parties to the CBD. It follows that the CBD's provisions constituted "relevant rules of international law" applicable in relations between them.[8] This approach to treaty interpretation, known as the "systemic approach," makes it possible to take into account as the context of the treaty all rules that spring from any of the formal sources of international law (treaties, customary law, general principles of law).[9] The rules in question may be the rules in force at the time the treaty was concluded or the rules in force at the time the treaty is being interpreted. It is the terms of the treaty being interpreted that indicates whether its provisions may be affected by the evolution of international law.[10] As we have seen in the previous chapter, the Philippines invoked the CBD, together with CITES, because it believed that the concepts contained in these two treaties were appropriate to describe the conduct of Chinese nationals and of China.[11] It is doubtful whether the Philippine government officials who drafted the communications with China had in mind Article 31(3)(c) of the Vienna Convention on the Law of Treaties or the "systemic approach" to treaty interpretation.[12]

The Philippines also cited Article 293(1) of the Convention, under the terms of which "[a] court or tribunal having jurisdiction under this

[7] *Vienna Convention on the Law of Treaties*, done at Vienna on 23 May 1969, entered into force on 27 January 1980, http://legal.un.org/ilc/texts/instruments/english/conventions/1_1_1969.pdf, accessed 9 April 2019.

[8] *MP*, 56, para. 11.4.

[9] Ulf Linderfalk, *On the Interpretation of Treaties: The Modern International Law as Expressed in the 1969 Vienna Convention on the Law of Treaties* (Dordrecht: Springer, 2007), 177.

[10] Mustafa Kamal Yasseen, "L'interprétation des traités d'après la Convention de Vienne sur le droit des traités [Treaty Interpretation According to the Vienna Convention on the Law of Treaties]," *Recueil des Cours de l'Académie de Droit International de La Haye* [Collected Courses of the Hague Academy of International Law] 151 (1976–III): 66–67. Yasseen was chair of the Drafting Committee of the Vienna Conference on the Law of Treaties, which drafted the Vienna Convention on the Law of Treaties.

[11] *Convention on International Trade in Endangered Species of Wild Fauna and Flora*, signed at Washington, DC on 3 March 1973, amended at Bonn, on 22 June 1979, amended at Gaborone, on 30 April 1983, https://www.cites.org/eng/disc/text.php, accessed 26 March 2019.

[12] On Article 31(3)(c) and the systemic approach to treaty interpretation, see also Oliver Dörr, "Article 31. General Rule of Interpretation," in Oliver Dörr and Kirsten Schmalenbach (eds.), *Vienna Convention on the Law of Treaties: A Commentary* (2nd ed.; Berlin: Springer

section shall apply this Convention and other rules of international law not incompatible with this Convention." The Philippines concluded that the Tribunal had jurisdiction to look to the CBD to inform the Tribunal's interpretation and application of Articles 192 and 194 of the Convention.[13]

Following the Philippines' explanation, the Tribunal was satisfied

> that the incidents alleged by the Philippines, in particular as to the use of dangerous substances such as dynamite or cyanide to extract fish, clams, or corals at and around Scarborough Shoal and Second Thomas Shoal, could involve violations of obligations under Article 194 of the Convention, read in conjunction with Article 192 of the Convention, to take measures to prevent, reduce and control pollution of the marine environment.[14]

It is likely that the Tribunal referred to "measures to prevent, produce and control pollution of the marine environment" because that was the title of Article 194, paragraph 5 of which contained the reference to "rare or fragile ecosystems" and "the habitat of depleted threatened or endangered species." The Tribunal accepted the Philippine assertions that while the Philippines considered China's actions and failures to be inconsistent with provisions of the CBD, the Philippines had not presented a claim under the CBD as such. The Tribunal was also satisfied that Article 293(1) of the Convention, together with Article 31(3)(c) of the Vienna Convention on the Law of Treaties, enabled it to consider provisions of the CBD for purposes of interpreting the content and standard of Articles 192 and 194 of the Convention.[15]

The Tribunal acknowledged that the Philippines' factual allegations could potentially give rise to disputes under both the Convention and

---

Verlag GmbH Germany, 2018), 603–12; Adolfo Maresca, *Il diritto dei trattati. La Convenzione codificatrice di Viena del 23 maggio 1969* [The Law of Treaties. The Codifying Convention of Vienna, 23 May 1969] (Milano: Dott. A. Giuffrè Editore, 1971), 358; Mark E. Villiger, *Commentary on the 1969 Vienna Convention on the Law of Treaties* (Leiden: Martinus Nijhoff Publishers, 2009), 433

[13] *SWSP*, vol. I, 58, para. 11.13.

[14] Award on Jurisdiction, 69, paras. 174–75.

[15] Ibid., 69, para. 176.

the CBD, but it was not convinced that this possibility could necessarily exclude jurisdiction to consider Philippine Submissions No. 11 and 12(b). The Tribunal pointed out that treaties often mirror each other in substantive content, and that as a result, it was not uncommon for more than one treaty to bear on a dispute. Nevertheless, the rights and obligations of a State under one treaty, whether they are similar to or identical with the rights and obligations under another convention, have a separate existence.[16]

The finding that a dispute existed between the Philippines and China relating to China's alleged violations of the obligation to preserve and protect the marine environment did not exhaust the question of the Tribunal's jurisdiction over the Philippine claims that China had tolerated illegal harvesting of endangered species and cyanide and dynamite fishing by its nationals and had failed to protect and preserve fragile marine ecosystems by its construction activities in the South China Sea. While the dispute settlement mechanism of the Convention has been described as comprehensive, Section 1 of Part XV allows States Parties to the Convention to resolve their disputes regarding the interpretation and application of the Convention through means of settlement previously agreed upon outside the Convention framework. The Tribunal called the conditions under which such agreements may preclude its jurisdiction as the "preconditions to the Tribunal's jurisdiction," an expression that does not appear in the Convention.[17] The fact that the preconditions have been fulfilled does not automatically mean that the Tribunal has jurisdiction to hear the case. The Tribunal also had to examine the question whether the limitations and optional exceptions to jurisdiction provided for in Section 3 of Part XV were applicable to the Philippine claims. The difference between the two categories is that the limitations apply automatically—no declaration is necessary in order to accept them—whereas the optional exceptions have to be activated by means of a declaration. China had made such a declaration in 2006 in which it excluded from compulsory dispute settlement entailing binding decisions all the categories of disputes referred to in Article 298(1)(a), (b), and (c) of the Convention. Should none of the limitations and exceptions be applicable, then the Tribunal would have jurisdiction to hear the case.

---

[16] Ibid., 69–70, para. 177.
[17] Ibid., 75.

It may be recalled that China refused to appear before the Arbitral Tribunal and that its 2014 Position Paper did not refer to environmental issues at all.[18] Given China's silence on the issues, the determination whether there were any possible objections to its jurisdiction in respect of Submissions No. 11 and 12(b) was undertaken by means of questions that the Tribunal posed to the Philippines on three occasions: in December 2014, following the publication of China's Position Paper[19]; in June 2015, prior to the Hearing on Jurisdiction[20]; and in July 2015, during the Hearing on Jurisdiction.[21] The Philippines gave written answers in March 2015 to the first set of questions and oral answers to the second and third set of questions during the Hearing on Jurisdiction in July 2015. The Philippine responses formed the basis of the Tribunal's decisions relating to the preconditions to jurisdiction, on the one hand, and the limitations and exceptions to jurisdiction, on the other.

The Tribunal's decisions on these two issues will be the object of this chapter. The first part will discuss the Tribunal's consideration of the question whether the Philippines and China had agreed to resolve their disputes concerning the protection and preservation of the marine environment through the mechanisms set out in the CBD and whether the Philippines had fulfilled the obligation to exchange views with China regarding the means of resolving the dispute relating to the protection and preservation of the marine environment. The second part will analyze the Tribunal's reasoning on the question whether the coastal State's

---

[18] *SWSP*, Annex 467, People's Republic of China, Position Paper of the Government of the People's Republic of China on the Matter of Jurisdiction in the South China Sea Arbitration Initiated by the Republic of the Philippines (7 December 2014), vol. VIII, 19–33 ("Position Paper"), https://files.pca-cpa.org/pcadocs/The%20Philippines%27%20Supplemental%20Written%20Submission%20-%20Volume%20VIII%20%28Annexes%20466-499%29.pdf, accessed 4 April 2019.

[19] The text of the Tribunal's request is not available. The reader can only deduce its contents by reading the Philippine submissions. *SWSP*, 55–58, paras. 11.1–11.13.

[20] The text of the Tribunal's list of issues is not available. The reader can only deduce its contents by reading the transcripts of the Hearing itself. *South China Sea Arbitration*, Hearing on Jurisdiction and Admissibility, Transcript, Day 2 (8 July 2015), 74–86 ("Hearing on Jurisdiction"), https://pcacases.com/web/sendAttach/1400, accessed 7 April 2019.

[21] The text of the Tribunal's questions is not available. The reader can only deduce its contents by reading the transcripts of the Hearing itself. Hearing on Jurisdiction, Transcript, Day 3 (13 July 2015), 42–47, https://pcacases.com/web/sendAttach/1401, accessed 7 April 2019.

sovereign rights in the exclusive economic zone ("EEZ") constituted a limitation to the Tribunal's jurisdiction and whether the military activities exception as well as the law enforcement activities exception referred to in China's 2006 Declaration were applicable to the disputes relating to the protection and preservation of the marine environment.

I. The Preconditions to the Tribunal's Jurisdiction

Under the heading of "preconditions," the Tribunal considered whether there were any circumstances that would preclude access to the compulsory dispute settlement procedures in Section 2 of Part XV of the Convention.[22] The first set of preconditions is that no other agreement on dispute settlement exists between the Philippines and China that excludes recourse to Section 2. The agreement that the Tribunal had in mind in relation to the Philippine Submissions No. 11 and 12(b) was the CBD. The second precondition that had to be fulfilled was an exchange of views between the parties to a dispute on the settlement of their dispute.

A. The CBD as an Agreement Excluding Recourse to Arbitration

The first and most obvious question relating to jurisdiction arose from the Philippine allegation made in its *Memorial* that China had breached the obligation to protect and not pollute the marine environment by violating the CBD.[23] If the dispute concerned interpretation and application of the CBD, then the Philippines admitted that it should have submitted the dispute to compulsory conciliation.[24] The Philippines clarified during the Hearing on Jurisdiction that it had made no attempt to invoke Article 27(4) of the CBD because it had no dispute with China relating to compliance with the CBD and that its dispute concerned interpretation of Part XII of the Convention and related provisions.[25]

The Tribunal responded to the Philippine clarification in two distinct ways. On the one hand, it accepted the assertion. The Tribunal

---

[22] Award on Jurisdiction, 75, para. 189.

[23] *MP*, vol. I, 190–91, paras. 6.82–6.89.

[24] Hearing on Jurisdiction, Transcript, Day 3, 42.

[25] Ibid.

acknowledged that the Philippine claims could give rise to a dispute under both the CBD and the Convention, but that this did not necessarily exclude its jurisdiction to consider the environmental claims of the Philippines. The Tribunal explained that although two treaties may have a bearing on the same dispute, the rights and obligations under the separate agreements continued to have separate existence. For the Tribunal, the CBD could be considered for purposes of interpreting the content and standard of Articles 192 and 194 of the Convention.[26] On the other hand, the Tribunal was obliged to consider whether the CBD could constitute a bar to its jurisdiction under Section 1 of Part XV the Convention.

To be more specific, the question that the Tribunal had to answer was the following: Did the Philippines and China, in ratifying the CBD, agree to settle disputes concerning Articles 192 and 194 of the Convention—insofar as those disputes concern the protection of marine biological diversity—using procedures set out in Article 27 of the CBD?[27] A response to this question necessitated interpretation and application of Articles 281 and 282 of the Convention. Under the former, the CBD could be considered as "a peaceful means of their [the parties to the dispute] own choice." Under the latter, the CBD could be considered as a "general, regional or bilateral agreement" through which the Philippines and China have agreed to settle their dispute by a procedure entailing a binding decision.

1. The CBD as an Agreement Within the Meaning of Article 281

States Parties to the Convention have the right to settle disputes by any peaceful means of their choice. In the event they have decided to do so outside the Convention framework, Article 281 of the Convention regulates the relationship between these means and the Convention's dispute settlement system.[28]

---

[26] Award on Jurisdiction, 69–70, paras. 176–78.

[27] Ibid., 103, para. 283.

[28] The text of Article 281, Procedure where No Settlement Has Been Reached by the Parties, is as follows:

1. If the States Parties which are parties to a dispute concerning the interpretation or application of this Convention have agreed to seek settlement of the dispute by a peaceful means of their own choice, the procedures provided for in this Part apply

Article 281 allows parties to a dispute relating to the interpretation and application of the Convention to resort to means of settlement outside the Convention.[29] Assuming they have chosen other means of settlement, they must first attempt to solve their dispute through the means that they have chosen. As the Tribunal explained, three preconditions must be fulfilled before they can subsequently resort to compulsory procedures of Section 2 of Part XV: No settlement has been reached through the agreed means; the agreement does not exclude any further procedure, such as compulsory dispute settlement entailing binding decisions; and any agreed time limits have expired.[30] Once these three conditions have been fulfilled, any one of the parties to the dispute may submit the dispute to the compulsory procedure that the two have chosen in common from the list in Article 287 or to an arbitral tribunal, if they have chosen different means, or no means, of compulsory dispute settlement entailing binding decisions.

The Tribunal's concern stemmed from the Philippine claim that China had violated the CBD as well as Articles 192 and 194 of the Convention.[31] To the extent that both the CBD and the Convention protect marine biodiversity, it could be argued that the Philippines and China had agreed to settle their disputes concerning the protection and conservation of biodiversity in accordance with Article 27 of the CBD. Furthermore, if the CBD excluded the recourse to further proceedings, then the Tribunal's jurisdiction to decide the Philippine claims would be barred. Concretely, this would have meant an obligation for the Philippines to attempt compulsory conciliation provided for in Article 27(4) of the CBD, as described in Chapter 3.[32]

only where no settlement has been reached by recourse to such means and the agreement between the parties does not exclude any further procedure.
2. If the parties have also agreed on a time-limit, paragraph 1 applies only upon the expiration of that time-limit.

[29] Myron H. Nordquist et al. (eds.), *United Nations Convention on the Law of the Sea 1982 Commentary*, vol. V, *Settlement of Disputes, General and Final Provisions and Related Annexes and Resolutions* (Leiden: Martinus Nijhoff Publishers, 1989), 22.

[30] Award on Jurisdiction, 76, para. 195.

[31] *MP*, vol. I, 190–91, paras. 6.82–6.89.

[32] Award on Jurisdiction, 102, paras. 274–76.

The Philippines was conscious that in the *Southern Bluefin Tuna Arbitration*, which had also been initiated under Annex VII, the Tribunal had declined jurisdiction because a previous agreement excluded the resort to compulsory dispute settlement under Section 2 of Part XV.[33] In that case, Australia and New Zealand had alleged that Japan had breached Articles 64 and 116 to 119 of the Convention in relation to the conservation and management of the Southern Bluefin Tuna stock. Japan had responded that recourse to arbitration under Annex VII was excluded because the 1993 Convention for the Conservation of Southern Bluefin Tuna, to which the three States were Parties, provided for a dispute settlement procedure.[34] The Tribunal agreed with Japan, despite the fact that the 1993 Convention did not explicitly exclude the resort to further proceedings under the Convention.[35]

Adopting an approach opposite to that of the Tribunal in the *Southern Bluefin Tuna* case, the Philippines denied that the CBD constituted an agreement within the meaning of Article 281, adducing four reasons. First, for Article 281 to apply, there should have been an agreement between the Philippines and China that the CBD's dispute compulsory dispute mechanisms were mechanisms to settle disputes relating to interpretation and application of the Convention.[36] The Philippines contended that the CBD's dispute settlement procedures applied exclusively to disputes relating to the interpretation and application of the CBD and not to disputes relating to the interpretation and application of the Convention. To convert the CBD's dispute settlement mechanisms into an agreement to settle disputes regarding the interpretation and

---

[33] Hearing on Jurisdiction, Transcript, Day 2, 114–16. *Southern Bluefin Tuna (New Zealand-Japan, Australia-Japan)*, Award on Jurisdiction and Admissibility, Decision of 4 August 2000, *Reports of International Arbitral Awards*, vol. 23 (New York: United Nations, 2006), 1–57, http://legal.un.org/riaa/cases/vol_XXIII/1-57.pdf, accessed 5 April 2019. On this point see Nilufer Oral, "The South China Sea Arbitral Award, Part XII of UNCLOS, and the Protection and Preservation of the Marine Environment," in S. Jayakumar et al. (eds.), *The South China Sea Arbitration: The Legal Dimension* (Cheltenham: Edward Elgar, 2018), 229.

[34] *Convention for the Conservation of Southern Bluefin Tuna*, done at Canberra on 10 May 1993, entered into force on 20 May 1994, https://www.ccsbt.org/sites/ccsbt.org/files/userfiles/file/docs_english/basic_documents/convention.pdf, accessed 12 May 2019.

[35] *Southern Bluefin Tuna Arbitration*, Award on Jurisdiction, 43, paras. 56–57.

[36] *SWSP*, vol. I, 57–58, para. 11.10.

application of the Convention, clear and unambiguous wording would have been required. In support of this argument, the Philippines quoted a statement made by an ITLOS judge, Judge Wolfrum, who happened to be one of the members of the Arbitral Tribunal, in the context of a Provisional Order issued by the ITLOS in a 2001 case:

> such agreement among the parties to a conflict cannot be presumed. An intention to entrust the settlement of disputes concerning the interpretation and application of the Convention to other institutions must be expressed explicitly in respective agreements.[37]

Second, neither the text of Article 27 of the CBD nor the CBD annexes relating to conciliation and arbitration expressly excluded further proceedings under the Convention. Third, the dispute between the Philippines and China centered on the protection and preservation of the marine environment and not at all on the conservation and sustainable use of biodiversity under the CBD. Fourth, the Tribunal should adopt the characterization of the dispute presented by the party bringing the case, i.e., the Philippines.[38]

The Tribunal could not but admit that there was some overlap between the subject matter of Part XII of the Convention and that of the CBD. A broad interpretation of Article 192 of the Convention could encompass the obligation to preserve and protect marine biodiversity. By the same token, the obligation under Article 194 could include the protection and preservation of biodiversity represented by coral reefs. From this perspective, the alleged violations committed by China could constitute breaches of both the Convention and the CBD. In the Tribunal's view, the two treaties did not overlap to such an extent that the CBD could be considered to be an agreement within the meaning of Article 281 of the Convention. The reason was that the scope of the two Conventions differed markedly. CBD's definition of biodiversity was not limited to marine biodiversity but extended to biodiversity from all

---

[37] Award on Jurisdiction, 102, para. 278. The quotation was taken from *MOX Plant (Ireland v. United Kingdom), Provisional Measures, Order of 3 December 2001, Separate Opinion of Judge Wolfrum, ITLOS Reports 2001*, 131, https://www.itlos.org/fileadmin/itlos/documents/cases/case_no_10/published/C10-O-3_dec_01-SO_W.pdf, accessed 5 April 2019.

[38] Award on Jurisdiction, 102–3, paras. 277–80.

sources. The CBD's aim is to protect biodiversity in general, not just biodiversity in the marine environment. "The objective of the CBD potentially overlaps with, but also goes well beyond, the scope of Articles 192 and 194 of the Convention."[39] For its part, the Convention was not confined to the protection and preservation of the marine environment: "the Convention's scope goes well beyond the obligation to protect and conserve the marine environment."[40] Indeed, the Convention also regulates, among many other matters, navigation, the exploration, and exploitation of the resources of the Area (the "Common Heritage of Mankind"), marine scientific research, and transfer of technology, not to mention the legal regimes of a host of maritime areas (the territorial sea, archipelagic waters, the contiguous zone, straits used for international navigation, the EEZ, the continental shelf, and the high seas).

As a result of the (partial) overlap in the subject matter of the two treaties, the same facts could give rise to violations of both. Despite this possibility, the Tribunal stressed that the two regimes that are parallel in some respects remain just that—parallel regimes. The violation of the Convention does not necessarily give rise to violations of the CBD. The Tribunal agreed with the Philippines that disputes under the Convention do not become disputes under the CBD merely because there is some overlap between the two treaties. This reasoning was supported by two circumstances. First, Article 27 of the CBD does not expressly exclude dispute settlement under Section 2 of Part XV of the Convention. Second, and perhaps more importantly, the CBD implicitly recognizes the possibility of overlap between the two treaties and stipulated that with respect to the marine environment, it should be implemented by CBD Contracting Parties consistently with the rights and obligations of States under the law of the sea. Those rights and obligations included, in the Tribunal's view, those under Section 2 of Part XV. The Tribunal concluded that the CBD's dispute settlement provisions could not, by virtue of Article 281, preclude its jurisdiction.[41]

The CBD could still have precluded the Tribunal's jurisdiction, this time under Article 282 of the Convention.

[39] Ibid., 105, para. 285.
[40] Ibid.
[41] Ibid., 103–5, paras. 284–89.

2. The CBD as an Agreement Within the Meaning of Article 282

Like Article 281, Article 282 refers to agreements concluded by States to settle their disputes concerning interpretation and application of the Convention outside the Convention framework. Unlike Article 281, Article 282 refers to compulsory procedures entailing binding decisions. Such procedures could be general, regional, or bilateral agreements, or "otherwise," this last could refer to the declaration of acceptance of compulsory jurisdiction of the ICJ under Article 38 of its Statute.[42] According to an authoritative commentary, Article 282 reflected the assumption that the parties to a dispute would normally prefer to have their dispute settled though procedures previously agreed by them. Agreements of this type could be general bilateral treaties with dispute settlement procedures; multilateral agreements for dispute settlement; bilateral agreements on particular subjects or broad categories of subjects with dispute settlement provisions; multilateral agreements with dispute settlement provisions; regional agreements; and special agreements for particular disputes or groups of disputes.[43] Such agreements could bar the Tribunal's jurisdiction if they allowed any one party to the dispute to submit the dispute to a procedure that entailed binding decisions; the procedure would then take the place of the procedures applicable under Section 2 of Part XV of the Convention, unless the Parties have agreed to retain access to the procedures under Section 2 of Part XV.[44]

The question that the Tribunal had to decide was the following: Did the CBD constitute an agreement to refer a dispute relating to interpretation and application of the Convention to compulsory procedures entailing binding decisions within the meaning of Article 282? If so, then Article 27 of the CBD should apply "in lieu" of Section 2 of Part XV.[45]

---

[42] Nordquist et al., vol. V, 26. International Court of Justice, *Statute of the International Court of Justice*, https://www.icj-cij.org/en/statute, accessed 5 May 2019.

[43] Nordquist et al., vol. V, 26. Article 282 (Obligations under general, regional or bilateral agreements) reads as follows:
If the States Parties which are parties to a dispute concerning the interpretation or application of this Convention have agreed, through a general, regional or bilateral agreement or otherwise, that such dispute shall, at the request of any party to the dispute, be submitted to a procedure that entails a binding decision, that procedure shall apply in lieu of the procedures provided for in this Part, unless the parties to the dispute otherwise agree.

[44] Award on Jurisdiction, 106, paras. 290–91.

[45] Ibid., 109, para. 312.

In response, the Philippines reiterated its argument that the CBD constituted an agreement only for the settlement of disputes concerning the interpretation or application of the CBD and that its dispute with China regarding China's violations of Articles 192 and 194 of the Convention did not concern the interpretation or application of the CBD. The Philippines added that even if Article 27 of the CBD could be deemed to constitute an agreement to submit disputes concerning the interpretation or application of the Convention to CBD procedures, neither the Philippines nor China had accepted a compulsory procedure under the CBD, be it arbitration or the ICJ. Without acceptance of such a compulsory procedure, the only compulsory procedure applicable to a dispute between the two States under the CBD would have been conciliation under Article 27 of the CBD.[46] In these circumstances, the CBD's compulsory dispute settlement procedure could not be considered as "entailing binding decisions," as required by Article 282.

There can of course be no doubt in anyone's mind that the CBD is a legally binding agreement, to which both the Philippines and China are Parties. The Tribunal identified three conditions that had to be fulfilled if the CBD were to constitute a bar to its jurisdiction by virtue of Article 282. First, Article 27 of the CBD must constitute an agreement for the settlement of disputes concerning the interpretation or application of the Convention. Second, there must be agreement to submit the dispute to a compulsory procedure, in the sense that the dispute is capable of being unilaterally initiated (at the request of any party to the dispute, even without the consent of the other). Third, the compulsory procedure must entail binding decisions. The Tribunal had already found that Article 27 of the CBD did not fulfill the first criterion. Now it found that Article 27 did not fulfill the second and third criteria either. The first step in dispute settlement under Article 27 involves diplomatic means. States must first seek to settle a dispute through negotiation, which is not a compulsory procedure that entails binding decisions. If Contracting Parties to the CBD fail to reach a settlement through negotiation, they may then jointly resort to good offices or mediation. Since the recourse to good offices or mediation is joint, neither procedure is a compulsory procedure. And by definition, neither good offices nor mediation results in a binding decision. The parties to a dispute are not bound to enter

---

[46] Hearing on Jurisdiction, Transcript, Day 2, 112.

into negotiations as the outcome of the good offices mission, nor are they bound by the proposals that the mediator puts forward. All of these procedures can only be initiated by mutual consent; none entails binding decisions.

It is true that Article 27(3) of the CBD provides for the possibility that States accept through a declaration the compulsory jurisdiction of the ICJ or arbitration, but neither the Philippines nor China had made such a declaration. If parties to a dispute have not accepted the same or any procedure, then the only compulsory procedure available under Article 27 of the CBD, following the failure of negotiation, good offices, and/or mediation, is conciliation—any one of the two parties to the dispute may submit the dispute to a conciliation commission. The conciliation procedure is compulsory but it does not entail a binding decision. Understandably, the Tribunal concluded that the CBD's dispute settlement provisions could not preclude its jurisdiction over the Philippine submissions.[47]

The second precondition to the recourse to compulsory dispute settlement procedures entailing binding decisions is that the parties to the dispute must have exchanged views on the settlement of the dispute.

B. The Obligation to Exchange Views

China seemed to believe that the scope of the obligation to exchange views required negotiations. Once this misunderstanding is dispelled, the conclusion is obvious—the Philippines had complied with this precondition.

1. The Scope of the Obligation to Exchange Views

In its Notification and Statement of Claim, the Philippines, citing Article 283 of the Convention, declared that on numerous occasions since 1995, the Philippines and China had exchanged views regarding the settlement of the disputes that the Philippines was submitting to arbitration.[48] China responded in its 2014 Position Paper that the Philippines

---

[47] Award on Jurisdiction, 110–12, paras. 317–21.

[48] *South China Sea Arbitration*, Notification and Statement of Claim, 22 January 2013, 10, para. 25, https://seasresearch.files.wordpress.com/2014/12/

and China had never engaged in negotiations concerning the subject matter of the arbitration and that general exchanges of views, without having the purpose of settling a given dispute, did not constitute negotiations. China contended that the exchanges of views that had occurred pertained to responses to incidents at sea in the disputed areas and the measures to prevent conflicts, reduce frictions, maintain stability in the region, and promote cooperation.[49] It is not clear if China's objection extended to Philippine allegations of the violations of Articles 192 and 194 of the Convention. These allegations, which had not been articulated in detailed fashion at the time the Philippines initiated the arbitration in January 2013, had been explained in detailed fashion in the *Memorial*, a copy of which had been transmitted to China.

Interestingly, China did not cite any Convention provision as the legal basis for this objection to jurisdiction, but the Philippines and the Tribunal took this to be an argument based on Article 283 of the Convention. During the Hearing on Jurisdiction, the Philippines stressed that Article 283 was not a requirement to negotiate as such; rather it was only an obligation to exchange views on the means by which the dispute may be settled and the substance of the disputes.[50] The Tribunal implicitly accepted the view that Article 283 imposed an obligation to exchange views on the means of settling the dispute as well as the substance of the disputes but did not actually require negotiations.[51] The voluminous records of Philippine-China consultations submitted by the Philippines convinced the Tribunal that the two countries had exchanged views on the means of resolving the dispute and the substance of the

notification-and-statement-of-claim-on-west-philippine-sea.pdf, accessed 5 April 2019. Article 283 (Obligation to exchange views) reads as follows:

1. When a dispute arises between States Parties concerning the interpretation or application of this Convention, the parties to the dispute shall proceed expeditiously to an exchange of views regarding its settlement by negotiation or other peaceful means.

2. The parties shall also proceed expeditiously to an exchange of views where a procedure for the settlement of such a dispute has been terminated without a settlement or where a settlement has been reached and the circumstances require consultation regarding the manner of implementing the settlement.

[49] Position Paper, paras. 45–47.

[50] Hearing on Jurisdiction, Transcript, Day 2, 25.

[51] Award on Jurisdiction, 119, para. 342.

disputes. The inevitable conclusion was that the requirement of Article 283 had been satisfied.[52]

2. *The Content of the Exchanges of Views*

The Philippines argued, not without reason, that it had had over many years extensive exchanges of views with China, in the course of which it had repeatedly protested against the harvesting of endangered species and cyanide and dynamite fishing by Chinese nationals and against China's construction activities on several maritime features in the South China Sea. In support of this contention, the Philippines gave as examples taken from a long list of communications two Notes Verbales addressed to the Embassy of the People's Republic of China in Manila in 2000 and 2012.[53]

The Philippines had not confined itself to protests in the diplomatic exchanges with China. Over the years, it had suggested that China instruct Chinese fishermen to fish closer to China's coast[54]; stop using illegal fishing methods that are harmful to the environment[55]; respect Philippine sovereignty over Philippine territorial waters and prevent

---

[52] Ibid., 115, para. 334; 117, para. 337; 119, para. 342; 120, para. 343; 121, para. 347.

[53] *MP*, Annex 205, Note Verbale from the Department of Foreign Affairs of the Republic of the Philippines to the Embassy of the People's Republic of China in Manila, No. 12-0894 (11 April 2012), vol. VI, 377–80, https://files.pca-cpa.org/pcadocs/The%20 Philippines%27%20Memorial%20-%20Volume%20VI%20%28Annexes%20158-221%29.pdf, accessed 26 March 2019; *MP*, Annex 186, Note Verbale from the Department of Foreign Affairs of the Republic of the Philippines to the Embassy of the People's Republic of China in Manila, No. 2000100 (14 January 2000), vol. VI, 259–61, https://files.pca-cpa.org/ pcadocs/The%20Philippines%27%20Memorial%20-%20Volume%20VI%20%28Annexes%20 158-221%29.pdf, accessed 26 March 2019; Hearing on Jurisdiction, Transcript, Day 2, 30.

[54] *MP*, Annex 19, Memorandum from Erlinda F. Basilio, Acting Assistant Secretary, Office of Asian and Pacific Affairs, Department of Foreign Affairs, Republic of the Philippines, to the Secretary of Foreign Affairs of the Republic of the Philippines (29 March 1995), vol. III, 217, https://files.pca-cpa.org/pcadocs/The%20Philippines%27%20 Memorial%20-%20Volume%20III%20%28Annexes%201-60%29.pdf, accessed 26 March 2019.

[55] *MP*, Annex 20, Memorandum from Lauro L. Baja Jr., Assistant Secretary, Office of Asian and Pacific Affairs, Department of Foreign Affairs, Republic of the Philippines, to the Secretary of Foreign Affairs of the Republic of the Philippines (7 April 1995), vol. III, 222, https://files.pca-cpa.org/pcadocs/The%20Philippines%27%20Memorial%20-%20 Volume%20III%20%28Annexes%201-60%29.pdf, accessed 26 March 2019.

the destruction of the marine environment[56]; and avoid the area of Scarborough Shoal in order to prevent tensions.[57] China, for its part, proposed in April 1995 that the Ministries of Foreign Affairs of the two States and their fishing authorities should meet and cooperate to reach agreement on an arrangement to avoid the detention of Chinese fishermen.[58] A few months later, China proposed a bilateral fishing agreement that would also cover the safety of fishermen in the South China Sea.[59] In 1999, China proposed that the two States should deal with territorial disputes and fishing incidents separately.[60]

After it had been decided that the CBD did not preclude recourse to Section 2 of Part XV and that the obligation to exchange views had been fulfilled, it remained for the Tribunal to determine whether any of the limitations and exceptions to its jurisdiction were applicable.

[56] MP, Annex 43, Memorandum from the Embassy of the Republic of the Philippines in Beijing to the Secretary of Foreign Affairs of the Republic of the Philippines, No. ZPE-06-2001-S (13 February 2001), vol. III, 334, https://files.pca-cpa.org/pcadocs/The%20 Philippines%27%20Memorial%20-%20Volume%20III%20%28Annexes%201-60%29.pdf, accessed 26 March 2019.

[57] MP, Annex 45, Memorandum from Willy C. Gaa, Assistant Secretary of Foreign Affairs, Republic of the Philippines to Secretary of Foreign Affairs, Republic of the Philippines (14 February 2001), vol. III, 366, https://files.pca-cpa.org/pcadocs/The%20 Philippines%27%20Memorial%20-%20Volume%20III%20%28Annexes%201-60%29.pdf, accessed 26 March 2019.

[58] MP, Annex 21, Memorandum from the Ambassador of the Republic of the Philippines in Beijing to the Undersecretary of Foreign Affairs of the Republic of the Philippines (10 April 1995), vol. III, 231, https://files.pca-cpa.org/pcadocs/The%20Philippines%27%20 Memorial%20-%20Volume%20III%20%28Annexes%201-60%29.pdf, accessed 26 March 2019.

[59] MP, Annex 23, Memorandum from the Secretary of Foreign Affairs of the Republic of the Philippines to the President of the Republic of the Philippines (31 July 1995), vol. III, 239–40, https://files.pca-cpa.org/pcadocs/The%20Philippines%27%20Memorial%20-%20 Volume%20III%20%28Annexes%201-60%29.pdf, accessed 26 March 2019.

[60] MP, Annex 37, Memorandum from the Embassy of the Republic of the Philippines in Beijing to the Secretary of Foreign Affairs of the Republic of the Philippines, No. ZPE-85-98-S (4 December 1998), vol. III, 318, https://files.pca-cpa.org/pcadocs/The%20 Philippines%27%20Memorial%20-%20Volume%20III%20%28Annexes%201-60%29.pdf, accessed 26 March 2019.

II. LIMITATIONS AND EXCEPTIONS TO THE TRIBUNAL'S JURISDICTION

China had not referred to the limitations in its Position Paper, but the Tribunal decided that as part of its duty to satisfy itself that it had jurisdiction over the dispute, it should not neglect other possible issues of jurisdiction and admissibility not raised in China's Position Paper.[61]

Section 3 of Part XV of the Convention distinguishes between limitations and exceptions. Limitations, which are enumerated in Article 297, apply automatically to any dispute between States. Exceptions, which are listed in Article 298, are optional and need to be activated by means of a declaration made by a State party to the Convention. In the event limitations or exceptions are applicable, disputes excluded under Article 297 or excepted by declaration under Article 298 may be submitted to compulsory procedures entailing binding decisions only by agreement between the parties to the dispute (Article 299(1)).[62]

In relation to the protection and the preservation of the marine environment, the Tribunal had to consider two limitations and two exceptions, any one of which could have precluded its jurisdiction. The automatic limitations related to disputes over the coastal State's sovereign rights in the EEZ and over fisheries conservation and management laws in the EEZ. The optional exceptions concerned the military or non-military character of China's construction activities and the law enforcement character of China's actions in relation to the harvesting of endangered species and dynamite and cyanide fishing by Chinese nationals in the South China Sea.

---

[61] *South China Sea Arbitration*, Procedural Order No. 4, Doc. PCA 1435452 (1 April 2015), https://pcacases.com/web/sendAttach/1807, accessed 5 April 2019.

[62] Award on Jurisdiction, 125, para. 354. The text of paragraph 1 of Article 299 (Right of the parties to agree upon a procedure) is as follows:

A dispute excluded under Article 297 or excepted by a declaration made under Article 298 from the dispute settlement procedures provided for in section 2 may be submitted to such procedures only by agreement of the parties to the dispute.

A. The Limitations to Jurisdiction Arising from the Coastal State's Sovereign Rights in the EEZ and China's Activities in the South China Sea

The EEZ constitutes one of the major innovations of the Convention. Within the EEZ, which extends to 200 nautical miles from the baselines from which the breadth of the territorial sea is measured, the coastal State enjoys sovereign rights for the purpose of exploring and exploiting, conserving and managing the living and non-living natural resources of the waters superjacent to the seabed and of the seabed and its subsoil and jurisdiction with regard to, among others, the protection and preservation of the marine environment.[63] At the Third United Nations Conference on the Law of the Sea (UNCLOS III, 1973–1982), a number of coastal States were of the view that the exclusive jurisdiction of the coastal State in the EEZ should not be jeopardized by submission to third-party adjudication, whereas other States sought safeguards against a coastal State's abuse of power.[64] Article 297 (Limitations on applicability of section 2) seeks to balance the interests of coastal States and other States in three areas: navigation, overflight, or laying of submarine cables and pipelines as well as the protection and preservation of the marine environment[65]; marine scientific research[66]; and fisheries.[67] Only the first and third limitations were potentially relevant in the *South China Sea Arbitration*. The first limitation raised the question whether Article 297(1) limited the jurisdiction of the Arbitral Tribunal to disputes concerning compliance by the coastal State with specified international rules and standards for the protection and preservation of the marine environment. The third limitation would have barred the jurisdiction of the Tribunal over disputes relating to terms and conditions of fisheries conservation and management laws and regulations.

---

[63]Convention, Article 56.

[64]Nordquist et al., vol. V, 98. United Nations Office of Legal Affairs, Codification Division, *United Nations Diplomatic Conferences. Third United Nations Conference on the Law of the Sea, 1973–1982*, 2019, http://legal.un.org/diplomaticconferences/1973_los/, accessed 4 April 2019.

[65]Article 297(1)(a), (b) and (c).

[66]Article 297(2)(a) and (b).

[67]Article 297(3).

1. The Protection and Preservation of the Marine Environment in the EEZ and China's Activities in the South China Sea

Article 297(1)(c) stipulates that disputes may be submitted to compulsory settlement entailing binding decisions if they involve claims that a coastal State has

> acted in contravention of specified international rules and standards for the protection and preservation of the marine environment which are applicable to the coastal State and which have been established by this Convention or through a competent international organization or diplomatic conference in accordance with this Convention.

The Tribunal reminded the Philippines that Article 297 could be understood as implicitly limiting its jurisdiction to the cases specifically identified in the article.[68] The question was the application of this paragraph to Philippine claims relating to the protection and preservation of the marine environment.[69]

In the event that the jurisdiction of the Tribunal was limited to disputes over rules and standards for the protection of the marine environment that are applicable to the coastal State and that have been established by the Convention or through a competent international organization or diplomatic conference, then presumably the Philippines should first have identified such international rules and standards and demonstrated that these had been established by the Convention or through a competent international organization or diplomatic conference.

The Philippines responded to the Tribunal's question by affirming that Article 297(1) supports, rather than limits, jurisdiction over environmental disputes within the territorial sea and the continental shelf, even assuming that China were the coastal State.[70] Indeed, Article 297(1)(c) represents an affirmation of compulsory jurisdiction with respect to the whole of the marine environment.[71] In support of this contention, the Philippines referred to an award issued by another Tribunal constituted under Annex VII of the Convention in the *Chagos*

---

[68] Award on Jurisdiction, 127, para. 359.
[69] Ibid., 128, para. 360.
[70] Hearing on Jurisdiction, Transcript, Day 2, 103–4.
[71] Ibid., 103.

Marine Protected Area Arbitration case. The latter dispute concerned a decision by the UK on 1 April 2010 to establish a Marine Protected Area ("MPA") around the Chagos Archipelago, which had been separated from Mauritius upon the independence of the latter from the UK in 1965 and had been administered by the UK since that time. Mauritius initiated arbitration alleging that the establishment of the MPA violated the Convention and other rules of international law. The UK objected to the *Chagos* Tribunal's jurisdiction on the grounds that fishing and the management of living resources fell outside the context of Article 297(1)(c) and that the expression "international rules and standards" referred to the prevention of marine pollution from ships, aircraft, and seabed activities.[72] The *Chagos* Tribunal rejected the UK's objection, ruling that Article 297(1) did not state that disputes relating to the exercise of sovereign rights and jurisdiction by the coastal State were only subject to compulsory settlement in the enumerated cases. In the *Chagos* Tribunal's view, Article 297(1) reaffirmed, but did not limit, the Tribunal's jurisdiction.[73] The *South China Sea* Tribunal tacitly accepted the reasoning of the Philippines, as the Tribunal refrained from considering Article 297(1)(c) as a limitation to its jurisdiction.

The question of the application of Article 297(3) to the Philippine Submissions was more complex.

2. Fisheries Conservation and Management Laws in the EEZ and China's Activities in the South China Sea

Article 297(3)(a) stipulates that disputes with a coastal State regarding fisheries are to be settled by means of compulsory procedures entailing binding decisions, except that

> the coastal State shall not be obliged to accept the submission to such settlement of any dispute relating to its sovereign rights with respect to the living resources in the exclusive economic zone or their exercise, including…the terms and conditions established in its conservation and management laws and regulations.

---

[72] *Chagos Marine Protected Area Arbitration (Mauritius v. United Kingdom)*, 18 March 2015, 94–95, paras. 234–36, https://files.pca-cpa.org/pcadocs/MU-UK%20 20150318%20Award.pdf, accessed 5 April 2019.

[73] Ibid., 119, para. 308.

Disputes arising out of allegations that a coastal State "has manifestly failed to comply with its obligations to ensure through proper conservation and management measures that the maintenance of the living resources in the exclusive economic zone is not seriously endangered" may be submitted only to compulsory conciliation, which is provided for in Section 2 of Annex V of the Convention.[74] In the Tribunal's view, Article 297(3) could potentially bar its jurisdiction over Philippine claims in relation to fisheries, to the extent that the events cited by the Philippines took place in China's EEZ or in an area where the Chinese and Philippine entitlements to an EEZ overlapped.[75]

To understand this scenario, it must be recalled that the Spratly Islands and Scarborough Shoal are subject to competing claims of sovereignty put forward by China and the Philippines (not to mention other littoral States). China claims sovereignty over the Spratly Islands as a whole, while the Philippines claims sovereignty over a number of features in the group. To complicate matters further, China contends that the Spratly Islands and Scarborough Shoal are entitled to an EEZ and a continental shelf. If China were indeed the sovereign over these maritime features and they were entitled to an EEZ, then the incidents of which the Philippines complained took place in China's EEZ or in areas in which China's EEZ overlapped with that of the Philippines. As the Tribunal did not have jurisdiction to consider questions of maritime delimitation, which had been excluded by a declaration made by China in 2006, the dispute regarding the harvesting of endangered species and cyanide and dynamite fishing would be excluded from the Tribunal's jurisdiction.

In its response to the Tribunal's question, the Philippines made a distinction between Scarborough Shoal and the Spratly Islands. In the waters around Scarborough Shoal, the harvesting of endangered species and dynamite and cyanide fishing occurred in the territorial sea of the Shoal, within 12 nautical miles from any putative baselines from which the territorial sea could be measured. In the Philippine view, the limitation to jurisdiction in Article 297(3) did not apply to the territorial sea. It applied to that part of the EEZ beyond the outer limit of the territorial sea up to a limit of 200 nautical miles (although the Philippines

---

[74] Convention, Article 297(3)(b)(i).
[75] Award on Jurisdiction, 127, para. 359.

denied that Scarborough Shoal was entitled to an EEZ or a continental shelf). As regards Second Thomas Shoal in the Spratly Islands, the Philippines claimed that it was a low-tide elevation and as such formed part of the seabed and subsoil of the Philippine continental shelf and EEZ.[76] Article 297(3) would only limit the jurisdiction of a tribunal if a foreign State were attempting to bring a case against the coastal State relating to the terms and conditions for fishing in the EEZ that are established by the latter in its conservation and management laws and regulations. The context of the *South China Sea Arbitration* was different. It was the coastal State, the Philippines, that was alleging that a foreign State, China, had violated obligations under the Convention to protect and preserve the marine environment. In this situation, Article 297(3) could not prevent the Philippines, as the coastal State, from resorting to compulsory dispute settlement in order to protect its marine environment.[77] For good measure, the Philippines added that Article 297(3) excluded only disputes concerning the living resources of the EEZ from compulsory jurisdiction; it did not apply to disputes relating to the protection of corals and giant clams, which are sedentary species subject to the regime of the continental shelf.[78]

The Tribunal agreed with the Philippines that Article 297 of the Convention did not apply in the territorial sea and consequently could not exclude jurisdiction over disputes that occurred in the territorial sea. The Tribunal decided at the time it issued the Award on Jurisdiction and Admissibility that it did not have to determine which of the two States—China or the Philippines—had sovereignty over Scarborough Shoal. Nor did it have to decide at that stage whether Second Thomas

---

[76]Article 13(1) of the Convention defines a low-tide elevation as "a naturally formed area of land which is surrounded by and above water at low tide but submerged at high tide." In its Submission No. 4 to the Tribunal, the Philippines claimed that Second Thomas Shoal, Mischief Reef, and Subi Reef were low-tide elevations and as such were not entitled to an EEZ and a continental shelf. In its Submission No. 5, the Philippines claimed that Second Thomas Shoal and Mischief Reef were part of the Philippine EEZ and continental shelf.

[77]Hearing on Jurisdiction, Transcript, Day 2, 103–104.

[78]Ibid., 104. Sedentary species are defined by Article 77 (Rights of the coastal State over the continental shelf) of the Convention as "organisms which, at the harvestable stage, either are immobile on or under the sea-bed or are unable to move except in constant physical contact with the sea-bed or the subsoil."

Shoal was an island or a low-tide elevation.[79] Whether the coastal State was the Philippines or China, a dispute relating to the marine environment could be submitted to a compulsory procedure entailing a binding decision. Even assuming that the harvesting of endangered species and cyanide and dynamite fishing occurred in the EEZ of the Philippines, or that of China, or in an area where the EEZs of the two countries overlapped, the Tribunal agreed with the Philippines that Article 297(1)(c) "expressly affirms the Tribunal's jurisdiction over disputes concerning the alleged violation of 'specified international rules and standards for the protection and preservation of the marine environment' in the exclusive economic zone."[80] The Tribunal did not find it necessary to rule on the argument based on the description of corals and giant clams as sedentary species.

The optional exceptions that could have precluded the Tribunal's jurisdiction, the military and law enforcement activities exceptions, had not been invoked by China in its Position Paper. This silence did not prevent the Tribunal from taking them into consideration.

B. The Optional Exceptions to Jurisdiction and China's Activities in the South China Sea

Pursuant to Article 298, a State party to the Convention is permitted three optional exceptions to the applicability of Section 2 of Part XV. Through a simple declaration that may be made when a State signs, ratifies or accedes to the Convention or at any time thereafter, a State may exclude from the compulsory procedures entailing binding decisions (a) disputes concerning the interpretation or application of Articles 15, 74 and 83 relating respectively to the delimitation of the territorial sea, EEZ, and continental shelf between States with opposite or adjacent coasts; or disputes involving historic bays or titles; (b) disputes concerning military activities, including military activities by government vessels and aircraft engaged in non-commercial service, and disputes concerning

---

[79] Award on Jurisdiction, 145, para. 408(a). In the Award of 12 July 2016, the Tribunal found that Second Thomas Shoal was a low-tide elevation that was situated within the Exclusive Economic Zone and the continental shelf of the Philippines. Award of 12 July 2016, 259–60, para. 646, https://pcacases.com/web/sendAttach/2086, accessed 26 March 2019.

[80] Award on Jurisdiction, 145, para. 408.

law enforcement activities in regard to the exercise of sovereign rights or jurisdiction excluded from the jurisdiction of a court or tribunal under Article 297, paragraph 2 or 3; and (c) disputes in respect of which the Security Council of the United Nations is exercising its functions. These categories of disputes were considered by a majority of States at UNCLOS III as so sensitive that they should not be subject to compulsory dispute settlement.[81]

On 25 August 2006, China filed a declaration by the terms of which it "[did] not accept any of the procedures provided for in Section 2 of Part XV of the Convention with respect to all the categories of disputes referred to in paragraph 1 (a) (b) and (c) of Article 298 of the Convention."[82] China's Position Paper invoked only Article 298(1)(a) to challenge the Arbitral Tribunal's Jurisdiction.[83] No reference was made to Article 298(1)(b). China's silence on the matter did not prevent the Tribunal from envisaging two hypotheses. First, the military activities exception could be applicable to China's island-building.[84] When the Philippines amended its submissions to cover China's island-building at Cuarteron Reef, Fiery Cross Reef, Johnson Reef, Hughes Reef, Gaven Reef (North), and Subi Reef, the Tribunal also had to consider the applicability of the military activities exception to Chinese construction activities on these reefs.[85] Second, the law enforcement activities exception could apply to China's toleration of harvesting of endangered species and dynamite and cyanide fishing at Scarborough Shoal and Second Thomas Shoal.[86]

The Tribunal's questions compelled the Philippines to devote time and energy to refuting challenges to jurisdiction that China had not raised and to provide the information that China had not supplied.

---

[81] Nordquist et al., vol. V, 109–10.

[82] United Nations Office of Legal Affairs, Treaty Section, Treaty Collection, *Multilateral Treaties Deposited with the Secretary-General, Chapter XXI: Law of the Sea*, https://treaties.un.org/Pages/ViewDetailsIII.aspx?src=TREATY&mtdsg_no=XXI-6&chapter=21&Temp=mtdsg3&clang=_en, accessed 4 April 2019.

[83] Position Paper, paras. 57–75.

[84] *SWSP*, vol. I, 51.

[85] See the Philippine answer to the Tribunal's Question 20, *South China Sea Arbitration*, Hearing on the Merits and the Remaining Issues of Admissibility, Transcript, Day 4 (30 November 2015), 169–70 ("Hearing on the Merits"), https://pcacases.com/web/sendAttach/1550, accessed 5 April 2019.

[86] *SWSP*, vol. I, 43.

As the Philippines pointed out, normally it would have been the duty of the respondent (China) to assert the objection to jurisdiction and provide the facts necessary to sustain the objection, particularly if the objection related to its own activities.[87] China's failure to appear before the Tribunal meant that it fell to the Philippines to supply information about China's activities. The burden was all the heavier because the Philippines would have to prove the negative—it would have to prove that China's activities were not covered by the military and law enforcement activities exception. The Philippines appealed to the Tribunal to take into account China's failure to invoke the military and law enforcement activities exception as well as its failure to supply any evidence in this respect.[88]

The question of the applicability of the military activities exception was resolved on the basis of official Chinese declarations, while a close reading of the text of Article 298(1)(b) made it clear that the law enforcement activities exception did not apply to China's toleration of the harvesting of endangered species and of destructive fishing methods.

1. The Military Activities Exception and China's Construction Activities in the South China Sea

Article 298 does not define "military activities," simply referring to "military activities by government vessels and aircraft engaged in non-commercial service."[89] The Philippines began by explaining that military activities are ordinarily conducted at sea by vessels and aircraft operated by armed forces. It observed that many States used naval vessels for law enforcement, at least some of the time. For example, the People's Liberation Army Navy ("PLAN") of China provides security support for maritime law enforcement, fisheries, and oil and gas exploration; it coordinates and cooperates with law enforcement organs of marine surveillance and fishery administration.[90] Conversely, if a dock were used by naval and other types of vessels, the dock's activity would not be considered military for purposes of Article 298(1)(b).[91]

---

[87] Hearing on Jurisdiction, Transcript, Day 3, 5.

[88] Ibid., Day 2, 76–77; *SWSP*, vol. I, 44, paras. 9.1–9.2.

[89] Hearing on Jurisdiction, Transcript, Day 2, 77.

[90] Ibid., 84.

[91] Ibid., Transcript, Day 3, 53.

At the time of the Hearing on Jurisdiction in July 2015, the Philippine submission on China's construction activities covered only Mischief Reef (Submission No. 12(b)). The Philippines had no choice but to admit that Mischief Reef was occupied by Chinese military personnel. In its view, this fact was insufficient to exclude the Tribunal's jurisdiction.[92] Since the mid-1990s, China had repeatedly told the Philippines that the facilities being built at Mischief Reef were for civilian use.[93] In 1995, no less than China's Foreign Minister declared that the structures were typhoon shelters for fishermen constructed by local fishing authorities.[94] If one were to believe China, Filipino fishermen would be welcome to use these shelters "in case of need," provided they obtained Chinese consent.[95] A few months later, China backtracked when the Philippines asked if Filipino fishermen could use the "wind shelters" after China's earlier statement. China set as a condition some kind of understanding between the two governments before the Filipino fishermen would be able to use the shelters.[96] In 1998, in response to a Philippine protest, China described the construction taking place on Mischief Reef as repair and renovation work being carried out on the

[92] *SWSP*, vol. I, 52, para. 10.3.

[93] Hearing on Jurisdiction, Transcript, Day 2, 75, 88; Award on Jurisdiction, 134, para. 377; Hearing on Jurisdiction, Transcript, Day 3, 53; Award on Jurisdiction, 134, para. 377.

[94] *MP*, Annex 18, Memorandum from the Ambassador of the Republic of the Philippines in Beijing to the Undersecretary of Foreign Affairs of the Republic of the Philippines (10 March 1995), vol. III, 211, https://files.pca-cpa.org/pcadocs/The%20Philippines%27%20 Memorial%20-%20Volume%20III%20%28Annexes%201-60%29.pdf, accessed 26 March 2019; *MP*, Annex 22, Memorandum from the Ambassador of the Republic of the Philippines in Beijing to the Secretary of Foreign Affairs of the Republic of the Philippines, No. ZPE-231-95 (20 April 1995), vol. III, 235, https://files.pca-cpa.org/pcadocs/ The%20Philippines%27%20Memorial%20-%20Volume%20III%20%28Annexes%201-60%29. pdf, accessed 26 March 2019.

[95] *MP*, Annex 180, Government of the Republic of the Philippines and Government of the People's Republic of China, Agreed Minutes on the First Philippines-China Bilateral Consultations on the South China Sea Issue (10 August 1995), vol. VI, 189, https://files. pca-cpa.org/pcadocs/The%20Philippines%27%20Memorial%20-%20Volume%20VI%20 %28Annexes%20158-221%29.pdf, accessed 26 March 2019.

[96] *MP*, Annex 181, Government of the Republic of the Philippines, Transcript of Proceedings Republic of the Philippines-People's Republic of China Bilateral Talks (10 August 1995), vol. VI, 196, https://files.pca-cpa.org/pcadocs/The%20 Philippines%27%20Memorial%20-%20Volume%20VI%20%28Annexes%20158-221%29.pdf, accessed 26 March 2019.

structures erected in 1995 which had "exceeded their service life" and had "deteriorated through natural causes." China denied that any PLAN vessels had been dispatched to the area.[97] In 1999, China informed the Philippines that it was Chinese fishermen who had requested that the government construct shelters for their use in inclement weather. The new structures were said to be very simple and not very big. The so-called radar facilities were dish-type television satellite antennae to enable the personnel on the reef to watch ordinary TV programs. China even attempted to convince the Philippines that the construction was proof that their purpose was not military: If China had allowed them to rot, it would have been evidence that China was responding to foreign criticism to the effect that these were military structures. Repair of the structures would demonstrate China's intention to use them as shelters for fishermen.[98]

In its Award on Jurisdiction, the Tribunal decided that it needed to have the specifics of China's activities on Mischief Reef before it could determine whether they were military in nature or not. An even more important consideration was the disputed status of Mischief Reef. The Philippines claimed that it was a low-tide elevation that formed part of its EEZ and continental shelf. China's position, on the other hand, is that the Spratly Islands as a whole, of which Mischief Reef is a part, are entitled to an EEZ and a continental shelf. One of the Philippine claims that the Tribunal had to adjudicate was precisely the status of Mischief Reef. If the Tribunal found that Mischief Reef in the Spratly Islands were an island, or even a rock, which is a type of island, it would constitute land territory.[99] Mischief Reef would come within the scope of the

---

[97] MP, Annex 33, Memorandum from Ambassador of the Republic of Philippines in Beijing to the Secretary of Foreign Affairs of the Republic of the Philippines, No. ZPE-76-98-S (6 November 1998), vol. III, 297, https://files.pca-cpa.org/pcadocs/The%20 Philippines%27%20Memorial%20-%20Volume%20III%20%28Annexes%201-60%29.pdf, accessed 26 March 2019.

[98] MP, Annex 38, Memorandum from Ambassador of the Republic of Philippines in Beijing to the Secretary of Foreign Affairs of the Republic of the Philippines, No. ZPE-18-99-S (15 March 1999), vol. III, 321, https://files.pca-cpa.org/pcadocs/The%20 Philippines%27%20Memorial%20-%20Volume%20III%20%28Annexes%201-60%29.pdf, accessed 26 March 2019.

[99] Article 121(1) of the Convention defines an island as "a naturally formed area of land, surrounded by water, which is above water at high tide." A definition of a "rock" is provided in paragraph 3 of the same article: "[r]ocks which cannot sustain human habitation or economic life of their own shall have no exclusive economic zone or continental shelf."

sovereignty dispute between the Philippines and China, over which the Tribunal had no jurisdiction. The Tribunal thus decided it was necessary to hear the arguments on the merits before it could determine whether it had jurisdiction to hear the Philippine submissions concerning Chinese construction activities on Mischief Reef (Submission No. 12(b)).[100]

At the merits stage, the Philippines amended its Submission No. 11 to encompass the marine environment at Cuarteron Reef, Fiery Cross Reef, Johnson Reef, Hughes Reef, Gaven Reef (North), and Subi Reef. The Tribunal's task was to determine whether its jurisdiction over Submissions No. 11 and 12(b) as amended were constrained by the military activities exception.[101] Once it had heard the arguments on the merits put forward by the Philippines concerning the status of Mischief Reef, the Tribunal concluded, for reasons that cannot be discussed here, that Mischief Reef was not an island or a rock but was a low-tide elevation within the continental shelf and the EEZ of the Philippines.[102] As such, Mischief Reef was not land territory and could not come within the scope of the sovereignty dispute between the Philippines and China.

At the merits stage, the Tribunal considered the statements repeatedly made by Chinese representatives to the effect that the installations and island-building were intended to fulfill civilian purposes. On 9 September 2014, a Chinese Foreign Ministry Spokesperson stated that "the construction work China is undertaking on relevant islands is mainly for the purpose of improving the working and living conditions of people stationed on these islands."[103] In April 2015, the spokesperson repeated that the "main purposes" of the "maintenance and construction work" on the Spratly Islands and Reefs were:

optimizing their functions, improving the living and working conditions of personnel stationed there, better safeguarding territorial sovereignty and

---

[100] Award on Jurisdiction, 146, para. 409.

[101] Award of 12 July 2016, 371–72, paras. 932–34.

[102] Ibid., 119–75, paras. 279–384.

[103] *Supplemental Documents of the Philippines* (19 November 2015), Annex 619, Ministry of Foreign Affairs, People's Republic of China, Foreign Ministry Spokesperson Hua Chunying's Regular Press Conference (9 September 2014), vol. I, 137–42 (*"SDP"*). https://files.pca-cpa.org/pcadocs/The%20Philippines%27%20Supplemental%20 Documents%20-%20Volume%20I%20%28Annexes%20607-667%29.pdf, accessed 9 April 2019; Award of 12 July 2016, 372, para. 836.

maritime rights and interests, as well as better performing China's international responsibility and obligation in maritime search and rescue, disaster prevention and mitigation, marine science and research, meteorological observation, environmental protection, navigation safety, fishery production service and other areas.[104]

At the meeting of the States Parties to the Convention held in June 2015, China once more explained the civilian purposes of the construction activities.[105] In September 2015, no less than China's President Xi Jinping stated during a state visit to the United States that the "[r]elevant construction activities that China [is] undertaking in the island of South – Nansha Islands do not target or impact any country, and China does not intend to pursue militarization."[106] In view of these statements, the Tribunal believed that it could not hold China's construction activities to be military in nature "when China itself has consistently and officially resisted such classifications and affirmed the opposite at the highest levels."[107] The Tribunal felt itself bound to accept China's "repeatedly affirmed position" that civilian use was "the primary (if not the only) motivation underlying the extensive construction on the seven reefs in the Spratly Islands." As civilian activity fell outside the scope of Article 298(1)(b)—it was not excluded by Article 298(1)(b)—the Tribunal

---

[104] SDP, Annex 624, Ministry of Foreign Affairs, People's Republic of China, Foreign Ministry Spokesperson Hua Chunying's Regular Press Conference (9 April 2015), vol. I, 165–70; *South China Sea Arbitration, Annexes cited during Hearing on Jurisdiction* (13 July 2015), Annex 579, Ministry of Foreign Affairs, People's Republic of China, Foreign Ministry Spokesperson Lu Kang's Remarks on Issues Relating to China's Construction Activities on the Nansha Islands and Reefs (16 June 2015), 159, https://files.pca-cpa.org/pcadocs/Annexes%20cited%20during%20Hearing%20on%20Jurisdiction%20%28Annexes%20574-583%29.pdf, accessed 9 April 2019; Award of 12 July 2016, 372, para. 936.

[105] There seems to be no copy of this statement available in the records of the proceedings. The Tribunal cited in footnote no. 1090 the statement made by China at the Meeting of the States Parties held in 2014. Award of 12 July 2016, 372, para. 937.

[106] "China Not to Pursue Militarization of Nansha Islands in South China Sea: Xi," *Xinhua* (25 September 2015), news.xinhuanet.com/english/2015-09/26/c_13466C930.htm, accessed 21 April 2017, quoted in Award of 12 July 2016, 412, para. 1027, note 1262; SDP, Annex 664, United States, The White House, Office of the Press Secretary, "Press Release: Remarks by President Obama and President Xi of the People's Republic of China in Joint Press Conference" (25 September 2015), 481–98; Award of 12 July 2016, 372, para. 937.

[107] Award of 12 July 2016, 373, para. 938.

concluded that it had jurisdiction to consider the Philippine claims concerning China's construction activities in the South China Sea.[108] As Stephens put it, the Tribunal's power to subject to scrutiny these activities was the legal price that China had to pay for its refusal to acknowledge their military character.[109] Had China openly admitted the military character of its activities, the Tribunal would have been prevented from examining them. China's repeated affirmations paved the way for an impartial, third-party examination of China's island-building in the South China Sea.

The only other possible exception that could have constrained the Tribunal's jurisdiction was the law enforcement activities exception.

2. The Law Enforcement Activities Exception and China's Toleration of the Harvesting of Endangered Species in the South China Sea

The Philippines argued that the law enforcement activities exception was irrelevant for the Philippine submission concerning China's toleration of the harvesting of endangered species in the South China Sea.[110] The law enforcement activities to which the exception could conceivably apply were those carried out by Chinese maritime law enforcement vessels while Chinese fishing vessels were engaged in harvesting of endangered species, particularly in April and May 2012. In passing, the Philippines noted that it was ironic to characterize such activities as "law enforcement," because they consisted in "standing by while [Chinese] fishing boats engage in environmentally destructive practice."[111]

Be that as it may, the Philippines argued convincingly that even if it were assumed that China were the coastal State, the law enforcement activities exception in Article 298(1)(b) was not a blanket exception for all disputes over law enforcement activities carried out by the coastal State.[112] The law enforcement activities exception did not exclude from jurisdiction any and all law enforcement activities by the coastal State in

---

[108] Ibid.

[109] Tim Stephens, "The Collateral Damage from China's 'Great Wall of Sand': The Environmental Dimensions of the South China Sea Case," *Australian Yearbook of International Law* 24 (2017): 44.

[110] Hearing on Jurisdiction, Transcript, Day 2, 88.

[111] Ibid., 87.

[112] Ibid., 77–78.

its EEZ. Article 298(1)(b) specifies that the disputes must concern "law enforcement activities in regard to the exercise of sovereign rights or jurisdiction excluded from the jurisdiction of a court or tribunal under Article 297, paragraph 2 or 3." The scope of the law enforcement activities exception is anchored to Articles 297(2) and (3). When the latter exceptions do not apply, neither does the former.[113]

Articles 297(2)(a) and (b) lay down automatic limitations on the applicability of Section 2. Article 297(2)(a) and (b) in turn point the reader to Articles 246 (Marine Scientific Research in the Exclusive Economic Zone and on the Continental Shelf), 253 (Suspension or Cessation of Marine Scientific Research Activities). Article 297(3)(a)–(e) lead us to Articles 62 (Utilization of the Living Resources of the EEZ), 69 (Right of Land-Locked States in the EEZ), and 70 (Right of Geographically Disadvantaged States).

The Philippines explained that the exclusion from jurisdiction in Article 297(2) relates only to certain questions relating to marine scientific research.[114] Under Article 246(3), a coastal State should, in normal circumstances, grant its consent for marine scientific research projects to be carried out by other States or by international organizations in its EEZ or on its continental shelf. Article 246(5) preserves the discretion of the coastal State to refuse consent in certain circumstances: (1) The project is of direct significance for the exploration and exploitation of living or non-living natural resources; (2) it involves drilling into the continental shelf, the use of explosives or the introduction of harmful substances into the marine environment; (3) it involves the construction, operation or use of artificial islands, installations and structures; or (4) contains inaccurate information submitted to the coastal State; or the researching State or international organization has outstanding obligations to the coastal State from a prior research project. Article 253 recognizes the right of the coastal State to suspend a marine scientific research project if the research is not carried out in accordance with the project submitted to it, or if the project is substantially modified, or if the coastal State's rights (e.g., the right to participate in the project) are not respected. Should a dispute arise between the coastal State and the researching State or international organization over the coastal State's

---

[113] *SWSP*, vol. I, 45, para. 9.6.
[114] Hearing on Jurisdiction, Transcript, Day 2, 78.

exercise of its discretionary powers under Articles 246 and 253, these disputes may not be submitted to a compulsory procedure entailing binding decisions (Section 2 of Part XV). The law enforcement activities that may be undertaken by the coastal State in connection with these disputes are the law enforcement activities that may be excluded by a coastal State in conformity with Article 298(1)(b). Once the coastal State has declared that it excludes these law enforcement activities from compulsory jurisdiction, the coastal State cannot be obliged to be a party to a compulsory procedure entailing a binding decision. Clearly, Philippine submissions relating to illegal harvesting of endangered species and dynamite and cyanide fishing had nothing to do with marine scientific research. The Philippines concluded that Article 297(2)(a) could not preclude the Tribunal's jurisdiction.[115]

Article 297(3) concerns the automatic limitation to jurisdiction over disputes regarding the sovereign rights of the coastal State over fisheries in the EEZ. These sovereign rights, listed in Article 62, include the right to determine its capacity to harvest the fish in the EEZ; if the coastal State does not have the capacity to harvest the entire allowable catch, it is obliged to give other States access to the surplus of the allowable catch. The coastal State may establish laws and regulations on such matters as (1) the licensing of fishermen, fishing vessels and equipment, (2) determining the species which may be caught, and fixing quotas of catch, (3) regulating seasons and areas of fishing, the types, sizes and amount of gear, and the types, sizes and number of fishing vessels that may be used and (4) fixing the age and size of fish and other species that may be caught. Article 69 recognizes the right of land-locked States to participate in fishing of part of the surplus of the living resources of the EEZs of coastal States of the same subregion or region. Article 70 grants geographically disadvantaged States the same right. If a dispute arises between a coastal State and other States over fisheries in the EEZ, that dispute may not be submitted to compulsory settlement entailing binding decisions, under Article 297(3)(a); instead, the dispute may be submitted by either State to conciliation, the outcome of which is not binding (Article 297(3)(b)). Even then, the conciliation commission is not permitted to substitute its discretion for that of the coastal State (Article 297(3)(c)). It is possible that in the course of events, a coastal

[115] Ibid.

State undertakes law enforcement activities relating to fisheries in its EEZ and that its activities give rise to a dispute with other States wishing to fish in its EEZ. This type of dispute may be excluded by the coastal State from compulsory dispute settlement by a declaration under Article 298(1).

With this in mind, the Philippines argued that the law enforcement activities exception did not apply to China's toleration of harvesting of endangered species and cyanide and dynamite fishing at Scarborough Shoal and at Second Thomas Shoal for four reasons. First, at Scarborough Shoal, the activities took place in the territorial sea, whereas the law enforcement activities exception covered only applies to activities in the EEZ. Second, the Philippine claims were not directed at China's law enforcement activities, but rather to the absence of such activities— China tolerated the harvesting of endangered species and cyanide and dynamite fishing. Third, the exception did not apply to the duties of the flag State. Fourth, none of the high-tide features within 200 nautical miles of Second Thomas Shoal and over which China claimed sovereignty generated entitlement to an EEZ.[116]

In the Award on Jurisdiction, the Tribunal merely noted the Philippine arguments, without expressing any views on them.[117] It may be assumed that the Philippine arguments convinced the Tribunal, as it refrained in the subsequent proceedings from referring to the law enforcement activities exception as a bar to its jurisdiction.

Once its jurisdiction in relation to Submission No. 11 and 12(b) was established, the Tribunal was in a position to adjudicate the question whether China, by tolerating Chinese fishermen's harvesting of endangered species and the use of dynamite and cyanide in fishing and by undertaking construction activities that damaged fragile marine ecosystems, had violated its obligation under the Convention to protect and preserve the marine environment.

References

Chagos Marine Protected Area Arbitration (Mauritius v. United Kingdom), Award of 18 March 2015, https://files.pca-cpa.org/pcadocs/MU-UK%20 20150318%20Award.pdf, accessed 5 April 2019.

[116]Ibid., 86.
[117]Award on Jurisdiction, 134, para. 378.

Convention for the Conservation of Southern Bluefin Tuna, done at Canberra on 10 May 1993, entered into force on 20 May 1994, https://www.ccsbt.org/sites/ccsbt.org/files/userfiles/file/docs_english/basic_documents/convention.pdf, accessed 12 May 2019.

Convention for the Pacific Settlement of International Disputes, done at The Hague on 18 October 1907, https://pca-cpa.org/wp-content/uploads/sites/6/2016/01/1907-Convention-for-the-Pacific-Settlement-of-International-Disputes.pdf, accessed 4 April 2019.

Convention on Biological Diversity, signed at Rio de Janeiro on 5 June 1992, entered into force on 29 December 1993, https://www.cbd.int/doc/legal/cbd-en.pdf, accessed 26 March 2019.

Convention on International Trade in Endangered Species of Wild Fauna and Flora, signed at Washington, DC., on 3 March 1973, amended at Bonn, on 22 June 1979, amended at Gaborone, on 30 April 1983, https://www.cites.org/eng/disc/text.php, accessed 26 March 2019.

Dörr, Oliver. "Article 31. General Rule of Interpretation." *Vienna Convention on the Law of Treaties: A Commentary*, 2nd ed., 559–616. Eds. Oliver Dörr and Kirsten Schmalenbach. Berlin: Springer Verlag GmbH Germany, 2018.

International Court of Justice. *Statute of the International Court of Justice*, https://www.icj-cij.org/en/statute, accessed 5 May 2019.

Linderfalk, Ulf. *On the Interpretation of Treaties: The Modern International Law as Expressed in the 1969 Vienna Convention on the Law of Treaties*. Dordrecht: Springer, 2007.

Maresca, Adolfo. *Il diritto dei trattati. La Convenzione codificatrice di Viena del 23 maggio 1969* [The Law of Treaties. The Codifying Convention of Vienna, 23 May 1969]. Milano: Dott. A. Giuffrè Editore, 1971.

MOX Plant (Ireland v. United Kingdom), Provisional Measures, Order of 3 December 2001, Separate Opinion of Judge Wolfrum, ITLOS Reports 2001, 131, https://www.itlos.org/fileadmin/itlos/documents/cases/case_no_10/published/C10-O-3_dec_01-SO_W.pdf, accessed 5 April 2019.

Nordquist, Myron H. et al. (eds.). *United Nations Convention on the Law of the Sea 1982 Commentary*. Vol. V. *Settlement of Disputes, General and Final Provisions and Related Annexes and Resolutions*. Leiden: Martinus Nijhoff Publishers, 1989.

Oral, Nilufer. "The South China Sea Arbitral Award, Part XII of UNCLOS, and the Protection and Preservation of the Marine Environment." *The South China Sea Arbitration: The Legal Dimension*, 223–46. Eds. S. Jayakumar et al. Cheltenham: Edward Elgar, 2018.

Shihata, Ibrahim F. I. *The Power of the International Court to Determine Its Own Jurisdiction: Compétence de la Compétence*. Dordrecht: Springer Science+Business Media, 1965.

South China Sea Arbitration. Annexes Cited During Hearing on Jurisdiction (13 July 2015). Annex 579. Ministry of Foreign Affairs, People's Republic of China, Foreign Ministry Spokesperson Lu Kang's Remarks on Issues Relating to China's Construction Activities on the Nansha Islands and Reefs (16 June 2015), 159, https://files.pca-cpa.org/pcadocs/Annexes%20cited%20during%20Hearing%20on%20Jurisdiction%20%28Annexes%20574–583%29.pdf, accessed 9 April 2019.

―――. Award of 12 July 2016, https://pcacases.com/web/sendAttach/2086, accessed 26 March 2019.

―――. Award on Jurisdiction and Admissibility, 29 October 2015, https://pcacases.com/web/sendAttach/1506, accessed 26 March 2019.

―――. Hearing on Jurisdiction and Admissibility. Transcript, Day 2 (8 July 2015), https://pcacases.com/web/sendAttach/1400, accessed 3 April 2019.

―――. Transcript, Day 3 (13 July 2015), https://pcacases.com/web/sendAttach/1401, accessed 5 April 2019.

―――. Hearing on the Merits and the Remaining Issues of Admissibility. Transcript, Day 4 (30 November 2015), https://pcacases.com/web/sendAttach/1550, accessed 5 April 2019.

―――. *Memorial of the Philippines* (30 March 2014). Vol. I, https://files.pca-cpa.org/pcadocs/Memorial%20of%20the%20Philippines%20Volume%20I.pdf, accessed 27 March 2019.

―――. Annex 18. Memorandum from the Ambassador of the Republic of the Philippines in Beijing to the Undersecretary of Foreign Affairs of the Republic of the Philippines (10 March 1995). Vol. III, 209–11, https://files.pca-cpa.org/pcadocs/The%20Philippines%27%20Memorial%20-%20Volume%20III%20%28Annexes%201-60%29.pdf, accessed 26 March 2019.

―――. Annex 19. Memorandum from Erlinda F. Basilio, Acting Assistant Secretary, Office of Asian and Pacific Affairs, Department of Foreign Affairs, Republic of the Philippines, to the Secretary of Foreign Affairs of the Republic of the Philippines (29 March 1995). Vol. III, 213–17, https://files.pca-cpa.org/pcadocs/The%20Philippines%27%20Memorial%20-%20Volume%20III%20%28Annexes%201-60%29.pdf, accessed 26 March 2019

―――. Annex 20. Memorandum from Lauro L. Baja Jr., Assistant Secretary, Office of Asian and Pacific Affairs, Department of Foreign Affairs, Republic of the Philippines, to the Secretary of Foreign Affairs of the Republic of the Philippines (7 April 1995). Vol. III, 219–24, https://files.pca-cpa.org/pcadocs/The%20Philippines%27%20Memorial%20-%20Volume%20III%20%28Annexes%201-60%29.pdf, accessed 26 March 2019.

―――. Annex 21. Memorandum from the Ambassador of the Republic of the Philippines in Beijing to the Undersecretary of Foreign Affairs of the Republic of the Philippines (10 April 1995). Vol. III, 219–24, https://files.pca-cpa.

org/pcadocs/The%20Philippines%27%20Memorial%20-%20Volume%20 III%20%28Annexes%201-60%29.pdf, accessed 26 March 2019.

―――. Annex 22. Memorandum from the Ambassador of the Republic of the Philippines in Beijing to the Secretary of Foreign Affairs of the Republic of the Philippines, No. ZPE-231-95 (20 April 1995). Vol. III, 233–35, https:// files.pca-cpa.org/pcadocs/The%20Philippines%27%20Memorial%20-%20 Volume%20III%20%28Annexes%201-60%29.pdf, accessed 26 March 2019.

―――. Annex 33. Memorandum from Ambassador of the Republic of Philippines in Beijing to the Secretary of Foreign Affairs of the Republic of the Philippines, No. ZPE-76-98-S (6 November 1998). Vol. III, 295–300, https://files.pca-cpa.org/pcadocs/The%20Philippines%27%20Memorial%20 -%20Volume%20III%20%28Annexes%201-60%29.pdf, accessed 26 March 2019.

―――. Annex 23. Memorandum from the Secretary of Foreign Affairs of the Republic of the Philippines to the President of the Republic of the Philippines (31 July 1995). Vol. III, 237–42, https://files.pca-cpa.org/pcadocs/The%20 Philippines%27%20Memorial%20-%20Volume%20III%20%28Annexes%20 1-60%29.pdf, accessed 26 March 2019.

―――. Annex 37. Memorandum from the Embassy of the Republic of the Philippines in Beijing to the Secretary of Foreign Affairs of the Republic of the Philippines, No. ZPE-85-98-S (4 December 1998). Vol. III, 315–19, https://files.pca-cpa.org/pcadocs/The%20Philippines%27%20Memorial%20 -%20Volume%20III%20%28Annexes%201-60%29.pdf, accessed 26 March 2019.

―――. Annex 43. Memorandum from the Embassy of the Republic of the Philippines in Beijing to the Secretary of Foreign Affairs of the Republic of the Philippines, No. ZPE-06-2001-S (13 February 2001). Vol. III, 351–55, https://files.pca-cpa.org/pcadocs/The%20Philippines%27%20Memorial%20 -%20Volume%20III%20%28Annexes%201-60%29.pdf, accessed 26 March 2019.

―――. Annex 45. Memorandum from Willy C. Gaa, Assistant Secretary of Foreign Affairs, Republic of the Philippines to Secretary of Foreign Affairs, Republic of the Philippines (14 February 2001). Vol. III, 363–67, https:// files.pca-cpa.org/pcadocs/The%20Philippines%27%20Memorial%20-%20 Volume%20III%20%28Annexes%201-60%29.pdf, accessed 26 March 2019.

―――. Annex 180. Government of the Republic of the Philippines and Government of the People's Republic of China. Agreed Minutes on the First Philippines-China Bilateral Consultations on the South China Sea Issue (10 August 1995). Vol. VI, 187–91, https://files.pca-cpa.org/pcadocs/The%20 Philippines%27%20Memorial%20-%20Volume%20VI%20%28Annexes%20158- 221%29.pdf, accessed 26 March 2019.

———. Annex 181. Government of the Republic of the Philippines. Transcript of Proceedings Republic of the Philippines-People's Republic of China Bilateral Talks (10 August 1995). Vol. VI, 193–217, https://files.pca-cpa. org/pcadocs/The%20Philippines%27%20Memorial%20-%20Volume%20 VI%20%28Annexes%20158-221%29.pdf, accessed 26 March 2019.

———. Annex 186. Note Verbale from the Department of Foreign Affairs of the Republic of the Philippines to the Embassy of the People's Republic of China in Manila, No. 2000100 (14 January 2000). Vol. VI, 259–61, https:// files.pca-cpa.org/pcadocs/The%20Philippines%27%20Memorial%20-%20 Volume%20VI%20%28Annexes%20158-221%29.pdf, accessed 26 March 2019.

———. Annex 205. Note Verbale from the Department of Foreign Affairs of the Republic of the Philippines to the Embassy of the People's Republic of China in Manila, No. 12-0894 (11 April 2012). Vol. VI, 377–80, https://files.pca-cpa.org/pcadocs/The%20Philippines%27%20Memorial%20-%20Volume%20 VI%20%28Annexes%20158-221%29.pdf, accessed 26 March 2019.

———. Notification and Statement of Claim, 22 January 2013, https://seasre-search.files.wordpress.com/2014/12/notification-and-statement-of-claim-on-west-philippine-sea.pdf, accessed 5 April 2019.

———. Procedural Order No. 4, Doc. PCA 1435452 (1 April 2015), https:// pcacases.com/web/sendAttach/1807, accessed 5 April 2019.

———. Supplemental Documents of the Philippines (19 November 2015). Annex 619. Ministry of Foreign Affairs, People's Republic of China, Foreign Ministry Spokesperson Hua Chunying's Regular Press Conference (9 September 2014). Vol. I, 137–42, https://files.pca-cpa.org/pcadocs/ The%20Philippines%27%20Supplemental%20Documents%20-%20Volume%20 I%20%28Annexes%20607-667%29.pdf, accessed 9 April 2019.

———. Annex 624. Ministry of Foreign Affairs, People's Republic of China, Foreign Ministry Spokesperson Hua Chunying's Regular Press Conference (9 April 2015). Vol. I, 165–70, https://files.pca-cpa.org/pcadocs/The%20 Philippines%27%20Supplemental%20Documents%20-%20Volume%20I%20 %28Annexes%20607-667%29.pdf, accessed 9 April 2019.

———. Supplemental Written Submission of the Philippines (16 March 2015). Vol. I, http://www.pcacases.com/pcadocs/Supplemental%20Written%20 Submission%20Volume%20I.pdf, accessed 5 April 2019.

———. Annex 467. People's Republic of China, Position Paper of the Government of the People's Republic of China on the Matter of Jurisdiction in the South China Sea Arbitration Initiated by the Republic of the Philippines (7 December 2014). Vol. VIII, 19–33, https://files.pca-cpa. org/pcadocs/The%20Philippines%27%20Supplemental%20Written%20 Submission%20-%20Volume%20VIII%20%28Annexes%20466-499%29.pdf, accessed 4 April 2019.

Stephens, Tim. "The Collateral Damage from China's 'Great Wall of Sand': The Environmental Dimensions of the South China Sea Case." *Australian Yearbook of International Law* 24 (2017): 41–56.

United Nations Convention on the Law of the Sea, Concluded at Montego Bay on 10 December, entered into force on 16 November 1994, http://www.un.org/Depts/los/convention_agreements/texts/unclos/closindx.htm, accessed 21 March 2019.

United Nations Office of Legal Affairs. Codification Division. *United Nations Diplomatic Conferences. Third United Nations Conference on the Law of the Sea, 1973–1982*, 2019, http://legal.un.org/diplomaticconferences/1973_los/, accessed 4 April 2019.

United Nations Office of Legal Affairs. Treaty Section. Treaty Collection. *Multilateral Treaties Deposited with the Secretary-General, Chapter XXI: Law of the Sea*, https://treaties.un.org/Pages/ViewDetailsIII.aspx?src=TREATY&mtdsg_no=XXI-6&chapter=21&Temp=mtdsg3&clang=_en, accessed 4 April 2019.

Vienna Convention on the Law of Treaties, done at Vienna on 23 May 1969, entered into force on 27 January 1980, http://legal.un.org/ilc/texts/instruments/english/conventions/1_1_1969.pdf, accessed 9 April 2019.

Villiger, Mark E. *Commentary on the 1969 Vienna Convention on the Law of Treaties*. Leiden: Martinus Nijhoff Publishers, 2009.

Yasseen, Mustafa Kamal. "L'interprétation des traités d'après la Convention de Vienne sur le droit des traités [Treaty Interpretation According to the Vienna Convention on the Law of Treaties]." *Recueil des Cours de l'Académie de Droit International de La Haye* [Collected Courses of the Hague Academy of International Law]. Vol. 151 (1976-III): 1–114.

CHAPTER 5

Endangered Species, Fragile Ecosystems, and the Obligation to Protect and Preserve the Marine Environment

After the Arbitral Tribunal had affirmed that it had jurisdiction over Philippine Submissions No. 11 and 12(b), the challenge for the Philippines was to demonstrate that China had breached its obligation to protect and preserve the marine environment under the United Nations Convention on the Law of the Sea ("Convention") by tolerating the harvesting of endangered species and dynamite and cyanide fishing by Chinese nationals (Submission No. 11) as well as by engaging in construction activities in the South China Sea (Submissions No. 11 and 12(b)). The scope of Submission No. 11, which in the *Memorial* had been limited to Scarborough Shoal and Second Thomas Shoal, was broadened in the Final Submissions to include Cuarteron Reef, Fiery Cross Reef, Gaven Reef, Johnson Reef, Hughes Reef, and Subi Reef, in response to the expansion of China's construction activities. Submission No. 12(b) was unaltered, focusing exclusively on Mischief Reef.[1]

---

[1] *South China Sea Arbitration, Memorial of the Philippines* (30 March 2014), vol. I, 156, para. 6.68 (*"MP"*), https://files.pca-cpa.org/pcadocs/Memorial%20of%20the%20Philippines%20Volume%20I.pdf, accessed 26 March 2019; *South China Sea Arbitration*, Award on Jurisdiction and Admissibility (29 October 2015), 35, para. 101 ("Award on Jurisdiction"), https://pcacases.com/web/sendAttach/1506, accessed 26 March 2019; Award of 12 July 2016, 41–42, para. 112, https://pcacases.com/web/sendAttach/2086, accessed 26 March 2019. *United Nations Convention on the Law of the Sea*, concluded at Montego Bay on 10 December, entered into force on 16 November 1994, http://www.un.org/Depts/los/convention_agreements/texts/unclos/closindx.htm, accessed 21 March 2019.

© The Author(s) 2020
A. C. Robles Jr., *Endangered Species and Fragile Ecosystems in the South China Sea*,
https://doi.org/10.1007/978-981-13-9813-1_5

The far-reaching scope of Article 192, which simply declares that "States have the obligation to protect and preserve the marine environment," is stressed by scholarly commentary. Article 192 is described as the first explicit statement, in a general international treaty of comprehensive and universal scope, of a general obligation to protect and preserve the marine environment.[2] The Philippines and the Tribunal both emphasized that the obligation applies to all States, and not just to States Parties to the Convention, with respect to the marine environment, in all maritime areas, inside the national jurisdiction of States and beyond it.[3] In the Philippine view, It is a general obligation to protect not only the marine environment of other States and that of areas beyond national jurisdiction, but also to protect their own marine environment, including their territorial seas, their EEZs, and their continental shelves.[4] Consistent with this view, the Tribunal deemed questions of sovereignty to be irrelevant to the application of Part XII of the Convention. The Tribunal explicitly declared that its findings would in no way be dependent upon which State is sovereign over the maritime features in

[2] *South China Sea Arbitration*, Hearing on the Merits and Remaining Issues of Admissibility, Transcript, Day 4 (30 November 2015), 173 ("Hearing on the Merits"), https://pcacases.com/web/sendAttach/1550, accessed 10 April 2019; Myron H. Nordquist et al. (eds.), *United Nations Convention on the Law of the Sea 1982 Commentary*, vol. IV, *Third Committee: Protection and Preservation of the Marine Environment, Marine Scientific Research, and Development and Transfer of Marine Technology* (Leiden: Martinus Nijhoff Publishers, 1991), 36, 40; Nilufer Oral, "Implementing Part XII of the 1982 UN Law of the Sea Convention and the Role of International Courts," in Nerina ·Boschiero et al. (eds.), *International Courts and the Development of International Law. Essays in Honour of Tullio Treves* (The Hague: T.M.C. Asser Press, 2013), 403, 405. Czybulka is one of the few to doubt that the general obligation is binding even on States that are not Parties to the Convention, on the grounds that the Convention does not make a systematic distinction between "States" and "States Parties." Detlef Czybulka, "Article 192. General Obligation," in Alexander Proelss (ed.), *The United Nations Convention on the Law of the Sea. A Commentary* (München: Verlag C.H. Beck oHG, 2017), 1283.

[3] See for the Philippine view, *South China Sea Arbitration*, Hearing on Jurisdiction and Admissibility, Transcript, Day 2 (8 July 2015), 95 ("Hearing on Jurisdiction"), https://pcacases.com/web/sendAttach/1400, accessed 3 April 2019; Hearing on the Merits, Transcript, Day 3 (26 November 2015), 23, https://pcacases.com/web/sendAttach/1549, accessed 26 March 2019; Award of 12 July 2016, 373, para. 940, https://pcacases.com/web/sendAttach/2086, accessed 26 March 2019.

[4] Hearing on the Merits, Transcript, Day 4, 172–73.

the South China Sea, nor would they have any bearing on the issue of sovereignty.[5]

The Tribunal agreed with the Philippines and scholarly commentators that the "general obligation" established in Article 192 extends both to the "protection" of the marine environment from future damage and its "preservation," in the sense of "maintaining or improving the present condition of the marine environment."[6] The general obligation entails a positive obligation to take active measures to protect and preserve the marine environment, which logically implies a negative obligation not to degrade it.[7]

A rapid perusal of Part XII of the Convention reveals that almost all of its provisions regulate marine pollution. Nevertheless, the Philippines argued that the general formulation of the obligation and indeed the very title of Part XII ("Protection and Preservation of the Marine Environment") justified the assertion that Part XII is broad and comprehensive.[8] Part XII of the Convention, in its view, applies to the marine environment as a whole, in all of its ecological dimensions, and not simply to the pollution of the sea.[9] The Philippines drew the Tribunal's attention to Article 194(5), which stipulates that "[t]he measures taken in accordance with this Part shall include those necessary to protect and preserve rare or fragile ecosystems as well as the habitat of depleted, threatened or endangered species and other forms of marine life." The presence of this paragraph in an article devoted to preventing, reducing, and controlling pollution of the marine environment proved that Article 194 and by extension Part XII are not limited to measures aimed at controlling pollution and extends to measures focused primarily on the conservation and preservation of ecosystems.[10] The Tribunal agreed with this interpretation of Article 194(5), which was first expressed by another

---

[5] Ibid., Transcript, Day 3, 13; Award on Jurisdiction, 145, para. 408; Award of 12 July 2016, 373, para. 940.

[6] *MP*, vol. I, 156, para. 6.68; Award of 12 July 2016, 373, para. 941. For scholarly commentators, "protection" entails measures relating to imminent or existing danger or injury: "preservation" involves "conserving natural resources and retaining the quality of the marine environment." Nordquist et al., vol. IV, 11–12, 40.

[7] Award of 12 July 2016, 373–74, para. 941.

[8] Hearing on Jurisdiction, Transcript, Day 2, 95.

[9] Hearing on the Merits, Transcript, Day 4, 174–75.

[10] Hearing on Jurisdiction, Transcript, Day 2, 96.

Arbitral Tribunal constituted under Annex VII of the Convention.[11] As the *South China Sea* Tribunal put it, marine pollution is certainly an aspect of environmental protection, but it is certainly not the only one. Article 194(5) covers all measures under Part XII taken by States or those acting under its jurisdiction and control to protect and preserve "rare or fragile ecosystems" and habitats of endangered species.[12]

The Philippines, recognizing that notwithstanding these explanations the formulation of Article 192 remained very general, sought to define the content of the obligation by reference to other treaties, notably the Convention on International Trade in Endangered Species of Wildlife and Fauna ("CITES"), 1972, and the Convention on Biological Diversity ("CBD"), 1992.[13] The CBD had been concluded a decade after the adoption of the Convention in 1982, but the Philippines pointed out that when the Convention was adopted, it was anticipated that its normative content would be developed in other international instruments. Thus to use the CBD to inform the scope and content of obligations under the Convention would be entirely consistent with the Convention scheme.[14] As we have seen, the Philippines, in order to justify this approach, invoked Article 31(3)(c) of the Vienna Convention on the Law of the Treaties and Article 293(1) of the Convention.[15] The Tribunal agreed that the content of the general duty to protect and preserve the marine environment

---

[11] *Chagos Marine Protected Area Arbitration (Mauritius v. UK)*, Award of 18 March 2015, 129, para. 320; 211, para. 538, https://files.pca-cpa.org/pcadocs/MU-UK%20 20150318%20Award.pdf, accessed 5 April 2019.

[12] Award of 12 July 2016, 376, para. 945 For the Philippine argument, see Hearing on Jurisdiction, Transcript, Day 2, 96; Hearing on the Merits, Transcript, Day 3, 24.

[13] Hearing on the Merits, Transcript, Day 3, 27, 29. *Convention on International Trade in Endangered Species of Wild Fauna and Flora*, signed at Washington, DC, on 3 March 1973, amended at Bonn, on 22 June 1979, amended at Gaborone, on 30 April 1983, https://www.cites.org/eng/disc/text.php, accessed 26 March 2019. *Convention on Biological Diversity*, signed at Rio de Janeiro on 5 June 1992, entered into force on 29 December 1993, https://www.cbd.int/doc/legal/cbd-en.pdf, accessed 26 March 2019.

[14] *Supplemental Written Submission of the Philippines* (16 March 2015), vol. I, 56, paras. 11.6, 11.8 ("*SWSP*"), http://www.pcacases.com/pcadocs/Supplemental%20Written%20 Submission%20Volume%20I.pdf, accessed 5 April 2019.

[15] *MP*, vol. I, 190, para. 6.82; *SWSP*, vol. I, 56, para. 11.4. *Vienna Convention on the Law of Treaties*, done at Vienna on 23 May 1969, entered into force on 27 January 1980, http://legal.un.org/ilc/texts/instruments/english/conventions/1_1_1969.pdf, accessed 9 April 2019.

was to be informed by reference to other international instruments, but it invoked Article 237 of the Convention:

> The content of the general obligation in Article 192 is further detailed in the subsequent provisions of Part XII, including Article 194, as well as by reference to specific obligations set out in other international agreements, as envisaged in Article 237 of the Convention.[16]

Article 237 has been described as providing a mechanism for integrating the detailed substantive provisions of existing and future instruments on the protection and preservation of the marine environment within the overall framework of Part XII. Article 237 accords priority to the application of obligations under existing and future agreements, but establishes that these obligations must be carried out in a manner consistent with the Convention.[17] The rationale for this approach may be found in the desire to avoid duplicating the work of existing or future conventions. This approach considers the Convention as a framework, many of whose provisions can only be applied by relying on other, more specific, international agreements, not to mention national laws and regulations.[18]

As we shall see in Part II below, the other provisions of the Convention that the Tribunal had in mind, in addition to Article 194, were Articles 197, 123, 204, 205, and 206. The obligations under these last five are obligations imposed on States, but Articles 192 and 194 set forth obligations not only in relation to activities directly undertaken by States and their organs, but also in relation to activities carried out by

---

[16] Award of 12 July 2016, 373, para. 941. See also Award on Jurisdiction, 69, para. 176.

[17] Nordquist et al., vol. IV, 425. The text of Article 237. Obligations under other Conventions on the Protection and Preservation of the Marine Environment are as follows:
1. The provisions of this Part are without prejudice to the specific obligations assumed by States under special conventions and agreements concluded previously which relate to the protection and preservation of the marine environment and to agreements which may be concluded in furtherance of the general principles set forth in this Convention.
2. Specific obligations assumed by States under special conventions, with respect to the protection and preservation of the marine environment, should be carried out in a manner consistent with the general principles and objectives of this Convention.

[18] Pierre-Marie Dupuy and Martine Rémond-Guilloud, "La préservation du milieu marin [The Preservation of the Marine Environment]," in René-Jean Dupuy and Daniel Vignes (eds.), *Traité du Nouveau Droit de la mer* [Treatise on the New Law of the Sea] (Paris: Éditions Economica, 1985), 1005.

agents within their jurisdiction and control, for which the duty of States is to ensure that such activities do not harm the marine environment.[19] By definition, a State is responsible for activities directly undertaken by the State and its organs. Beyond this, the State is also responsible for ensuring that activities within their jurisdiction and control do not harm the environment. To explain the nature of this second type of obligation, the Philippines and the Tribunal both referred to a 2015 advisory opinion of the International Tribunal for the Law of the Sea ("ITLOS"), which clarified that the obligation to "ensure" is an obligation of conduct. It requires "due diligence" in the sense that a flag State (the State of registration of a ship) should not only adopt appropriate rules and measures that are applicable to the ships flying its flag but also a "certain level of vigilance in their enforcement and the exercise of administrative control," in relation to the ships flying its flag.[20] During the Hearing on the Merits, the Philippines reiterated the widely held view that in general, the international environmental obligations of States are obligations of "due diligence": "The obligations created by Articles 192 and 194 are not absolute. States are only required to take appropriate measures and must act with due diligence."[21]

The latter simply means that a State must make every effort to reach a desired result, in the knowledge that the achievement of the result depends on the circumstances.[22] A due diligence obligation

---

[19] Award of 12 July 2016, 375, para. 944.

[20] Hearing on the Merits, Transcript, Day 3, 32–33; Award of 12 July 2016, 375–76, para. 944.

[21] Hearing on the Merits, Transcript, Day 2, 30.

[22] Malgosia A. Fitzmaurice, "International Protection of the Environment," *Recueil des Cours de l'Académie de Droit International de La Haye* [Collected Courses of the Hague Academy of International Law], vol. 293 (2001-VI), 288. On the origins of due diligence, see Awalou Ouedraogo, "La due diligence en droit international: de la règle de la neutralité au principe général [Due Diligence in International Law: From the Rule of Neutrality to the General Principle]," *Revue générale de droit* [General Journal of Law] 42 (2012): 642–83, https://www.erudit.org/fr/revues/rgd/2012-v42-n2-rgd01542/1026909ar.pdf, accessed 11 April 2019. For recent surveys, see Thiago Braz Jardim Oliveira, "La diligence due dans la prévention des dommages à l'environnement [Due Diligence in the Prevention of Damage to the Environment]," *Anuário Brasileiro de Direito Internacional Annuaire Brésilien de Droit International* [Brazilian Yearbook of International Law] 7 (2012): 205–42, https://papers.ssrn.com/sol3/papers.cfm?abstract_id=2253408, accessed 11 April 2019; Duncan French and Tim Stephens, *ILA Study Group on Due Diligence in International Law. First Report* (7 March 2014), https://olympereseauinternational.files.

is the obligation to take all measures that a State could reasonably be expected to take.[23] The due diligence standard leaves States discretion in deciding which measures to take in order to comply with an international obligation; in this manner autonomy and flexibility in discharging its international obligations are preserved for a State.[24] The standard of due diligence applied to the conduct of a State plays a key role in international environmental law for two reasons. First, the violation of an international obligation might be due to a private actor or entity, rather than to the State of residence or origin of that actor or entity. This being the case, the State cannot be held responsible for the act itself; it will be responsible for its failure to prevent the damage.[25] Second, it is not always possible to totally prevent environmental damage or harm. The State will only be responsible for its failure to act diligently.[26] The Philippines, basing itself on the case law of the International Court of Justice ("ICJ"), identified three elements of the obligation of due diligence: the adoption of reasonably appropriate rules and measures; a certain level of vigilance in their enforcement; and the exercise of administrative control applicable to public and private operators.[27] Agreeing with the Philippines, the Tribunal added that upon receipt of reports of non-compliance, the flag State is under an obligation to investigate the

---

wordpress.com/2015/07/due_diligence_-_first_report_2014.pdf, accessed 11 April 2019; Doris König, "The Elaboration of Due Diligence Obligations as a Mechanism to Ensure Compliance with International Legal Obligations by Private Actors," in *La contribution du Tribunal international du droit de la mer à l'état de droit: 1996–2016* [The Contribution of the International Tribunal for the Law of the Sea to the Rule of Law: 1996–2016] (Leiden: Martinus Nijhoff Publishers, 2017), 83–95; and Tim Stephens and Duncan French, *ILA Study Group on Due Diligence in International Law. Second Report* (July 2016), https://ila.vettoreweb.com/Storage/Download.aspx?DbStorageId=1427...4796, accessed 11 April 2019.

[23] Stephens and French, 8.

[24] Ibid., 2, 7.

[25] French and Stephens, 25–26.

[26] Patricia Birnie, Alan Boyle and Catherine Redgewell, *International Law and the Environment* (3rd ed.; Oxford: Oxford University Press, 2009), 147–48. Boyle was one of the counsels of the Philippines in the *South China Sea Arbitration*.

[27] Hearing on the Merits, Transcript, Day 3, 30.

matter and, if appropriate, to take any action necessary to remedy the situation as well as to inform the reporting State of that action.[28]

In its decision on the merits, the Tribunal assessed China's compliance with its obligation to protect and preserve the marine environment on the basis of a distinction between two types of activities that harmed the marine environment: Those carried out by Chinese nationals (the harvesting of endangered species and cyanide and dynamite fishing) and those undertaken by the Chinese State itself (construction activities and island-building). In the first scenario, China was under a due diligence obligation, which was identified by the Tribunal with reference to other international environmental treaties, to prevent and punish such acts. China's compliance with its due diligence obligation with respect to the activities of Chinese fishermen will be discussed in Part I. In the second scenario, China had a due diligence obligation to prevent damage to fragile ecosystems and the habitat of depleted, threatened, or endangered species. To this obligation may be added two further obligations: the duty to cooperate with other States in the region and the duty to monitor and assess, notably by carrying out an environmental impact assessment ("EIA") and communicating the results of the assessment. China's failure to comply with all these obligations will be discussed in Part II.

Before any further discussion, it should be recalled that China refused to appear before the Tribunal. As in the jurisdictional phase, the Tribunal went to extraordinary lengths to ascertain China's views and to give the Philippines the opportunity to address issues that in the Tribunal's view the Philippines may have inadequately addressed or failed to address at all. As already mentioned, during the Hearing on the Merits, the Tribunal posed 22 questions to the coral reef expert presented by the Philippines, Dr. Kent E. Carpenter, and 6 questions to the counsel that had presented the Philippine arguments on Submissions No. 11 and 12(b).[29] One of the six questions obliged the Philippines to search over the weekend for Chinese statements on the environmental impact of its construction activities in the South China Sea.[30] After the Hearing, the Philippines submitted a written response to a supplemental question

---

[28] Award of 12 July 2016, 376, para. 944.
[29] Hearing on the Merits, Transcript, Day 4, 138–62, 166–87.
[30] Ibid., 181–82.

from Judge Wolfrum.[31] In a departure from the practice of other inter-national courts and tribunals, the Tribunal, even after the oral phase of the proceedings had concluded and as it was already carrying out its deliberations, took further steps to ensure itself that the Philippine sub-missions were well founded in fact and in law. First, it undertook its own research to determine whether China had indeed carried out an EIA. It then made available to the Philippines and China the documents it found, including statements made by Chinese officials on at least five occasions (in April, May, and June 2015, and February and May 2016) concerning China's construction activities, and requested their com-ments on them.[32] Second, the Tribunal appointed an independent coral reef expert in February 2016 and two additional coral reef experts in April 2016.[33] Third, the Tribunal communicated directly with China. It transmitted to China two reports on the damage to coral reefs caused by harvesting of giant clams. The Tribunal asked China whether it had car-ried out an EIA, and if it had done so, it requested a copy.[34] China did not respond.

I. China's Toleration of the Harvesting Endangered Species and of Dynamite and Cyanide Fishing and the Obligation to Protect and Preserve the Marine Environment

The task of the Tribunal was to determine whether China had indeed breached the Convention by failing to fulfill its due diligence obligations to prevent the harvesting of endangered species and the use of dynamite and cyanide in fishing. Since the obligation under Article 192 is a due dil-igence obligation, the occurrence of harm or damage is not the object of the obligation. Rather the occurrence of harm is "the teleological reference

---

[31] *The Philippines' Written Responses to the Tribunal's November 2015 Question* (18 December 2015), https://pcacases.com/web/sendAttach/1848, accessed 11 April 2019.

[32] The list of documents may be found in *South China Sea Arbitration, The Philippines' Written Responses to the Tribunal's 5 February 2016 Request for Comments (11 March 2016) (Annexes 864–892)*, https://files.pca-cpa.org/pcadocs/The%20Philippines%27%20 Written%20Responses%20%2811%20March%202016%29%20%28Annexes%20864-892%29.pdf, accessed 7 April 2016.

[33] Award of 12 July 2016, 29, para. 84; 30, para. 85; 31, para. 89.

[34] Ibid., 364, para. 915; 365, paras. 917–20; 369, para. 924.

making it possible to evaluate if the means provided [by the holder of the obligation]...were such as to enable the execution of the obligation."[35] To put it more prosaically, the harm to the environment reveals the breach of the obligation by China.[36] It still remained for the Philippines to prove that China had failed to take the measures necessary to prevent and punish the harvesting of endangered species and dynamite and cyanide fishing.

A. The Harvesting of Endangered Species, Dynamite and Cyanide Fishing, and the Harm to the Marine Environment

The Philippines, the expert presented by the Philippines, and the independent experts appointed by the Tribunal provided ample evidence that the harvesting of endangered species and dynamite and cyanide fishing caused damage to the marine environment

1. The Harm to the Marine Environment Caused by the Harvesting of Endangered Species

The evidence provided by the Philippines consisted of contemporaneous reports of Philippine naval, coast guard and fisheries authorities, diplomatic exchanges, and photographic evidence.[37] The Philippine allegations of illegal harvesting of endangered species and cyanide and dynamite fishing referred to Scarborough Shoal and Second Thomas Shoal. The Tribunal accepted Philippine evidence that Chinese fishing vessels had been involved in the following incidents at or in the waters of Scarborough Shoal:

1. In 1998:
   a. January—22 Chinese fishermen were found with several sacks of precious and semi-precious corals and dead sea turtles.[38]

---

[35] Jean Combacau, "Obligation de résultat et obligation de comportement: Quelques questions et pas de réponse [Obligation of Result and Obligation of Conduct: Some Questions and No Response]," in Daniel Bardonnet et al. (eds.), Mélanges offerts à Paul Reuter. Le droit international: Unité et diversité [Essays in Honor of Paul Reuter. International Law: Unity and Diversity] (Paris: Editions A. Pedone, 1981), 186.

[36] Pierre-Marie Dupuy, "Reviewing the Difficulties of Codification: On Ago's Classification of Obligations of Means and Obligations of Result in relation to State Responsibility," European Journal of International Law 10 (1999): 381.

[37] MP, vol. I, 179–81, para. 6.55; Award of 12 July 2016, 322–29, paras. 827–51.

[38] MP, Annex 29, Memorandum from the Assistant Secretary for Asian and Pacific Affairs, Department of Foreign Affairs, Republic of the Philippines, to the Secretary of

    b. March—29 fishermen were found in possession of semi-precious corals and 2 sacks of dynamite.[39]

2. In January 2000—3 vessels were found loaded with corals, cyanide, blasting caps, detonating cord, and dynamite.[40]

3. In January 2001—4 vessels' catch of endangered sharks, eels, sea turtles, and corals were confiscated.[41]

4. In 2002:

    a. February—3 fishing vessels were found to have between them 4 tons of seaweed and sea turtle, 1 live sea turtle, 4 tons of seaweed, and 4 sea clams.[42]

Foreign Affairs, Republic of the Philippines (23 March 1998), vol. III, 277–82, https://files.pca-cpa.org/pcadocs/The%20Philippines%27%20Memorial%20-%20Volume%20III%20%28Annexes%201-60%29.pdf, accessed 26 March 2019; Award of 12 July 2016, 322, para. 827.

[39] *MP*, Annex 29, vol. III, 277–82; Award of 12 July 2016, 322–23, para. 827.

[40] *MP*, Annex 41, Situation Report the Philippine Navy to the Chief of Staff, Armed Forces of the Philippines, No. 004-18074 (18 April 2000), vol. III, 341–44, https://files.pca-cpa.org/pcadocs/The%20Philippines%27%20Memorial%20-%20Volume%20III%20%28Annexes%201-60%29.pdf, accessed 26 March 2019; *MP*, Annex 42, Letter from Vice Admiral, Armed Forces of the Philippines, to Secretary of National Defense of the Republic of the Philippines (27 May 2000), vol. III, 345–50, https://files.pca-cpa.org/pcadocs/The%20Philippines%27%20Memorial%20-%20Volume%20III%20%28Annexes%201-60%29.pdf, accessed 26 March 2019 (the photographs of sticks of dynamite and corals being transferred to *BRP Republika ng Pilipinas Lanao del Norte* [LT-504] are at 350); Award of 12 July 2016, 323, para. 828.

[41] *MP*, Annex 44, Memorandum from Acting Secretary of Foreign Affairs of the Republic of the Philippines to the President of the Republic of the Philippines (5 February 2001), vol. III, 357–62, https://files.pca-cpa.org/pcadocs/The%20Philippines%27%20Memorial%20-%20Volume%20III%20%28Annexes%201-60%29.pdf, accessed 26 March 2019; *MP*, Annex 46, Office of Asian and Pacific Affairs, Department of Foreign Affairs, Republic of the Philippines, Apprehension of Four Chinese Fishing Vessels in the Scarborough Shoal (23 February 2001), vol. III, 369–74, https://files.pca-cpa.org/pcadocs/The%20Philippines%27%20Memorial%20-%20Volume%20III%20%28Annexes%201-60%29.pdf, accessed 26 March 2019; Award of 12 July 2016, 323–24, para. 830.

[42] *MP*, Annex 43, Memorandum from the Embassy of the Republic of the Philippines in Beijing to the Secretary of Foreign Affairs of the Republic of the Philippines, No. ZPE-06-2001-S (13 February 2001), vol. III, 395–98, https://files.pca-cpa.org/pcadocs/The%20Philippines%27%20Memorial%20-%20Volume%20III%20%28Annexes%201-60%29.pdf, accessed 26 March 2019; Award of 12 July 2016, 324, para. 831.

b. March—2 vessels were found to have a total of 3 tons of corals and 1 sack of seashells.[43]

c. September—the Philippine Navy rescued 14 Chinese fishermen onboard 4 dinghies; they had diving gear for gathering sea clams and other marine products.[44]

5. In October 2004—the Philippines intercepted Chinese vessels laden with giant clams.[45]

6. In December 2005—the Philippine Navy intercepted 4 Chinese vessels with assorted corals, 15 tons of giant clams, 1 ton of live clamshells, and illegal fishing gear.[46]

7. In April 2006—A Philippine naval patrol located 3 Chinese vessels, one with 20 tons of assorted shells, a second with 1 ton of live fish, and a third with 10 tons of corals and shells.[47]

8. In 2012:

---

[43] *MP*, Annex 50, Letter from Vice Admiral, Philippine Navy, to the Assistant Secretary for Asian and Pacific Affairs, Department of Foreign Affairs, Republic of the Philippines (26 March 2002), vol. III, 399–404, https://files.pca-cpa.org/pcadocs/The%20 Philippines%27%20Memorial%20-%20Volume%20III%20%28Annexes%201-60%29.pdf, accessed 26 March 2019; Award of 12 July 2016, 324, para. 831.

[44] *MP*, Annex 52, Report from CNS to Flag Officer in Command, Philippine Navy, File No. N2D-0802-401 (1 September 2002), vol. III, 411–14, https://files.pca-cpa.org/ pcadocs/The%20Philippines%27%20Memorial%20-%20Volume%20III%20%28Annexes%20 1-60%29.pdf, accessed 26 March 2019; Award of 12 July 2016, 324, para. 831.

[45] *MP*, Annex 55, Report from Lt. Commander, Philippine Navy, to Flag Officer in Command, Philippine Navy, No. N2E-F-1104-012 (18 November 2004), vol. III, 435–54, https://files.pca-cpa.org/pcadocs/The%20Philippines%27%20Memorial%20-%20 Volume%20III%20%28Annexes%201-60%29.pdf, accessed 26 March 2019 (a photograph of confiscated giant clams is at 448); Award of 12 July 2016, 324, para, 832.

[46] *MP*, Annex 57, Letter from Rear Admiral, Armed Forces of the Philippines, to the Assistant Secretary for Asian and Pacific Affairs, Department of Foreign Affairs, Republic of the Philippines (2006), vol. III, 461–80, https://files.pca-cpa.org/pcadocs/The%20 Philippines%27%20Memorial%20-%20Volume%20III%20%28Annexes%201-60%29.pdf, accessed 26 March 2019; Award of 12 July 2016, 324, para. 833.

[47] *MP*, Annex 59, Report from the Commanding Officer, NAVSOU-2, Philippine Navy, to the Acting Commander, Naval Task Force 21, Philippine Navy, No. NTF21-0406- 011/NTF21 OPLAN (BANTAY AMIANAN) 01-05 (9 April 2006), vol. III, 485–502, https://files.pca-cpa.org/pcadocs/The%20Philippines%27%20Memorial%20-%20 Volume%20III%20%28Annexes%201-60%29.pdf, accessed 26 March 2019; Award of 12 July 2016, 325, para. 834.

a. On 10 April, the Philippine Navy conducted "Visit, Board, Search and Seizure Operations" inside Scarborough Shoal on eight Chinese vessels; onboard the vessels they found large amounts of corals and giant clams.[48] Two Chinese survey ships, Zhingguo Haijian 75 and Zhingguo Haijian 84, maneuvered themselves in such a way as to position themselves at the mouth of the entrance of Scarborough Shoal, between a Philippine Navy vessel and 8 Chinese fishing vessels that were gathering giant clams, assorted corals, and endangered species.[49]

b. On 23 April—1 Chinese fishing vessel was observed to have giant clams inside its cargo hold.[50]

c. On 26 April—a Chinese vessel was seen overloaded with clamshells heading out of the lagoon.[51]

At Second Thomas Shoal, the Tribunal accepted a Philippine report that in May 2013, two Philippine vessels sighted two Chinese fishing vessels that were gathering corals and clams and dredging in the shoal. They were accompanied by a Chinese Jianghu V Missile Frigate and China Marine Surveillance ("CMS") 84 and 167. In fact, since the beginning of that month, 33 Chinese fishing vessels had been fishing at Mischief Reef and nearby maritime features, escorted by a People's Liberation Army Navy ("PLAN") Ship and CMS vessels.[52] The Philippines also

---

[48] MP, Annex 77, Memorandum from Colonel, Philippine Navy, to Chief of Staff, Armed Forces of the Philippines, No. N2E-0412-008 (11 April 2012), vol. IV, 131–47, https://files.pca-cpa.org/pcadocs/The%20Philippines%27%20Memorial%20-%20Volume%20IV%20%28Annexes%2061-102%29.pdf, accessed 26 March 2019; Award of 12 July 2016, 324, para. 835.

[49] MP, Annex 77, vol. III, 135; Award of 12 July 2016, 324, para. 836.

[50] MP, Annex 78, Report from the Commanding Officer, SARV-003, Philippine Coast Guard, to Commander, Coast Guard District Northwestern Luzon, Philippine Coast Guard (28 April 2012), vol. IV, 147–59, https://files.pca-cpa.org/pcadocs/The%20Philippines%27%20Memorial%20-%20Volume%20IV%20%28Annexes%2061-102%29.pdf, accessed 26 March 2019; Award of 12 July 2016, 325–26, para. 837.

[51] MP, Annex 78, 154.

[52] MP, Annex 94, Armed Forces of the Philippines, Near-Occupation of Chinese Vessels of Second Thomas (Ayungin) Shoal in the Early Weeks of May 2012 (May 2013), vol. IV, 337, 339, https://files.pca-cpa.org/pcadocs/The%20Philippines%27%20Memorial%20-%20Volume%20IV%20%28Annexes%2061-102%29.pdf, accessed 26 March 2019; Award of 12 July 2016, 327, para. 846.

supplied evidence, in the form of a BBC report, that Chinese fishermen had been engaged in large-scale harvesting of endangered hawksbill turtles near Mischief Reef.[53]

It might seem obvious that the harvesting of endangered species will cause harm to the marine environment. The Philippines had already explained in its *Memorial* that the extraction of corals reduces the structural complexity of reefs, affecting the reef's ability to support fishes and other, and that the harvesting giant clams entails crushing and destruction of surrounding corals.[54] This did not prevent the Tribunal from asking the expert presented by the Philippines, Dr. Kent E. Carpenter, to present further details of such harm. Dr. Carpenter, the Philippine-appointed expert, assured the Tribunal that there were no turtle species in the region that were not considered endangered or threatened. The five species of marine turtles in the region were either critically endangered, endangered, or vulnerable according to the International Union for the Conservation of Nature ("IUCN") Red List of threatened species.[55] He added that marine turtles as a group had the highest percentage of threatened species of all major groups of threatened species.[56] In answer to a question about the detrimental effects of extraction of corals, Dr. Carpenter explained that corals are a source of food for some reef inhabitants. In his view, any extraction of corals will have a negative environmental impact, depending on the amount of coral extracted.[57] In answer to a question how the removal of giant clams can be detrimental to the functioning of an ecosystem, Dr. Carpenter explained that giant clams contribute to the overall growth and maintenance of the reef

---

[53] *Philippines Written Responses to the Tribunal's November 2015 Questions (Annexes 860–63)* (18 December 2015), Annex 862, "Why Are Chinese Fishermen Destroying Coral Reefs in the South China Sea?" by Rupert Wingfield-Hayes, BBC (15 December 2015), 14–24, http://www.pcacases.com/pcadocs/The%20Philippines%27%20Written%20Responses%20to%20the%20Tribunal%27s%20November%202015%20Question%20%28Annexes%20860-863%29.pdf, accessed 11 April 2019.

[54] *MP*, vol. I, 178, para. 6.56; 181, para. 6.57.

[55] The IUCN is a non-governmental organization founded in 1948 and with headquarters in Gland, Switzerland. Its members are both governments and civil society organizations. It describes itself as "the global authority on the status of the natural world and the measures needed to safeguard it." International Union for Conservation of Nature, *About* (Gland: IUCN, 2019), https://www.iucn.org/about, accessed 12 April 2019.

[56] Hearing on the Merits, Transcript, Day 4, 145–46.

[57] Ibid., 150–51.

structure itself. They contribute a large mass of calcium carbonate that has taken many years to form and provide mass and diversity of topography to the reef. Many species of fishes and invertebrates in turn rely on this topography as refuge.[58]

In their report submitted several months after the Hearing on the Merits, the independent experts drew the Tribunal's attention to the dangers associated with the harvesting of giant clams. The largest of these species, which grow slowly, can reach approximately 1.5 meters in size and weigh more than 100 kilograms. They are harvested widely in Southeast Asia and the rest of the world, mainly for their meat and shells, and as a result have become rare on most reefs. The techniques employed by Chinese fishermen to harvest giant clams compounded the damage to the marine environment. As the giant clams have become rare on most reefs, collectors have begun to target the fossil shells buried in reef flats, the shallow extensive habitat on top of reefs. The excavation of these fossil shells is highly destructive, resulting in a decline in coral cover by 95%. More recently, Chinese fishermen utilized boat propellers to excavate shells from the reef flats of the Spratly Islands on an industrial scale, a technique that causes nearly complete destruction of reef areas.[59] The experts' report surveyed the different reefs for damage resulting from extraction of giant clams and observed arc-shaped scars indicating extensive propeller damage on Cuarteron Reef, Fiery Cross Reef, Hughes Reef, and Mischief Reef.[60] Cuarteron Reef was subject to intense clamshell mining, carried out by no less than 89 boats using propellers to dig up substrata of shallow reef.[61] Propeller damage was reported on a total of 28 reefs in the Spratly Islands and the Paracel Islands.[62] Dr. John McManus, another coral reef expert whose testimony was submitted by the Philippines, initially estimated that

---

[58] Ibid., 145.

[59] *South China Sea Arbitration*, Independent Expert Report. Assessment of the Potential Environmental Consequence of Construction Activities on Seven Reefs in the Spratly Islands in the South China Sea, Sebastian C.A. Ferse, Peter Mumby, and Selina Ward, 26 April 2016, 10 ("Ferse Report"), https://pcacases.com/web/sendAttach/1809, accessed 24 March 2019; Award of 12 July 2016, 381, para. 957; *Philippines Written Responses to the Tribunal's November 2015 Questions*, Annex 862, 15–24.

[60] Award of 12 July 2016, 329, para. 851.

[61] Ferse Report, 17–19.

[62] Award of 12 July 2016, 328, para. 848.

Chinese nationals were responsible for approximately 70 square kilometers of coral reef damage from giant clam harvesting using propellers.[63] Through the Tribunal, Dr. Ferse, one of the coral reef experts appointed by the Tribunal, asked him in writing to clarify what percentage of his estimates on the extent of reef areas damaged could be attributed to dredging associated with extraction and to clam shell extraction. Dr. McManus responded through the Philippines that the dredgers generally used by China for the island-building have drafts too deep for the affected areas, which were very shallow (1–3 meters) and that the presence of masses of dead broken branching coral and abundant sand on one reef ruled out dredging as the cause of the damage.[64] His revised estimates were that China was responsible for 39 square kilometers of damage from shallow dredging and 69 square kilometers of damage from giant clam harvest using propellers to dig up the bottom within the Spratly Islands. Based on interviews, studies of satellite imagery and underwater inspection at clam inspection sites near Thitu Island, he concluded that the thoroughness of the damage exceeded anything he had seen in forty years.[65]

As for corals, the Ferse Report explained that stony corals are often harvested as construction material or for the curio trade, while branching corals are targeted mainly for the curio trade. The Tribunal accepted the Ferse Report's finding that the repeated targeted removal of coral colonies can modify the community structure and lead to overall loss of structural complexity. The reduction in live coral cover and structural complexity severely affects the reef fish community, as a large proportion of fishes on the reef utilize live corals, as food, or as refuges against predators, at some point in their life history.[66] In light of the evidence, the

---

[63] Ibid., 381, para. 958, *The Philippines' Annexes Cited During the Hearing on the Merits*, Annex 850, "Offshore Coral Reef Damage, Overfishing and Paths to Peace in the South China Sea", by John McManus, *The South China Sea: An International Law Perspective*, Conference Papers (6 March 2015), 578–608, https://pcacases.com/web/view/7, accessed 26 March 2019.

[64] Draft (draught) is defined as the depth of water that a ship requires to float freely. *Larousse Dictionary of Science and Technology* (Edinburgh: Larousse plc, 1995), 332.

[65] Award of 12 July 2016, 329, para. 850. See Letter from Professor John W. McManus to the Tribunal, 22 April 2016, https://pcacases.com/web/sendAttach/1917, accessed 17 April 2019.

[66] Ferse Report, 10; quoted in the Award of 12 July 2016, 380, para. 955; *MP*, vol. I, 181–82, para. 6.56; Carpenter, in Hearing on the Merits, Transcript, Day 4, 150.

Tribunal found that "harvesting of corals and giant clams has a harmful effect on the fragile marine environment."[67]

At least as harmful to the marine environment as the harvesting of endangered species is the use of dynamite and cyanide for fishing.

2. The Harm to the Marine Environment Caused by Dynamite and Cyanide Fishing

As regards cyanide and dynamite fishing at Scarborough Shoal and Second Thomas Shoal, the Tribunal was satisfied, based on Philippine Navy, Philippine Coast Guard and Philippine police reports and photographic evidence, that Chinese fishing vessels used dynamite or cyanide on the following occasions:

1. In 1995—62 Chinese fishermen were arrested after being found in possession of explosives and cyanide.[68]
2. In March 1998—29 Chinese fishermen were found in possession of dynamite.[69]
3. In April 2000—3 Chinese vessels were found with one-half liter of cyanide, twenty-one blasting caps (non-electric), eight feet detonating cord, and seven sticks of dynamite.[70]
4. In 2002:
   a. February—3 fishing vessels were found to have between them 1 box blasting caps, 2 meters of detonating cord, 20 bottles of cyanide, and 200 pieces of cyanide tube pumps.[71]

---

[67] Award of 12 July 2016. 382, para. 960.

[68] *MP*, Annex 21, Memorandum from the Ambassador of the Republic of the Philippines in Beijing to the Undersecretary of Foreign Affairs of the Republic of the Philippines (10 April 1995), vol. III, 230, https://files.pca-cpa.org/pcadocs/The%20Philippines%27%20Memorial%20-%20Volume%20III%20%28Annexes%201-60%29.pdf, accessed 26 March 2019.

[69] *MP*, Annex 29, vol. III, 277–82; Award of 12 July 2016, 322–23, para. 827.

[70] *MP*, Annex 41, vol. III, 341–44; Award of 12 July 2016, 323, para. 829.

[71] *MP*, Annex 49, Memorandum from Perfecto C. Pascual, Director, Naval Operation Center, Philippine Navy, to The Flag Officer in Command, Philippine Navy (11 February 2002), vol. III, 395–97, https://files.pca-cpa.org/pcadocs/The%20Philippines%27%20Memorial%20-%20Volume%20III%20%28Annexes%201-60%29.pdf, accessed 26 March 2019; Award of 12 July 2016, 324, para. 831.

    b. March—2 vessels were found to have a total of 25 pieces of cyanide tube pumps.[72]

    c. In September—the Philippine Navy rescued 14 Chinese fishermen onboard 4 dinghies; they used cyanide cakes to catch fish.[73]

5. In December 2005, the Philippine Navy intercepted 4 Chinese vessels with 4 cyanide fishing pumps.[74]

6. In April 2006—A Philippine naval patrol located 3 Chinese vessels with cyanide pumps.[75]

The Tribunal did not take into consideration a reference to Chinese vessels' alleged use of explosives on 12 May 2012. The reasons were that the provenance of the vessels was uncertain and the report was unsupported by contemporaneous reports, inventories, and photographs.[76]

In the *Memorial*, the Philippines explained that the use of explosives is particularly destructive of the surrounding ecosystem. When a bottle bomb containing half a kilogram of explosives is dropped on a reef, it will shatter all the coral reef structure within a meter radius from the reef. A gallon-sized drum filled with explosives will reduce a coral reef to rubble within a five-meter radius, but the killing area for fish and/or invertebrates will be much wider.[77] To make matters worse, the corals most vulnerable to being pulverized by explosion are the more delicate branching forms that are home to the smaller animals upon which larger animals depend. The destruction of the corals has a ripple effect on the entire ecosystem.[78] Exposure to cyanide is also destructive of the surrounding system. It results in the loss of coral cover and affects the vitality of all species that depend on the reef. The inefficiency of cyanide

---

[72] *MP*, Annex 50, vol. III, 388–404 (see a photograph of cyanide tube pumps at 403).

[73] *MP*, Annex 52, vol. III, 411–14.

[74] *MP*, Annex 57, Letter from Rear Admiral, Armed Forces of the Philippines, to the Assistant Secretary for Asian and Pacific Affairs, Department of Foreign Affairs, Republic of the Philippines (2006), vol. III, 466, https://files.pca-cpa.org/pcadocs/The%20 Philippines%27%20Memorial%20-%20Volume%20III%20%28Annexes%201-60%29.pdf, accessed 26 March 2019.

[75] *MP*, Annex 59, vol. III, 485–502; *MP*, vol. I, 181–84, paras. 6.58–6.61; Award of 12 July 2016, 385–86, paras. 9.68–9.69.

[76] Award of 12 July 2016, 386, para. 969.

[77] Hearing on the Merits, Transcript, Day 3, 20–21.

[78] *MP*, vol. I, 182, para. 6.59.

fishing exacerbates the problem as cyanide merely stuns fish, so that the fishermen have to pound apart and destroy corals to extract the fish.[79]

During the Hearing on the Merits, the expert presented by the Philippines gave more detailed accounts of the impact of blast fishing operations and the use of cyanide on the marine environment. Dr. Carpenter explained that cyanide is a very strong poison, which prevents cells from using oxygen for respiration and inhibits photosynthesis. It is sprayed into the water if the purpose is to capture ornamental fish for the aquarium fish industry. Fishing often involves breaking the corals to extract the hiding fish.[80] The area of coral destroyed is one square meter per fish.[81] If the purpose is to capture large reef fish, such as groupers, for the live reef fish or food trade, large quantities of cyanide are dumped on to the reef. When the large fishes swim to the surface in distress, they are scooped up with nets and taken to clean water to recuperate. The large release of cyanide kills non-target species, invertebrates, and corals.[82] The independent coral reef experts confirmed that cyanide and blast fishing are highly destructive methods that have been used in the Spratly Islands in the past.[83]

No State ever wishes to be caught admitting that it is failing in its obligation to protect and preserve the environment, and China is no exception. China's responses to the Philippines imply that the marine environment had not been damaged by the activities of Chinese fishermen. In March 1992, referring to an incident about which very

[79] Ibid., 183–84, para 6.61.

[80] Hearing on the Merits, Transcript, Day 4, 158–59.

[81] *MP*, Annex 240, "Eastern South China Sea Environmental Disturbances and Irresponsible Fishing Practices and their Effects on Coral Reefs and Fisheries." by Kent E. Carpenter, Ph.D., vol. VII, 407 ("First Carpenter Report"), https://files.pca-cpa.org/pca-docs/The%20Philippines%27%20Memorial%20-%20Volume%20VII%20%28Annexes%20 222-255%29.pdf, accessed 7 April 2019.

[82] Hearing on the Merits, Transcript, Day 4, 159; First Carpenter Report, 407; *MP*, vol. I, 182, para. 6.60–6.61; Award of 12 July 2016, 357, para. 898.

[83] Award of 12 July 2016, 386, para. 970; Ferse Report, 10. According to another scholar, in fishing for purposes of the restaurant and aquarium trade, cyanide is almost exclusively used, as it is deemed as the most "cost-effective" way of fishing. Herman S.J. Cesar, "Coral Reefs: Their Threats, Functions and Economic Value," in Herman S.J. Cesar (ed.), *Collected Essays on the Economics of Coral Reefs* (Kalmar: Linnaeus University, 2002), 17, http://www.reefbase.org/resource_center/publication/pub_12370.aspx, accessed 31 March 2019.

little information is available, China asserted that Chinese fishermen were "carrying out their traditional and normal fishing activities within China's territorial waters."[84] In March 1995, China pointed out that China also prohibited illegal methods of fishing and attached great importance to the protection of rare species and denied that the Chinese fishermen had violated Philippine laws.[85] The implication was that the fishermen were not violating Chinese laws either. In April 1995, China reiterated that Chinese fishermen detained by the Philippines were conducting normal fishing operations in traditional Chinese fishing grounds.[86] China insisted that it was "totally legal and irreproachable for Chinese fishermen to fish" in the Spratly Islands, which were their traditional fishing grounds.[87] In February 2001, China again declared that Chinese fishermen were conducting normal operations in China's traditional fishing grounds.[88] In April 2002, China protested the apprehension of Chinese fishing vessels which were "normally working" in the area of Scarborough Shoal.[89] In 2011, China protested that the

---

[84] MP, Annex 16, Memorandum to the Assistant Secretary, Office of Asian and Pacific Affairs, Department of Foreign Affairs, Republic of the Philippines (23 March 1992), vol. III, 201–4, https://files.pca-cpa.org/pcadocs/The%20Philippines%27%20Memorial%20-%20Volume%20III%20%28Annexes%201-60%29.pdf, accessed 26 March 2019.

[85] MP, Annex 19, Memorandum from Erlinda F. Basilio, Acting Assistant Secretary, Office of Asian and Pacific Affairs, Department of Foreign Affairs, Republic of the Philippines, to the Secretary of Foreign Affairs of the Republic of the Philippines (29 March 1995), vol. III, 217, https://files.pca-cpa.org/pcadocs/The%20Philippines%27%20Memorial%20-%20Volume%20III%20%28Annexes%201-60%29.pdf, accessed 26 March 2019.

[86] MP, Annex 20, Memorandum from Lauro L. Baja, Jr., Assistant Secretary, Office of Asian and Pacific Affairs, Department of Foreign Affairs, Republic of the Philippines, to the Secretary of Foreign Affairs of the Republic of the Philippines (7 April 1995), vol. III, 219–24, https://files.pca-cpa.org/pcadocs/The%20Philippines%27%20Memorial%20-%20Volume%20III%20%28Annexes%201-60%29.pdf, accessed 26 March 2019.

[87] MP, Annex 21, vol. III, 228.

[88] MP, Annex 43, vol. III, 353.

[89] MP, Annex 51, Memorandum from Josue L. Villa, Embassy of the Republic of the Philippines in Beijing, to the Secretary of Foreign Affairs of the Republic of the Philippines (19 August 2002), vol. III, 405–409, https://files.pca-cpa.org/pcadocs/The%20Philippines%27%20Memorial%20-%20Volume%20III%20%28Annexes%201-60%29.pdf, accessed 26 March 2019.

Philippines had shot at a Chinese fishing vessel conducting fishing in China's traditional fishing grounds.[90]

Now at the time practically all of these incidents occurred, China had already passed a series of laws and regulations to protect endangered species. The Fisheries Law of China, adopted in 1986 and amended in 2000, 2004, 2009, and 2013, prohibited fishing and killing or hunting of aquatic wild animals under State protection. The same Law prohibited methods such as catching fish by explosives, poison, or electricity. The Law on the Protection of Wildlife, adopted in 1989, amended in 2004 and 2009 and revised in 2016, prohibited hunting, catching, or killing of wildlife under special State protection as well as the sale and purchase of this wildlife or their products. The Law also prohibited techniques using explosives and other prohibited methods in hunting or catching wildlife. The Regulations on the Protection of Aquatic Wild Animalsprohibited any entity or individual from damaging the waters, places, and living conditions where aquatic wild animals under State priority protection and local priority protection live and breed. They prohibited the capture or killing of any aquatic wild animal under State priority protection, as well as the sale and purchase of such animals and their products.[91]

The implications of the Chinese government's responses to the Philippines are twofold. First, if the Chinese fishing vessels' operations were "normal" at a time when China's own laws prohibited the harvesting of endangered species and the use of cyanide and dynamite in fishing, then the Chinese fishing vessels were not engaged in any such activities and were consequently not causing any harm to the environment. Second, if the Chinese fishing vessels were not engaged in any illegal activity, then the activities of Chinese official vessels, whether they belonged to the Coast Guard or to Maritime Law Enforcement, could not be described as toleration of illegal activities. Indeed, in 2012, the

---

[90] *MP*, Annex 75, Memorandum from the Embassy of the Republic of the Philippines in Beijing to the Secretary of Foreign Affairs of the Republic of the Philippines, No. ZPE-121-2011-S (2 December 2011), vol. IV, 123, https://files.pca-cpa.org/pcadocs/The%20Philippines%27%20Memorial%20-%20Volume%20IV%20%28Annexes%2061-102%29.pdf, accessed 26 March 2019.

[91] Chinese Society of International Law ("CSIL"), "The South China Sea Arbitration Awards: A Critical Study," *Chinese Journal of International Law* 17 (2018): 580–82, paras. 791–95.

Chinese Foreign Ministry spokesperson declared that China had sent administrative vessels "to protect the safety and legitimate fishing activities of Chinese fishermen and fishing vessels."[92] To sum up, all Chinese activity in the area of Scarborough Shoal was legal: "activities by Chinese ships, including government public service ships and fishing boats, in Huangyan Island [Scarborough Shoal] and its waters are completely within China's sovereignty."[93]

On two occasions, Chinese officials displayed a markedly different attitude. In February 2001, a Chinese diplomat in Manila informed the Philippines that China had a law on the illegal catching of turtles and corals and that those who violated the law would be punished.[94] In March 2001, a Chinese Foreign Ministry official declared to the Philippine Ambassador in Beijing that the Chinese government was always against illegal fishing.[95]

As already mentioned, the Tribunal accepted most of the Philippine claims, which were accompanied not just by contemporaneous reports but also by photographs of the endangered species and the equipment for dynamite and cyanide fishing found onboard Chinese fishing vessels. The issue that the Philippines and the Tribunal had to address was whether China had breached the obligation to protect and preserve the

---

[92] MP, Annex 117, Ministry of Foreign Affairs of the People's Republic of China, Foreign Ministry Spokesperson Liu Weimin's Regular Press Conference on April 12, 2012 (12 April 2012), vol. V, 111, https://files.pca-cpa.org/pcadocs/The%20 Philippines%27%20Memorial%20-%20Volume%20V%20%28Annexes%20103-157%29.pdf, accessed 16 April 2019.

[93] MP, Annex 211, Note Verbale from the Embassy of the People's Republic of China in Manila to the Department of Foreign Affairs of the Republic of the Philippines, No. (12) PG-239 (25 May 2012), vol. VI, 401–4, https://files.pca-cpa.org/pcadocs/The%20 Philippines%27%20Memorial%20-%20Volume%20VI%20%28Annexes%20158-221%29.pdf, accessed 26 March 2019.

[94] MP, Annex 45, Memorandum from the Assistant Secretary for Asian and Pacific Affairs, Department of Foreign Affairs, Republic of the Philippines, to the Secretary of Foreign Affairs, Republic of the Philippines (14 February 2001), vol. III, 363–68, https://files.pca-cpa.org/pcadocs/The%20Philippines%27%20Memorial%20-%20Volume%20 III%20%28Annexes%201-60%29.pdf, accessed 26 March 2019.

[95] MP, Annex 47, Memorandum from the Embassy of the Republic of the Philippines in Beijing to the Secretary of Foreign Affairs, Republic of the Philippines, No. ZPE-09-2001-S (17 March 2001). Vol. III, 375–78, https://files.pca-cpa.org/pcadocs/The%20 Philippines%27%20Memorial%20-%20Volume%20III%20%28Annexes%201-60%29.pdf, accessed 26 March 2019.

marine environment under Article 192 by tolerating the harvesting of endangered species and dynamite and cyanide fishing.

B. The Toleration of the Harvesting of Endangered Species and of Dynamite and Cyanide Fishing and China's Obligation to Protect and Preserve the Marine Environment

The Tribunal's approach to the question whether China had fulfilled its obligation under Article 192 of the Convention to protect and preserve the marine environment involved interpretation of Article 192 in light of other provisions of Part XII of the Convention and international environmental treaties. On this basis, the Philippines and the Tribunal deduced China's due diligence obligations to prevent and punish the harvesting of endangered species and the use of dynamite and cyanide for fishing.

1. The Obligation to Protect and Preserve the Marine Environment in the Convention and in Other International Environmental Treaties

The Philippines argued, on the basis of Article 293(1) of the Convention, that the interpretation of Article 192 should be guided by reference to standards in other multilateral environmental instruments.[96] Four such instruments were identified in the *Memorial.*[97] Principle 21 of the 1972 Stockholm Declaration laid down the principle that States have "the responsibility to ensure that activities within their jurisdiction or control do not cause damage to the environment of other States or of areas beyond the limits of national jurisdiction."[98] Concretely, this meant that China was internationally responsible for the damage caused by the actions of fishermen under its control.[99] Principle 17 of Agenda 21, adopted by the 1992 UN Conference on Environment and Development, declares that it is "necessary" for States to, "[m]aintain or

---

[96] *MP*, vol. I, 187, paras. 6.71–6.72; 190–93, paras. 6.82–6.89; Award of 12 July 2016, 361, para. 907.

[97] *MP*, vol. I, 187, paras. 6.71–6.72; Hearing on the Merits, Transcript, Day 4, 171–72.

[98] *Stockholm Declaration of the United Nations Conference on the Human Environment*, 16 June 1972, https://www.jus.uio.no/english/services/library/treaties/06/6-01/stockholm_decl.xml, accessed 22 April 2019.

[99] *MP*, vol. I, 187, para. 6.71.

restore populations of marine species at levels that can produce the maximum sustainable yield," "[p]rotect and restore endangered marine species," and "[p]reserve rare or fragile ecosystems, as well as habitats and other ecologically sensitive areas."[100] CITES lists giant clams as under threat in its Appendix II, which together with its Appendix I constitutes generally accepted rules and standards that should inform the interpretation and application of Articles 192 and 194.[101]

The fourth instrument that the Philippines believed was relevant when interpreting the Convention was the CBD, to which the Philippines and China were both Parties.[102] For the Philippines, the protection and preservation of the marine environment should be interpreted to include the protection and preservation of biodiversity.[103] The CBD applied both to areas under State sovereignty and to activities under their control, within the area of its national jurisdiction or beyond the limits of national jurisdiction. Whether China was the sovereign or not over the contested maritime features, it had to fulfill its obligations under the CBD in the South China Sea.[104] China's obligations were to "regulate or manage biological resources important for the conservation of biodiversity, whether within or outside protected areas, with a view to ensuring their conservation and sustainable use" and to promote "the protection of ecosystems, natural habitats and the maintenance of viable populations of species in natural surroundings."[105] In order to comply with these obligations, China had to control all activities independent of location, which could affect biological resources, including direct use (e.g., hunting and harvesting) and indirect effects (e.g., pollution or tourism) on biological resources, and to prevent activities that "destroy ecosystems and natural habitats or which impede the maintenance of viable populations."[106]

---

[100] United Nations Conference on Environment and Development ("UNCED"), *Agenda 21*, Rio de Janeiro, Brazil, 3–14 June 1992, 183, paras. 17.74(d), (e), and (f), https://sustainabledevelopment.un.org/content/documents/Agenda21.pdf, accessed 11 April 2019.

[101] Hearing on the Merits, Transcript, Day 3, 29.

[102] Ibid., 28.

[103] Ibid., Transcript, Day 4, 178.

[104] *MP*, I, 191–92, paras. 6.84–6.85.

[105] CBD, Article 8(c) and (d).

[106] *MP*, vol. I, 192–93, paras. 6.85–6.89.

The Philippines denied that it was suggesting that the CBD had been wholly incorporated into the Convention or that there was a dispute about China's compliance with the CBD.[107] The reference to the CBD was necessary, in order to make clear that Article 192 included the protection and preservation of biodiversity, and that Article 192 was not limited to the categories identified in Article 194(5)—"rare or fragile ecosystems" and "habitat of depleted, threatened or endangered species."[108] This interpretation of Article 192, while intuitively appealing, does not seem to be indispensable to the Philippine case, to the extent that all the species that it identified were depleted, threatened, or endangered, and that coral reefs are rare or fragile ecosystems. Perhaps for this reason, this aspect of the Philippine argument did not figure in the Tribunal's reasoning. The Philippines further argued that Article 192, informed by the CBD, entails two obligations: The obligation to take measures to ensure the sustainable use of biological resources, which is undermined by wasteful and unsustainable blast fishing, and the obligation to protect and preserve endangered species.[109]

With respect to the use of dynamite and cyanide in fishing, the Philippines argued that China had violated its obligation to prevent pollution of the marine environment under Article 194.[110] The Convention defines pollution as the "introduction by man [sic], directly or indirectly, of substances or energy into the marine environment ... which results or is likely to result in such deleterious effects as harm to living resources and marine life."[111] Article 194(1) obliges States to take actions to prevent pollution of the marine environment.[112] The Philippines noted that the Convention did not specify the substances the dumping of which is

---

[107] *SWSP*, 56, para. 11.3; *MP*, vol. I, 191–92, paras. 6.82–6.85; Hearing on Jurisdiction, Transcript, Day 2, 107; Hearing on the Merits, Transcript, Day 4, 177–78.

[108] Hearing on the Merits, Transcript, Day 4, 178.

[109] Ibid., Day 3, 28; Award of 12 July 2016, 361, para. 910.

[110] *MP*, vol. I, 188–90, paras. 6.75–6.81.

[111] Convention, Article 1(4).

[112] Article 194(1) provides that

> States shall take, individually or jointly as appropriate, all measures consistent with this Convention that are necessary to prevent, reduce and control pollution of the marine environment from any source, using for this purpose the best practicable means at their disposal and in accordance with their capabilities, and they shall endeavour to harmonize their policies in this connection.

regulated by Article 194, but it believed that the use of cyanide, a highly toxic chemical, was a form of pollution.[113] Article 194(3) emphasizes that States are obliged to manage "all sources" of pollution, placing a special importance on "minimiz[ing] to the fullest possible extent ... the release of toxic, harmful or noxious substances ... by dumping."[114] Dumping is defined as the "deliberate disposal of wastes or other matter from vessels ..."[115] Article 194(5) further emphasizes that the obligations set out in Article 194 are heightened in areas of rare or fragile ecosystems and in places that provide habitats for vulnerable species.[116] The London Convention on the Prevention of Marine Pollution by Dumping of Wastes and Other Matter, 1972, which, according to scholarly commentary, guides the definition of "dumping" in the Convention, specifically requires a prior special permit to dump cyanide.[117] The environmental damage is aggravated by the fact that the dumping occurred at Scarborough Shoal and Second Thomas Shoal, which are fragile ecosystems or habitats of endangered species, within the meaning of Article 194(5) of the Convention.[118]

The Tribunal's interpretation of Article 192 was informed first of all by other provisions of Part XII of the Convention, and particularly by Article 194(5), under which States were required to take measures "necessary to protect and preserve fragile ecosystems, as well as the habitats of depleted, threatened or endangered species and other forms of marine life."[119] For the identification of "depleted, threatened or endangered species," the Tribunal turned to CITES, as a convention that was "the subject of nearly universal adherence, including by the Philippines and

---

[113] MP, vol. I, 189, para. 6.79.

[114] Convention, Article 194(3)(a).

[115] Ibid., Article 1(5)(a)(i).

[116] Article 194(5) stipulates that "[t]he measures taken in accordance with this Part shall include those necessary to protect and preserve rare or fragile ecosystems as well as the habitat of depleted, threatened or endangered species and other forms of marine life."

[117] MP, vol. I, 189–90, para. 6.79. *Protocol to the Convention on the Prevention of Marine Pollution by Dumping of Wastes and Other Matter*, 1972, done at London, 2 November 1996 (as amended in 2006), http://www.imo.org/en/OurWork/Environment/LCLP/Documents/PROTOCOLAmended2006.pdf, accessed 11 April 2019.

[118] MP, vol. I, 190, paras. 6.80–6.81.

[119] Award of 12 July 2016, 381, para. 959.

China."[120] In the Tribunal's view, CITES formed "part of the general corpus of international law that informs the content of Article 192 and Article 194(5) of the Convention."[121]

The Tribunal noted that all the sea turtles that the Philippines found onboard Chinese vessels on various occasions were listed in Appendix I of CITES. As sea turtles were species threatened with extinction, they were subject to the strictest levels of control on trade.[122] Having in mind the expert testimony provided by Dr. Carpenter, the Tribunal concluded forcefully that harvesting of sea turtles constituted harm to the environment as such.[123] The mere act of harvesting of the sea turtles, regardless of the methods through which they are harvested, harms the environment. For giant clams, the Tribunal, basing itself on the fact that giant clams are listed in Appendix II of CITES, concluded that they "were unequivocally threatened," even if they were not subject to the same level of international controls as species listed in Appendix I.[124] The Tribunal was "particularly troubled" by the evidence relating to the harvesting of giant clams (*Tridacnidae*).[125]

The Tribunal turned to another international treaty to fill in a significant lacuna in the Convention, the absence of a definition of "ecosystem."[126] The term is used in Article 194(5), referring to the "rare or fragile ecosystems." The negotiating history of the Article does not shed any light on its meaning. All that is known is that paragraph 5 was added upon a proposal of the United States at the seventh session of the Third UN Conference on the Law of the Sea ("UNCLOS III"), in 1978. There is no record of the exact motivation of the US proposal, as negotiations for the entire article were informal.[127] Perhaps the concept of ecosystem is related to that of "the ecological balance of the marine environment," which appeared in a 1970 UN General Assembly

---

[120] Ibid., 383, para. 956.

[121] Ibid.

[122] Ibid.

[123] Ibid., 382, para. 960.

[124] Ibid., 383, para. 956; 384, para. 957.

[125] Ibid., 381, para. 957.

[126] Ibid., 376, para. 945.

[127] Nordquist et al., vol. IV, 63–64.

Declaration.[128] Be that as it may, for the definition of "ecosystem," the Tribunal quoted the definition provided by Article 2 of the CBD, which the Tribunal characterized as an "internationally accepted definition," to wit, "a dynamic complex of plant, animal and micro-organism communities and their non-living environment interacting as a functional unit."[129] From the scientific evidence that was submitted to it, the Tribunal had no doubt that the marine environments where illegal harvesting of endangered species, cyanide and dynamite fishing and island-building took place were "rare or fragile ecosystems" and that they are also habitats of "depleted, threatened or endangered species."[130]

With respect to the use of dynamite and cyanide in fishing, the Tribunal considered, as did the Philippines, that such use was pollution

---

[128]Ibid., 53. United Nations General Assembly Resolution 2749 (XXV), Declaration of Principles Governing the Sea-Bed and the Ocean Floor, and the Subsoil Thereof, beyond the Limits of National Jurisdiction, para. 11(a) adopted on 17 December 1970, https://www.unescwa.org/declaration-principles-governing-seabed-and-ocean-floor-and-sub-soil-thereof-beyond-limits-national, accessed 12 April 2019.

[129]Award of 12 July 2016, 376, para. 945.

[130]Ibid. In the definition formulated by the Conference of the Parties (COP) of the CBD, the scientific criterion of "rarity" for "ecologically or biologically significant marine areas in need of protection in open-ocean waters and deep sea-habitats" means that that the "Area contains either (i) unique ('the only one of its kind'), rare (occurs only in few locations) or endemic species, populations or communities, and/or (ii) unique, rare or distinct, habitats or ecosystems; and/or (iii) unique or unusual geomorphological or oceanographic features." The scientific criterion of "vulnerability, fragility, sensitivity, or slow recovery" means that the area contains "a relatively high proportion of sensitive habitats, biotopes or species that are functionally fragile (highly susceptible to degradation or depletion by human activity or by natural events) or with slow recovery." The scientific criterion of "Importance for threatened, endangered or declining species and/or habitats" means that the area contains "habitat for the survival and recovery of endangered, threatened, declining species or area with significant assemblages of such species." Conference of the Parties to the CBDCOP 9, Decision IX/20, Marine and Coastal Biodiversity, Annex 1, Scientific Criteria for Identifying Ecologically or Biologically Significant Marine Areas in Need of Protection in Open-ocean Waters and Deep Sea-habitats 7–8, Doc. UNEP/CBD/COP/DEC/IX/20 (9 October 2008), https://www.cbd.int/doc/decisions/cop-09/cop-09-dec-20-en.pdf, accessed 6 May 2019. Aichi Biodiversity Target 10 identifies coral reefs as "vulnerable ecosystems." The goal is to reduce by 2020 "multiple anthropogenic pressures on coral reefs, and other vulnerable ecosystems impacted by climate change or ocean acidification are minimized, so as to maintain their integrity and functioning." CBD COP 10, Decision X/2, X/2. Strategic Plan for Biodiversity 2011–2020 and the Aichi Biodiversity Targets, 9, Doc. UNEP/CBD/COP/DEC/X/2 (29 October 2010), https://www.cbd.int/doc/decisions/cop-10/cop-10-dec-02-en.pdf, accessed 6 May 2019.

of the environment as defined by Article 1 of the Convention and threat-ened the fragile ecosystems of coral reefs and the habitats of endangered species at Scarborough Shoal. The failure to take measures against the use of dynamite and cyanide would constitute breaches of Articles 192, 194(2) and 194(3) of the Convention.[131]

It was on the basis of Article 192, interpreted in light of Article 194 and other international environmental instruments, that the Philippines and the Tribunal assessed China's fulfillment of its due diligence obliga-tions to protect and preserve the marine environment.

2. China's Compliance with Its Obligation to Protect and Preserve the Marine Environment

The Philippines readily recognized that China was not responsible for the actions of its fishermen. China's obligations under the Convention were due diligence obligations. As the flag State, China was obliged to monitor and enforce compliance with its laws by all ships flying its flag.[132] The Philippines added that if violations occur and are reported by other States, the flag State is obliged to investigate and, if appropri-ate, to take any action necessary to remedy the situation.[133] China was responsible for its own failure to control the illegal and damaging activ-ities of its nationals who harvested endangered species.[134] According to the Philippines, China had not even attempted to exercise such control. On the contrary, it had actively supported, protected, and facilitated their harmful practices.

Interpreting Article 192 in light of Article 194(5) and CITES, the Tribunal deduced two due diligence obligations that were incumbent on China. The first was to take measures to prevent "harms that would affect depleted, threatened or endangered species indirectly through the destruction of their habitat."[135] The second was "a due diligence obli-gation [under Article 192] to prevent the harvesting of species that are recognised internationally as being at risk of extinction and requiring

---

[131] Award of 12 July 2016, 386, para. 970.

[132] Hearing on the Merits, Transcript, Day 3, 32–34; Award of 12 July 2016, 361, para. 909.

[133] Hearing on the Merits, Transcript, Day 3, 33.

[134] Ibid., 32; Award of 12 July 2016, 361, para. 909.

[135] Award of 12 July 2016, 382, para. 959.

international protection."[136] The Tribunal then sought to determine China's responsibility for the failure to protect and preserve the marine environment that the occurrence of harm necessarily implied. Well-established principles of international law, particularly of international responsibility of States, were the context of the Tribunal's inquiry.

The notion of international responsibility "designates the legal consequences of the international wrongful act of a State, the obligations of the wrongdoer, on the one hand, and the rights and powers of any State affected by the wrongdoing, on the other."[137] One general principle is that "the conduct of private persons or entities is not attributable to the State under international law."[138] Another principle is that a State's failure to comply with its international obligations can consist of actions or omissions.[139] Responsibility of the State is entailed by the "inaction" of its authorities which "[fail] to take appropriate steps," in circumstances where such steps [are] evidently called for.[140] Such steps involve the duty to prevent the violation of the international obligation or failing that, to punish the individuals or entities that have committed the unlawful acts.[141] To this might be added the suggestion by an authoritative

---

[136] Ibid., 384, para. 956.

[137] Antonio Cassese, *International Law* (2nd ed.; Oxford: Oxford University Press, 2005), 241.

[138] United Nations International Law Commission, *Draft articles on Responsibility of States for Internationally Wrongful Acts, with Commentaries 2001*, in *Report of the International Law Commission on the Work of Its Fifty-Third Session, 23 April–1 June and 2 July–10 August 2001, Official Records of the General Assembly, Fifty-Sixth Session*, Supplement No. 10, Doc. A/56/10, 47 ("International Law Commission, 'Commentaries'"), legal.un.org/ilc/texts/instruments/english/commentaries/9_6_2001. pdf, accessed 12 April 2019. The commentary adds that "the conduct of private persons is not as such attributable to the State," 38, and that "the conduct of a person or group of persons not acting on behalf of the State is not considered as an act of the State under international law," 52.

[139] Ibid., 35.

[140] Ibid., quoting *United States Diplomatic and Consular Staff in Tehran, Judgment, I.C.J. Reports 1980*, 3, at 31–32, paras. 63 and 67, https://www.icj-cij.org/files/case-related/64/064-19800524-JUD-01-00-BI.pdf, accessed 12 April 2019.

[141] Eduardo Jiménez de Aréchaga, "International Responsibility," in Max Sørensen (ed.), *Manual of Public International Law* (London: Macmillan, 1968), 560; Eduardo Jiménez de Aréchaga and Atilla Tanzi, "International State Responsibility," in Mohammed Bedjaoui (ed.), *International Law: Achievements and Prospects* (Paris: UNESCO, 1991), 359–60. Jiménez de Aréchaga was member (1970–1979) and President (1976–1979) of the ICJ.

commentator that besides the duty to take sufficient care to prevent and punish, a duty of inquiring and explaining may be inferred from the case law of the ICJ. In the *Corfu Channel* Case, the ICJ had observed that a State in whose territory an act contrary to the rights of other States has taken place cannot evade a request to give an explanation "by limiting itself to a reply that it is ignorant of the circumstances of the act and of its authors."[142]

Consistent with these principles of international responsibility, the Tribunal declared that the responsibility of a State party to the Convention, under Articles 192 and 194, is to ensure that activities within its jurisdiction and control do not harm the environment.[143] The Tribunal held that it is an obligation under the Convention of a State that is aware that vessels flying its flag are harvesting endangered species or inflicting significant damage on rare or fragile ecosystems or the habitats of depleted, threatened, or endangered species "to adopt rules and measures to prevent such acts and to maintain a level of vigilance in enforcing those rules and measures."[144] Now China could not have claimed ignorance of the activities of its nationals. Philippine protests, as early as 2000, and at the latest by 2005, should have made China aware of the illegal activities of Chinese nationals and of Philippine concerns. The Tribunal acknowledged that China had become a State party to CITES in 1981 and that a Law of Protection of Wildlife had been in force in China since 1989. The latter prohibited the catching and killing of two classes of special State-protected wildlife, which included sea turtles and giant clams.[145] In short, the Tribunal acknowledged that through ratification of CITES and passage of the Law of Protection of Wildlife, China had adopted rules and measures to prohibit the harmful practices.

The Tribunal recalled that taking measures to prevent the harmful practices is only one component of due diligence.[146] The other

---

[142] Jiménez de Aréchaga, "International Responsibility," 560–61. The phrase cited is quoted from *Corfu Channel Case, Judgment of April 9th, 1949, I.C. J. Reports 1949* 18, https://www.icj-cij.org/files/case-related/1/001-19490409-JUD-01-00-BI.pdf, accessed 12 April 2019.

[143] Award of 12 July 2016, 375, para. 944.

[144] Ibid., 382, para. 961.

[145] Ibid., 383, para. 963.

[146] Ibid., para. 964.

component is that of punishing acts that are contrary to these measures. Notwithstanding the fact that CITES had been in force for China since 1981 and the fact that it had passed a Law of Protection of Wildlife in 1989, the Tribunal saw "no evidence that Chinese fishermen involved in poaching of endangered species have been prosecuted under Chinese law....there is no evidence that China has taken any steps to enforce its rules and measures against poachers."[147] On the contrary, China was aware of the harvesting of giant clams. It not only turned a blind eye to the practice; it provided armed government vessels to protect Chinese fishing vessels. In April 2012, the Chinese Foreign Ministry publicly admitted that it had dispatched administrative vessels to the area for the protection of the "legitimate" fishing activities of Chinese fishermen and vessels.[148] In May 2013, Chinese naval and CMS vessels escorted Chinese vessels gathering clams at Second Thomas Shoal. In the Tribunal's view, the photographic evidence indicates that China must have known of the harmful activities and deliberately tolerated and protected them. The Tribunal therefore had "no hesitation" in finding, as the Philippines had requested, that China had breached its obligations under Articles 192 and 194(5) of the Convention "to take the necessary measures to protect and preserve the marine environment, with respect to the harvesting of endangered species from fragile ecosystems at Scarborough Shoal and Second Thomas Shoal."[149]

The Tribunal having been "particularly troubled" by the evidence concerning the harvesting of giant clams by propeller chopping, it devoted a separate conclusion to the matter. It found that "fishermen from Chinese flagged vessels have engaged in the harvesting of giant clams in a manner that is severely destructive of the coral reef ecosystem."[150] Two circumstances were relevant to its conclusion. First, the areas in which the harvesting had taken place were under China's jurisdiction and control. At the time that the harvesting took place in these

---

[147] Ibid.

[148] *MP*, Annex 117, Ministry of Foreign Affairs, People's Republic of China, Foreign Ministry Spokesperson Liu Weimin's Regular Press Conference (12 April 2012), vol. V, 109,   https://files.pca-cpa.org/pcadocs/The%20Philippines%27%20Memorial%20-%20 Volume%20V%20%28Annexes%20103-157%29.pdf, accessed 12 April 2019; quoted in Award of 12 July 2016, 383, para. 964.

[149] Award of 12 July 2016, 384, para. 965.

[150] Ibid., 475, para. 1202, (12)(b)

areas, China was planning and implementing island-building. The small propeller vessels were within China's jurisdiction and control. Second, China was fully aware of the practice. It seemed to the Tribunal that China had actively tolerated the practice, giving Chinese fishermen one last opportunity to harvest giant clams before the reefs were permanently destroyed by China's island-building. In view of the circumstances the Tribunal could not but conclude that China had also breached its obligation to preserve and protect the marine environment by tolerating and protecting the harvesting of giant clams by the propeller chopping method.[151]

The Tribunal's separate inquiry into the use of cyanide and dynamite at Scarborough Shoal and Second Thomas Shoal arrived at a different conclusion concerning China's responsibility.[152] While being satisfied that Chinese fishing vessels used dynamite or cyanide in 1995, 1998, 2000, and 2002, the Tribunal doubted whether there was sufficient evidence that China should be held responsible for the failure to prevent the incidents.[153] The Tribunal noted that in the preceding decade (i.e., since 2005), there was "scant evidence" about the use of explosives and dynamite or Philippine complaints about its use.[154] In April 1995, the Philippines had raised the issue of 62 fishermen in possession of explosives and cyanide.[155] In 1999, China passed a Marine Environment Protection Law, and in October 2000, it updated its Fisheries Law. Article 30 of the latter prohibits the use of explosives, poisons, electricity, and any other means that impair fishing resources.[156] In March 2000, a Chinese Foreign Ministry official expressed to the Philippine ambassador her "particular concern about dynamite fishing" and informed him that she had requested that the Ministry of Agriculture take steps to

---

[151] Ibid., 384, paras. 967–68.

[152] Ibid., 385–88, paras. 968–75.

[153] Ibid., 386, para. 971.

[154] Ibid., 388, para. 975.

[155] *MP*, Annex 21, vol. III, 225–31.

[156] *Supplemental Documents of the Philippines* (19 November 2015), Annex 614, People's Republic of China, Marine Environment Protection Law of The People's Republic of China (25 December 1999), vol. I, 101–11 ("*SDP*"), https://files.pca-cpa.org/pca-docs/The%20Philippines%27%20Supplemental%20Documents%20-%20Volume%20I%20%20%28Annexes%20607-667%29.pdf, accessed 12 April 2019; Fisheries Law of the People's Republic of China, 2000, cited in Award of 12 July 2016, 387, para. 974.

remedy the situation.[157] Reports from the Philippine Embassy in Beijing show that the Chinese government was aware of the risks to the environment caused by cyanide and dynamite fishing. A Chinese diplomat had expressed particular concern about dynamite fishing in 2000.[158] In the same year, the Philippine Ambassador was informed that "the Chinese Government attaches great importance to environmental protection and violators are dealt with in accordance with Chinese laws and regulations." He reported that "local fishing authorities imposed a penalty on the fishermen caught blasting coral reefs near Scarborough Shoal in early 2000."[159] The Tribunal, recalling that the two components of due diligence were the adoption of rules and measures and the adoption of a certain level of vigilance in the enforcement of rules, concluded that there was little in the record to suggest that China had failed to enforce the rules regarding dynamite and cyanide fishing.[160] On the contrary, the circumstances suggested that China may have taken measures to prevent these practices in the Spratly Islands.[161] One may surmise that what was crucial for the Tribunal was the fact that even if incidents occurred after 1995, the Philippines did not seem to have complained to Beijing about them. After 2002, there were no further internal Philippine reports of incidents of dynamite and cyanide fishing, a circumstance that implied that China had taken steps to enforce the law to prevent this practice in the Spratly Islands.[162] With these considerations in mind, the Tribunal was "not prepared to make a finding on the evidence available, under Submission No. 11 with respect to cyanide and explosives."[163]

---

[157] MP, Annex 40, Memorandum from the Embassy of the Republic of the Philippines in Beijing to the Secretary of Foreign Affairs of the Republic of the Philippines, No. ZPE-24-2000-S (14 March 2000), vol. III, 337–40, https://files.pca-cpa.org/pcadocs/The%20 Philippines%27%20Memorial%20-%20Volume%20III%20%28Annexes%201-60%29.pdf, accessed 26 March 2019.

[158] Ibid.

[159] MP, Annex 43, vol. III, 3531–55

[160] Award of 12 July 2016, 387, para. 974.

[161] Ibid., 388, para. 975.

[162] Ibid.

[163] Ibid. Nilufer Oral observes that this finding of the Tribunal demonstrated its impartiality. Nilufer Oral, "The South China Sea Arbitral Award, Part XII of UNCLOS and the Protection and Preservation of the Marine Environment," in S. Jayakumar et al. (eds.), *The South China Sea Arbitration: The Legal Dimension* (Cheltenham, Gloucestershire: Edward Elgar, 2018), 234 ("Oral, 'The South China Sea Arbitral Award'").

Incidentally, the recapitulation of this part of the Award demonstrates that Mbengue errs in concluding that "when expert evidence was not convincing, the Tribunal did not make any finding." He believes, mistakenly, that the Tribunal declined to make a finding because of a lack of scientific evidence and that whenever experts were not able to provide the Tribunal with "hard facts," the Tribunal would infer that it was probably because China was acting consistently with Part XII of the Convention.[164] Mbengue's unjustified conclusion may perhaps be explained by a hasty reading of the Award. Carpenter, the expert presented by the Philippines, and Ferse, Mumby and Ward, the independent experts appointed by the Tribunal, provided evidence of the impact of dynamite and cyanide fishing on coral reefs. They did not provide evidence of the frequency of incidents involving such practices. As the basis for its conclusion, the Tribunal relied exclusively on the evidence (or the lack of it) submitted by the Philippines, and the evidence that it examined on this matter consisted almost entirely of Philippine official documents.

The Chinese Society of International Law ("CSIL") criticized the Tribunal's findings on several grounds. First, citing an advisory opinion of the ITLOS, it argued that a due diligence duty requires States to perform it "in accordance with their capacities." States are not to be unrealistically required to take measures beyond their capacities.[165] Second, it accused the Tribunal of deliberately distorting China's actions to affirm and safeguard its sovereignty. In April 2012, China sent official ships to Scarborough Shoal in order to affirm and safeguard its territorial sovereignty and to protect the life, safety, and property of Chinese fishermen.[166] Third, the Tribunal allegedly erred in applying Article 194(5) of the Convention. The title ("Measures to prevent, reduce and control pollution of the marine environment") and content of Article 194 allegedly indicate that States are to take measures to protect rare or fragile ecosystems and the habitat of depleted, threatened, and endangered species only when these suffer from pollution. The existence of pollution is said to be a precondition for applying paragraph 5 of Article 194.[167]

---

[164] Makane Moïse Mbengue, "The South China Sea Arbitration. Innovation in Marine Environmental Fact-finding and Due Diligence Obligations," *ASIL Unbound* (2016): 289.

[165] CSIL, 576, para. 784.

[166] Ibid., 578, para. 789.

[167] Ibid., 578–79, para. 790.

Fourth, the Tribunal was said to have disregarded the fact that China has actively taken measures to fulfill its due diligence obligations. It enacted the Law on the Protection of Wildlife, the Fisheries Law, and the Regulations on the Protection of Aquatic Wild Animals.[168] Moreover, China is said to have actively taken administrative measures to protect endangered species. In June 2003, it started a special law enforcement program to penalize the illegal hunting, killing, purchasing, selling, transporting, and importing and exporting of aquatic wild animals. In June 2012, it started a special law enforcement program to combat the illegal harvesting, trading, and utilizing and smuggling of aquatic wild animals. At the regional level, two provinces and one autonomous region have established natural reserves.[169] China has criminally prosecuted those engaged in illegal harvesting of endangered species, including giant clams, corals, and sea turtles, in the South China Sea. In December 2007, three individuals were convicted in Hainan Province for illegally purchasing, transporting, and selling sea turtles. In May 2014, an individual was convicted for the illegal sale of red coral products. In a third case, which the CSIL admitted was not related directly to the South China Sea, four individuals were convicted in 2013, for the illegal sale of red corals in Fujian province, situated opposite the island of Taiwan.[170]

It is not necessary to dwell too much on these criticisms. The first strains the credulity of the observer. Is the CSIL really asking us to believe that China lacks the capacity to enforce its laws on the protection of endangered species? The second assumes that if a State is taking steps that it deems to be required to safeguard its sovereignty it is not thereby violating international law. One could argue that States not infrequently attempt to excuse many violations of international law by claiming that they are taking steps to safeguard their sovereignty. The third misreads Article 194(5), the first part of which reads as follows: "The measures taken in accordance with *this Part....*" It does not say "The measures taken in accordance with this *Article.*"[171] The expression

---

[168] Ibid., 580–82, paras. 793–95.

[169] Ibid., 582, paras. 796–97.

[170] Ibid., 583–84, paras. 800–801.

[171] Underscoring supplied in the two sentences.

"This Part" refers to Part XII, whose title is "Protection and Preservation of the Marine Environment." Consequently, pollution as defined by the Convention is not a precondition for States to take measures to protect and preserve rare or fragile ecosystems as well as the habitat of depleted, threatened, or endangered species. The fourth is based on a misreading of the Award of 12 July 2016, while simultaneously providing evidence of China's failure to enforce its own laws. The Award recognizes that China has put in place laws to protect endangered species. That it did not enumerate all the laws that the CSIL listed does not prove that the Tribunal took no notice of all Chinese laws. Had the Tribunal listed all the relevant Chinese laws, the Tribunal's conclusion would probably have remained unchanged: China had breached the Convention for its failure to enforce the laws. The information given by the CSIL relating to China's enforcement only serves to highlight China's failure to enforce its laws and regulations. The incidents that triggered Philippine protests to China go back at least to the mid-1990s and continued to occur until six to eight months before the initiation of the arbitration. In that period of 17–18 years, there were only three prosecutions in China, according to the evidence that the CSIL itself provided. Two of the three cases were brought before Chinese courts in 2013 and 2014, after the arbitration had been initiated. Two of the three cases were apparently related to the South China Sea but they do not seem to have any connection with any incident reported by the Philippines to China. Now China's duty under the Convention was to investigate the incidents reported by the Philippines to it and to take any necessary action to remedy the situation, if appropriate. The CSIL provided no evidence at all that China had taken any action in response to the incidents reported by the Philippines in its protests. On the contrary, China consistently maintained throughout the years that the activities of the Chinese fishermen—harvesting of endangered species and the use of cyanide and dynamite in fishing—were "normal" and "legitimate." The third prosecution reported by the CSIL, undertaken in 2013, was not even related to the South China Sea. None of these facts suggests that China was taking its due diligence obligations seriously.

Having disposed of these criticisms, we can now turn to the Tribunal's assessment of the damage to fragile marine ecosystems caused by China's construction activities and island-building in the South China Sea.

II. China's Construction Activities and the Obligation to Preserve and Protect the Marine Environment

The Tribunal examined China's conduct in relation to the general obligation to protect and preserve the marine environment under Article 192 in conjunction with Article 194(5), the obligation to cooperate with other littoral States of the South China Sea under Articles 197 and 123, and the obligation to monitor and assess under Articles 205 and 206.

A. The General Obligation to Protect and Preserve the Marine Environment and the Obligation to Preserve and Protect Fragile Marine Ecosystems

The Philippines laid great stress on the harm to the environment resulting from China's construction activities. China publicly contested the Philippine arguments, notwithstanding the fact that it refused to appear before the Tribunal. The existence of such harm, confirmed by the independent experts, provided the basis for the assessment of China's compliance with its obligation to protect and preserve the marine environment.

1. The Harm to Fragile Marine Ecosystems Caused by China's Construction Activities

The Tribunal's detailed survey of Chinese island-building on seven reefs in the South China Sea is summarized in Tables 5.1 and 5.2. Here we shall simply present the Tribunal's overall assessment of the scale of the construction activities and its description of the methods used by China for dredging.

Between the 1990s and 2013, China undertook some construction and land reclamation, starting with basic aluminum, wooden or fiberglass structures and supported by steel bars with cement base. Over time, China installed more sophisticated structures, including concrete multi-story buildings, wharves, helipads, and weather communication instruments. By 2013, the largest project was the construction of an artificial island of 115 meters by 80 meters at Fiery Cross Reef. The Tribunal

Table 5.1 Overview of China's construction activities in the Spratly Islands

International name/Filipino name/Chinese name

Cuarteron Reef / *Calderon Reef* / *Huayang Jiao*	*Fiery Cross Reef* / *Kagitingan Reef* / *Yongshu Jiao*	*Gaven Reef (North)* / *Burgos Reef* / *Nanxun Jiao*	*Johnson Reef* / *Mabini Reef* / *Chigua Jiao*	*Hughes Reef* / *Chigua Reef* / *Dongmen Jiao*	*Subi Reef* / *Zamora Reef* / *Zhubi Jiao*	*Mischief Reef* / *Panganiban Reef* / *Meiji Jiao*
Coordinates						
08° 51′ 41″ N / 112° 50′ 08″ E	09° 33′ 00″ N / 112° 53′ 25″ E	10° 12′ 27″ N / 114° 13′ 21″ E	9° 43′ 00″ N / 114° 16′ 55″ E	09° 54′ 48″ / N 114°29′ 48″ E	10° 55′ 22″ N / 114° 05′ 04″ E	09° 54′ 17″ N / 115° 31′ 59″ E
Description						
Ellipse-shaped table-like reef, extending about 5 km. W to E, with shallow reef flat area, no lagoon in the center	Open spindle-shaped atoll, extending 25 km from NE to SW, width 6 km, extensive reef flat on SW end surrounds small closed lagoon in the center, with maximum depth of 12 m	Sits on W end of largely submerged atoll of Tizard Bank; reef flat extends 1.9 km from N to S and 1.2 km E to W; has no central lagoon	Large coral reef platform with shallow central lagoon, located at SW end of Union Bank atoll; 4.6 km by 2.4 km	Forms part of rim of Union Bank atoll; lies to NE of Johnson Reef; 2.1 km from N to S and 2 km from E to W; features natural lagoon meandering across the center and opening to adjacent deeper lagoon through narrow channel on E side of reef	Coral atoll enclosing large lagoon that lies to SW of Thitu; 5.75 km long, 3.25 km wide; originally closed atoll with no passages into lagoon	Large oval-shaped atoll, 6.5 km wide, with 3 natural entrances into lagoon; lagoon featured a number of well-developed patch reefs, with massive, foliose, and branching corals
Resources						
Surveys in the late 1990s—abundance of fish resources, including sharks, parrotfishes, and groupers	Surveys in the late 1990s—abundant fisheries; 2004–05 survey—highly biodiverse communities	Surveys in 1998 and 2005—fisheries resources were less than at Fiery Cross Reef	Fisheries resources were less than at other reefs; live coral covered about 15% of reef flat	Surveys in late 1990s—some productive value, less than at other reefs	Survey in late 1990s—it may have already suffered from overfishing; 2002 study—it was home to over 300 macrobenthic species; 2007 coral surveys—64–74 species of coral; live coral cover is highest in inner reef flat and lagoon areas	Fisheries surveys in late 1990s—some productive value; 2007 survey—94 species of stony corals; live coral cover of 51% on reef slope; by 2005, it was under pressure of increased fishing

(continued)

Table 5.1 (continued)

International name/Filipino name/Chinese name

Year of observation	Cuarteron Reef / Calderon Reef / Huayang Jiao	Fiery Cross Reef / Kagitingan Reef / Yongshu Jiao	Gaven Reef (North) / Burgos Reef / Nanxun Jiao	Johnson Reef / Mabini Reef / Chigua Jiao	Hughes Reef / Chigua Reef / Dongmen Jiao	Subi Reef / Zamora Reef / Zhubi Jiao	Mischief Reef / Panganiban Reef / Meiji Jiao
1988		Small naval post, followed by oceanographic observation post, pier and other buildings	Start of Chinese presence		Start of Chinese presence		
1989						Start of Chinese presence	
1992	3 small buildings			Heavily fortified area with observation tower			
1994						5 buildings, wharf, helipad	
1995							Start of Chinese presence, construction of typhoon shelters
1996			Outpost with barracks, 2 octagonal structures				

(continued)

Table 5.1   (continued)

International name/Filipino name/Chinese name

Cuarteron Reef *Calderon Reef* *Huayang Jiao*	*Fiery Cross Reef* *Kagitingan Reef* *Yongshu Jiao*	*Gaven Reef (North)* *Burgos Reef* *Nanxun Jiao*	*Johnson Reef* *Mabini Reef* *Chigua Jiao*	*Hughes Reef* *Chigua Reef* *Dongmen Jiao*	*Subi Reef* *Zamora Reef* *Zhubi Jiao*	*Mischief Reef* *Panganiban Reef* *Meiji Jiao*
Year of observation: 1997 More buildings, wharves, communication facilities						
Year of observation: 1999						Multi-storey structure; communications equipment; wharves; helipad
Year of observation: 2005		3-story concrete building with communication equipment				
Year of observation: 2006 3-story building, concrete platforms, helipad			3-story concrete building, communication equipment, solar panels, helipads	3-story concrete building, helipad	4-story concrete building	
Year of observation: 2011		Further enhancements				

(continued)

Table 5.1 (continued)

International name/Filipino name/Chinese name

Cuarteron Reef / Calderon Reef / Huayang Jiao	Fiery Cross Reef / Kagitingan Reef / Tongun Jiao	Gaven Reef (North) / Burgos Reef / Nanxun Jiao	Johnson Reef / Mabini Reef / Chigua Jiao	Hughes Reef / Chigua Reef / Dongmen Jiao	Subi Reef / Zamora Reef / Zhubi Jiao	Mischief Reef / Panganiban Reef / Meiji Jiao
Year of observation: 2013						
More concrete buildings, solar panels, weather and communication instruments, observation towers, temporary pier	Communication system, and lighthouses; complete complex of buildings with significant communication and defense and military features; greenhouse				More concrete structures, lighthouse, communication equipment	
Year of observation: 2014						
Start of more substantial land reclamation	Runway being built	Intense reclamation work began	Intense reclamation work began	Start of large-scale land reclamation	Start of large-scale dredging	
Month and year of observation: May 1015						
Land reclamation intensified; permanent pier; ceremony to mark the start of construction of a 50-m lighthouse	Land reclamation intensified		More buildings; solar panels; paved roads; piers ceremony to mark the start of construction of a lighthouse	Permanent pier; massive onshore construction of 6-storey building; large cargo vessels transporting sand sediments with newly reclaimed land		

(continued)

Table 5.1 (continued)

International name/Filipino name/Chinese name

Cuarteron Reef *Calderon Reef* *Huayang Jiao*	*Fiery Cross Reef* *Kagitingan Reef* *Tongshu Jiao*	*Gaven Reef (North)* *Burgos Reef* *Nanxun Jiao*	*Johnson Reef* *Mabini Reef* *Chigua Jiao*	*Hughes Reef* *Chigua Reef* *Dongmen Jiao*	*Subi Reef* *Zamora Reef* *Zhubi Jiao*	*Mischief Reef* *Panganiban Reef* *Meiji Jiao*
Month and year of observation: December 2015						
231,000 m² of new land, channel 125 m wide for large vessels to access and berth within a harbor cut out of the reef platform	2,740,000 m² of new land created, sand and rock dredged from seabed covered the entire platform of south-west reef flat; artificial island is 300 times larger than preexisting installations, which covered 11,000 m²; 3 km runway, 630,000 m² harbor; cement plants; support buildings; temporary loading pier; communication facilities; defense equipment, 2 lighthouses; green-house; 2 helipads; multi-level administrative facility	Gaven Reef was an artificial island 300 by 250 m created from 136,000 m² of materials dredged from the seabed	Johnson Reef was an artificial island, with area of 109,000 m², 1000 times larger than the previous structure; dredged harbor channel into reef center	Hughes Reef was an artificial island, with area of 75,000 m², coastal fortifications, defensive towers, multi-level facility; reef entrance enlarged to create 118 m wide access channel for larger vessels	Artificial island, 3,950,000 m² covering most of the reef; start of 3 km runway; multi-level facility; reinforced sea walls; towers; communication facilities; 230 m-wide access channel; nearly the entire atoll has been transformed into an artificial island	Total land area created was 5,580,000 m²; southern entrance of reef was widened from 110 to 275 m

Source Prepared by the author, based on the Award of 12 July 2016, 121–22, 337–55, paras. 285–90, 863–89

Table 5.2   The Tribunal's analysis of the aerial and satellite imagery of maritime features in the South China Sea prior to and after Chinese construction

Prior to construction	After construction
Cuarteron Reef	
The original Chinese installation is barely visible in the 2012 photograph (338, para. 866; Fig. 19, 341)	The artificial island is approximately 200 times larger than the original installation in 2012 (photograph on 23 August 2013; 338, para. 866; Fig. 19, 341)
Fiery Cross Reef	
In January 2012, the reef was in its nearly natural state. China's original installation is just visible at the southern end (340, para. 870; Fig. 20, 343)	The reef appears as an artificial island complex, complete with a large runway, covering the entire reef platform (photograph on 19 October 2015; 340, para. 870; Fig. 21, 343)
Gaven Reef (North)	
In 2012, China's original installation and the naturally occurring sand cay are barely visible at the north end of the reef (345, para. 874; Fig. 22, 347)	A large artificial island in the shape of a sideways "Y" dominates the reef in the image taken on 16 November 2015. (345, para. 874; Fig. 23, 347)
Johnson Reef	
China's original installation cannot even be seen without enlargement in satellite imagery from 20 March 2013 (346, para. 878; Fig. 24, 347)	A large artificial island, along with a dredged harbor channel into the center of the reef, is readily visible in satellite imagery from 4 November 2015 (346, para. 878; Fig. 25, 347)
Hughes Reef	
China's original installation cannot even be seen without enlargement in satellite imagery from February 2010 (349, para. 882; Fig. 26, 351)	A large artificial island, along with a dredged harbor channel into the center of the reef, is apparent in imagery from 22 September 2015 (349, para. 882; Fig. 27, 351)
Subi Reef	
Subi Reef was originally a closed atoll, with no passages into the lagoon (2012; 349–50, para. 883; Fig. 28, 353)	Nearly the entire atoll had been transformed into an artificial island (6 November 2015; 350, para. 886; Fig. 29, 353)
Mischief Reef	
Construction on Mischief Reef between 1999 and 2013 appears to have been relatively limited (402, para. 1004; Fig. 31, 405)	An artificial island covers the entire northern half of the reef (photograph on 19 October 2015; 403, para. 1007; Fig. 32, 405)

Source Prepared by the author, based on the Award of 12 July 2016

acknowledged that the Philippines, Vietnam, and other claimants undertook similar constructions on other reefs. From the end of 2013, China undertook massive island-building, far exceeding the scale of earlier projects. All in all, China created more than 12.8 million square meters of new land in less than three years.[172]

For this purpose, China deployed a large fleet of vessels to the seven reefs using heavy "cutter-suction dredge" equipment. In the "cutter-suction dredge" method, a shipborne drill is extended from the dredging vessel into the seabed.[173] The drill's rotating teeth act like picks that chisel away at the seabed or reef, breaking apart and extracting the soil, rock, and reef. The material is pumped up through a floating tube pipeline from the vessel's stern to the dredging area, kilometers away. China's largest suction cutter dredger is Tian Jing Hao, capable of extracting 4500 cubic meters of sand, rock, and other materials from the surrounding seabed per hour.[174] In 2013–2014, China was reported to have utilized it to navigate between five reefs in the Spratly Islands, where it dredged and blasted more than 10 million cubic meters of material onto the reefs.[175]

The Philippines protested to China as it obtained information about the expanding scale of Chinese activities at Johnson Reef,[176]

---

[172] Award of 12 July 2016, 329–30, paras. 853–54.

[173] See Fig. 12 in ibid., 331.

[174] See Fig. 15 in ibid., 335.

[175] *SDP* (19 November 2015), Annex 699, "Environmental Consequences of Land Reclamation Activities on Various Reefs in the South China Sea (14 November 2015)," by K.E. Carpenter and L.M. Chou, vol. II, 248–49 ("Second Carpenter Report"), https:// files.pca-cpa.org/pcadocs/The%20Philippines%27%20Supplemental%20Documents%20 -%20Volume%20II%20%28Annexes%20608-709%29.pdf, accessed 7 April 2019; Award of 12 July 2016, 330–31, paras. 855–56; Ferse Report, 22. The Ferse Report adds (24) that the cutter-suction dredger is capable of digging a 0.6 meter-deep-pit the size of a football field in one hour.

[176] *SDP*, Annex 670, Note Verbale from the Department of Foreign Affairs of the Republic of the Philippines to the Embassy of the People's Republic of China in Manila, No. 14-1180 (4 April 2014), vol. II, 9–12, https://files.pca-cpa.org/pcadocs/ The%20Philippines%27%20Supplemental%20Documents%20-%20Volume%20II%20 %28Annexes%20608-709%29.pdf, accessed 7 April 2019.

McKennan/Hughes Reef,[177] Cuarteron Reef,[178] Gaven Reef,[179] and Fiery Cross Reef.[180] The Philippines estimated that as of March 2015, the area of reef ecosystems destroyed in the Spratly Islands was 311 hectares.[181] It calculated that the destruction of the coral reef systems resulting from China's activities would lead to economic losses to the

---

[177] *SDP*, Annex 672, Note Verbale from the Department of Foreign Affairs of the Republic of the Philippines to the Embassy of the People's Republic of China in Manila, No. 14-2093 (6 June 2014), vol. II, 17–20, https://files.pca-cpa.org/pcadocs/The%20Philippines%27%20Supplemental%20Documents%20-%20Volume%20II%20%28Annexes%20608-709%29.pdf, accessed 7 April 2019.

[178] *SDP*, Annex 673, Note Verbale from the Department of Foreign Affairs of the Republic of the Philippines to the Embassy of the People's Republic of China in Manila, No. 14-2276 (23 June 2014), vol. II, 21–24, https://files.pca-cpa.org/pcadocs/The%20Philippines%27%20Supplemental%20Documents%20-%20Volume%20II%20%28Annexes%20608-709%29.pdf, accessed 7 April 2019.

[179] *SDP*, Annex 674, Note Verbale from the Department of Foreign Affairs of the Republic of the Philippines to the Embassy of the People's Republic of China in Manila, No. 14-2307 (24 June 2014), vol. II, 25–27, https://files.pca-cpa.org/pcadocs/The%20Philippines%27%20Supplemental%20Documents%20-%20Volume%20II%20%28Annexes%20608-709%29.pdf, accessed 7 April 2019.

[180] *SDP*, Annex 679, Note Verbale from the Department of Foreign Affairs of the Republic of the Philippines to the Embassy of the People's Republic of China in Manila, No. 14-3504 (10 October 2014), vol. II, 51–54, https://files.pca-cpa.org/pcadocs/The%20Philippines%27%20Supplemental%20Documents%20-%20Volume%20II%20%28Annexes%20608-709%29.pdf, accessed 7 April 2019; *SDP*, Annex 677, Note Verbale from the Department of Foreign Affairs of the Republic of the Philippines to the Embassy of the People's Republic of China in Manila, No. 14-2889 (18 August 2014), vol. II, 41–44, https://files.pca-cpa.org/pcadocs/The%20Philippines%27%20Supplemental%20Documents%20-%20Volume%20II%20%28Annexes%20608-709%29.pdf, accessed 7 April 2019.

[181] *SDP*, Annex 609, Republic of the Philippines, Bureau of Fisheries and Aquatic Resources, Press Release: DA-BFAR, National Scientist condemns the destruction of marine resources in the West Philippine Sea (23 April 2015), vol. I, 11–15, https://files.pca-cpa.org/pcadocs/The%20Philippines%27%20Supplemental%20Documents%20-%20Volume%20I%20%28Annexes%20607-667%29.pdf, accessed 17 April 2019.

coastal States of $100 million annually.[182] China rejected all these protests.[183]

In the arbitration proceedings, the Philippines initially submitted, in support of its claim that China's construction activities were causing harm to the marine environment, an expert report as an Annex to

[182] *SDP*, Annex 608, Department of Foreign Affairs of the Republic of the Philippines, Statement on China's Reclamation Activities and their Impact on the Region's Marine Environment (13 April 2015), vol. I, 7–9, https://files.pca-cpa.org/pcadocs/The%20 Philippines%27%20Supplemental%20Documents%20-%20Volume%20I%20%28Annexes%20 607-667%29.pdf, accessed 17 April 2019.

[183] See the following documents in *SDP*, vol. II:

1. Annex 671, Verbatim Text of Response by Deputy Chief of Mission, Embassy of the People's Republic of China in Manila, to Philippine Note Verbale No. 14-1180 dated 04 April 2014 (11 April 2014), 13–16;
2. Annex 675, Note Verbale from the Embassy of the People's Republic of China in Manila to the Department of Foreign Affairs, Republic of the Philippines, No. 14 (PG)-195 (30 June 2014), 29–32;
3. Annex 676, Note Verbale from the Embassy of the People's Republic of China in Manila to the Department of Foreign Affairs, Republic of the Philippines, No. 14 (PG)-197 (4 July 2014);
4. Annex 678, Note Verbale from the Embassy of the People's Republic of China in Manila to the Department of Foreign Affairs, Republic of the Philippines, No. 14 (PG)-264 (2 September 2014), 45–50;
5. Annex 680, Note Verbale from the Embassy of the People's Republic of China in Manila to the Department of Foreign Affairs, Republic of the Philippines, No. 14 (PG)-336 (28 October 2014), 55–59;
6. Annex 681, Note Verbale from the Ministry of Foreign Affairs, People's Republic of China to the Embassy of the Republic of the Philippines in Beijing, No. (2015) Bu Bian Zi No. 5 (20 January 2015), 61–66;
7. Annex 685, Note Verbale from the Embassy of the People's Republic of China to the Department of Foreign Affairs, Republic of the Philippines, No. 15 (PG)-068 (4 March 2015), 83–87;
8. Annex 686, Note Verbale from the Department of Boundary and Ocean Affairs, Ministry of Foreign Affairs, People's Republic of China, to the Embassy of the Republic of the Philippines in Beijing, No. 2015) Bu Bian Zi No. 22 (30 March 2015), 89–93;
9. Annex 687, Note Verbale from the Department of Boundary and Ocean Affairs, Ministry of Foreign Affairs, People's Republic of China, to the Embassy of the Republic of the Philippines in Beijing, No. (2015) Bu Bian Zi No. 23 (30 March 2015), 97–101.

its *Memorial* ("First Carpenter Report").[184] In response to China's expansion of its construction activities as the arbitral proceedings were ongoing, the Philippines submitted two further expert reports, the first transmitted to the Tribunal prior to the Hearing on the Merits ("Second Carpenter Report")[185] and the second during the Hearing itself.[186] The Philippines called on Professor Kent E. Carpenter as an expert witness during the Hearing on the Merits.

There was no doubt on the part of the Philippines and Dr. Carpenter that the coral reefs of the South China Sea constituted fragile marine ecosystems. Dr. Carpenter explained that Scarborough Shoal and Second Thomas Shoal are unlike other oceanic reefs in other parts of the world in that they represent assemblage of species found nowhere else.[187] Scarborough Shoal and the Spratly Islands were formed on top of seamounts, or uplifts in the sea beds, which are the result of the convergence of more tectonic plates than anywhere else in the world. It is the seamounts that serve as the attachment site for coral growth.[188] The present-day coral reef ecosystems around the numerous naturally shallow reefs and the low islands of Scarborough Shoal, the Spratly Islands and the Reed Bank have developed since the end of the last ice age 10,000–15,000 years ago.[189] Dr. Carpenter also stressed that coral reefs are fragile and depend on a symbiotic relationship of corals and algae. Many of the coral species in the South China Sea are threatened with extinction. In his view, Scarborough Shoal and Second Thomas Shoal belong to one of the most fragile marine ecosystems globally.[190]

Dr. Carpenter emphasized the harmful effects of China's island-building on coral reefs as fragile marine ecosystems. The abrupt man-made alteration to the shallow reef features had a direct impact on the functioning of delicate reefs and altered topography that had taken thousands of years to form. The millions of tons of rock and sand dredged from the seabed and deposited on the shallow reefs destroyed large swathes of

---

[184] *MP*, Annex 240, vol. VII, 389–437.

[185] Second Carpenter Report, 235–92.

[186] *The Philippines Annexes Cited During the Merits Hearing*, Annex 850.

[187] Hearing on the Merits, Transcript, Day 4, 185.

[188] First Carpenter Report, 395.

[189] Second Carpenter Report, 243.

[190] Hearing on the Merits, Transcript, Day 4, 155.

reefs, constituting a catastrophic disturbance of the reefs. The sedimentation caused by the construction smothers coral, deprives it of sunlight, and impedes its ability to grow. The wholesale removal and destruction of coral reef habitat by direct destruction and the replacement of the shallow portions of the reef ecosystem remove vital components of the available reef habitat that have functioned as a single ecosystem for many generations of reef inhabitants. Dr. Carpenter anticipated a dramatic decline in population and local extinction of prominent fishes and invertebrates. The direct ecosystem harm can be multiplied many times over by the wider effects of sediment plumes caused by island-building.[191]

Dr. Carpenter was pessimistic about the duration of the effects of China's construction activities and the prospects of recovery. The demolition, burial, and landfill resulted in total destruction of large swathes of reef structures, destabilizing the reef substrate and affecting negatively the potential for recovery. The reefs smothered by sedimentation are unlikely to ever recover if the unstable sediments remain in place, because reef-building requires a hard substrate to recruit and thrive.[192]

Counsel for the Philippines admitted during the Hearing on the Merits that the Philippines was unable to investigate conditions at Mischief Reef, to send scientists there to investigate and report on these conditions, or to draw on reports of independent observers. The Philippine conclusions on the harm that island-building could cause drew on "obvious" inferences and relied on satellite photos of sedimentation caused by island-building at Mischief Reef. In the view of the Philippines, the satellite photos showed "obvious" disturbance of the seabed and water column.[193] Judge Wolfrum then asked counsel for the Philippines:

> What hard facts do you have that this has been taking place? You said at the beginning there is a caveat: there was no fact-finding you could undertake in this region. But still you must present to us something, that we know that what you qualified as illegal fishing, illegally taking parts of the sea, marine biomass, and destroying the coral, so that we find a factual basis for invoking Articles 192 and 194...[194]

---

[191] Ibid., Transcript, Day 3, 51–52; Award of 12 July 2016, 359, para. 903.

[192] Hearing on the Merits, Transcript, Day 3, 53; Award of 12 July 2016, 359, para. 903.

[193] Hearing on the Merits, Transcript, Day 3, 13.

[194] Ibid., 46.

The significance of the term "hard facts" should not be exaggerated, as if "hard facts" were a new category of facts and "hard fact-finding" a more rigorous type of inquiry, as Mbengue thinks.[195] The transcripts of the Merits Hearing shed light on the proper meaning of the expression. The use of the word "something" and of the expression "factual basis" hardly suggests the existence of a new category of facts called "hard facts." The expression "hard facts" should properly be contrasted not with "facts" but with "inferences" and perhaps even "satellite photos," terms that the Philippines used. To mark the difference with these two terms, Judge Wolfrum could just as well as have used the term "facts."

While China did not appear before the Tribunal, Chinese officials made public statements on several occasions denying that the marine environment would be damaged by China's construction activities. In April 2015, a Foreign Ministry spokesperson categorically declared that the South China Sea "ecological environment [sic] will not be damaged."[196] This assertion was echoed a month later by the Director General of the Department of Boundary and Ocean Affairs of China's Foreign Ministry.[197] In June 2015, a Foreign Ministry spokesperson declared that one of the purposes of the construction was to enable China to "better perform" its international obligations in areas such as environmental conservation.[198] In May 2016, a Foreign Ministry

---

[195] Mbengue, 288.

[196] *SDP*, Annex 624, Ministry of Foreign Affairs, People's Republic of China, Foreign Ministry Spokesperson Hua Chunying's Regular Press Conference (9 April 2015), vol. I, 165–69, https://files.pca-cpa.org/pcadocs/The%20Philippines%27%20Supplemental%20Documents%20-%20Volume%20I%20%28Annexes%20607-667%29.pdf, accessed 9 April 2019; quoted in Award of 12 July 2016, 365, para. 917.

[197] *The Philippines Annexes Cited During the Merits Hearing (Annexes 820–59)* (30 November 2015), Annex 820, Embassy of the People's Republic of China in Canada, An Interview on China's Construction Activities on the Nansha Islands and Reefs 2015/05/27, 5–10, https://pcacases.com/web/view/7, accessed 26 March 2019; quoted in the Award of 12 July 2016, 365, para. 918.

[198] *The Philippines' Annexes Cited During Hearing on Jurisdiction (Annexes 574–583)*, Annex 579, Ministry of Foreign Affairs, People's Republic of China, Foreign Ministry Spokesperson Lu Kang's Remarks on Issues Relating to China's Construction Activities on the Nansha Islands and Reefs (16 June 2015), 158–59, https://files.pca-cpa.org/pcadocs/Annexes%20cited%20during%20Hearing%20on%20Jurisdiction%20%28Annexes%20574-583%29.pdf, accessed 17 April 2019; quoted in the Award of 12 July 2016, 365, para. 919.

spokesman declared that as owners of the Spratly Islands, China cared about protecting the environment of the relevant islands, reefs, and waters of the South China Sea more than any other State, organization or people in the world.[199]

China justified its denial of the occurrence of damage to the South China Sea marine environment resulting from its construction activities by claiming that it had taken the appropriate measures to prevent any such harm. It is therefore essential to examine the nature of such measures if one is to assess China's compliance with its obligation to preserve and protect the marine environment.

2. China's Compliance with Its Obligation to Preserve and Protect Fragile Marine Ecosystems

The Philippines argued that under Part XII, States are under an obligation to protect and preserve marine ecosystems. For the Philippines, the thrust of Article 192 is not limited to the prevention of prospective damage to the marine environment but extends to the preservation of the latter. Preservation requires active measures to maintain and improve the present condition of the marine environment.[200] This duty is a due diligence obligation, "to take the necessary measures to ensure that land creation, construction work and fishing practices do not destroy coral reefs or to pollute the marine environment or to alter the ecological balance."[201] Coral reefs are fragile and vitally important parts of marine ecosystems, so that creating artificial islands out of coral reefs is the worst possible way to treat these fundamental building blocks. [202]

The Tribunal undertook its interpretation of Article 192 in light of Article 194 of the Convention and the corpus of international law relating to the environment. This corpus included two obligations. First, States should ensure that activities within their jurisdiction and control respect the environment of other States or of areas beyond national

---

[199] Ministry of Foreign Affairs, People's Republic of China, Foreign Ministry Spokesperson Hong Lei's Regular Press Conference (6 May 2016), quoted in the Award of 12 July 2016, 365, para. 920.

[200] *MP*, vol. I, 156, para. 6.68.

[201] Hearing on the Merits, Transcript, Day 3, 32.

[202] Award of 12 July 2016, 361, para. 910.

control.[203] Second, States have a positive "duty to prevent or at least mitigate significant harm to the environment when pursuing large-scale construction activities."[204] This duty was reaffirmed in two international arbitral awards issued in the decade preceding the arbitration and cited by the Tribunal: the *Iron Rhine ("IJzeren Rijn") Railway Arbitration* between Belgium and the Netherlands (2005) and the *Indus Waters Kishenganga Arbitration (Pakistan v. India)* Arbitration (2013). In the first case, Belgium, which was planning to activate a railway that linked a Belgian port to the Rhine Basin in Germany via the Netherlands, objected to Dutch requirements of wildlife and nature protection measures. The *Iron Rhine ("IJzeren Rijn") Railway* Tribunal held that "where development [in this case, the restoration, adaptation and modernization of the railway line] may cause significant harm to the environment, there is a duty to prevent or at least mitigate such harm."[205] In the second case, Pakistan questioned the legality under the 1960 Indus Treaty, which regulates the use by India and Pakistan of the Indus system of rivers, of the construction and operation of an Indian hydroelectric project located in Indian-administered Jammu and Kashmir. The *Indus Waters Kishenganga* Tribunal held that the Treaty permitted the transfer of water by India from one tributary of the Jhelum to another to generate hydroelectric power but that Pakistan retained to right to receive a minimum flow of water from India in the Kishenganga/Neelum riverbed at all times.[206] In its Award, the *South China Sea* Tribunal recalled the acknowledgment by the Tribunal in the *Iron Rhine Arbitration* that

---

[203] *Legality of the Threat or Use of Nuclear Weapons, Advisory Opinion, I.C.J. Reports 1996*, 240–42, para. 29, https://www.icj-cij.org/files/case-related/95/095-19960708-ADV-01-00-BI.pdf, accessed 19 April 2019, quoted in Award of 12 July 2016, 374, para. 941.

[204] Award of 12 July 2016, 374, para. 941.

[205] *Iron Rhine Arbitration (Belgium/Netherlands)*, Award of 24 May 2005, 28, para. 59, https://pcacases.com/web/sendAttach/478, accessed 17 April 2019; quoted in the Award of 12 July 2016, 374, note 1095.

[206] Permanent Court of Arbitration, *Indus Waters Kishenganga Arbitration (Pakistan v. India). Court of Arbitration Issues Partial Award in First Arbitration Under the Indus Waters Treaty 1960*, Press Release, PCA Doc. 87220 (19 February 2013), https://pcacases.com/web/sendAttach/1684, accessed 17 April 2019.

States have a "duty to prevent or at least mitigate significant harm to the environment when pursuing large-scale construction activities."[207]

The Philippines strongly doubted that China had taken all necessary measures to protect and preserve the fragile marine ecosystems of the South China Sea. This conclusion was based on an examination of a 500-word statement by China's State Oceanic Administration (SOA), published on 18 June 2015, which it had located in response to a Tribunal question during the Hearing on the Merits,[208] statements by China's Ministry of Foreign Affairs, China's Embassy in Canada, and further publications of the SOA that the Tribunal itself had researched and transmitted to it during and after the Hearing on the Merits.[209] Apart from the fact that China did not undertake an EIA, which will be discussed separately, there was no evidence to back up China's claims that China had followed "strict environmental standards and requirements," which China did not even take the trouble to specify; that China had adopted "many effective measures to preserve the ecological environment (sic)"; that China had made efforts to monitor ecological effects; that such efforts, if any, were sufficient to prevent the environmental harm alleged by the Philippines; or that any measures to promote the recovery of coral reefs would actually be undertaken.[210] Satellite images showed no

---

[207] *Indus Waters Kishenganga Arbitration (Pakistan v. India)*, Partial Award, 18 February 2013, 170, para. 451, https://pcacases.com/web/sendAttach/1681, accessed 17 April 2019.

[208] *The Philippines' Written Responses (Annexes 864–892)* (11 March 2016), Annex 872, China State Ocean Administration, "Construction Activities at Nansha Reefs Did Not Affect the Coral Reef Ecosystem" (10 June 2015), 201–8, https://files.pca-cpa.org/pcadocs/The%20Philippines%27%20Written%20Responses%20%2811%20March%202016%29%20%28Annexes%20864-892%29.pdf, accessed 19 April 2019.

[209] Hearing on the Merits, Transcript, Day 4, 182. The document in question was included in the collection *The Philippines' Annexes Cited During the Merits Hearing 30 November 2015 (Annexes 820–859)* (30 November 2015), Annex 821, China State Oceanic Administration, "Construction Work at Nansha Reefs Will Not Harm Oceanic Ecosystems" (18 June 2015), 11–16, https://files.pca-cpa.org/pcadocs/The%20Philippines%27%20Annexes%20cited%20during%20Merits%20Hearing%20%28Annexes%20820-859%29.pdf, accessed 7 April 2019.

[210] *The Philippines Annexes Cited During the Merits Hearing*, Annex 820; *Responses of the Philippines to the Tribunal's 5 February 2016 Request for Comments* (11 March 2016), 5–6, paras. 13–15; 7–9, paras. 17–24, https://pcacases.com/web/sendAttach/1849, accessed 19 April 2019.

signs that remedial work had been or would be started. The Philippines observed that the existence of Chinese technical guidelines (Guidance for the Assessment of Coastal Marine Ecosystem Health, Code of Practice for Marine Monitoring Technology, and Technical Guidelines for Environmental Impact Assessment of Marine Engineering) did not prove that China had followed any of these guidelines in the areas where island-building took place.[211]

Of course, China did not contest the existence of a duty to protect and preserve marine ecosystems—it would have been inconceivable for China to do so. Its strategy may be recapitulated in four assertions. First, China had undertaken rigorous scientific tests before undertaking the construction. Second, China had taken all necessary measures to protect and preserve fragile marine ecosystems during the construction. Third, China denied strenuously that construction would damage the marine environment. Fourth, China was optimistic about the possibility of recovery of any coral reefs that may have been damaged.

On the whole, the independent experts' report (the Ferse Report) confirmed the findings of the expert reports submitted by the Philippines. The Tribunal noted at least nine specific claims of China that were contradicted by the Ferse Report. The details may be found in Table 5.3. The following paragraphs will present summaries of the major discrepancies between Chinese assessments and those of the independent experts.

One claim concerned the "comprehensive technical concept" of the construction activities, described as "nature simulation." The Ferse Report pointed out that the construction process, rather than stimulating the natural process of island development, increased erosion of the reefs by shifting the balance between carbonate accretion and erosion.[212] A second set of Chinese claims that were questioned by the Ferse Report related to the measures that China had allegedly taken in order to minimize the impact of construction on coral reefs. China claimed that the construction was timed to avoid the spawning periods of red snapper (mid-April), tuna (June–August), and bonito (March–August). Unfortunately for China, the satellite and aerial imagery analyzed by

---

[211] *Responses of the Philippines to the Tribunal's 5 February 2016 Request for Comments,* 15–16, paras. 39–41.

[212] Award of 12 July 2016, 393, para. 982(h).

Table 5.3 Construction methods, impact of construction on coral reefs, and prospects of recovery of coral reefs: assessments by Chinese authorities and by the independent experts

Chinese authorities	Independent experts (Ferse report)
Methods	
We put equal emphasis on construction and protection by following a high standard of environmental protection and taking into full consideration the protection of ecological environment and fishing resources. (MOFA, 9 April 2015)	No information on the Environmental Impact Assessment, the measures taken to prevent negative impacts, and the standards applied during construction was found online. (47–48)
China's construction projects on the islands and reefs have gone through scientific assessments and rigorous tests. We put equal emphasis on construction and protection by following a high standard of environmental protection and taking into full consideration the protection of ecological environment and fishing resources. (MOFA, 28 April 2015)	
We have taken into full account issues of ecological preservation and fishery protection, followed strict environmental protection standards and requirements in the construction process, and adopted many effective measures to preserve the ecological environment. (PRC Embassy in Canada, 27 May 2015)	Chinese-language documents specific to the construction activities that were found online dealt with the engineering aspects of the projects, but did not appear to include consideration of how to minimize negative ecological effects. (47–48)
The construction was undertaken with an emphasis on the protection of ecosystem and fishery resources, carried out after scientific assessment and feasibility studies. (SOA, 10 June 2015)	The available information on the construction activities does not indicate a particular emphasis on the protection of ecosystem and fishery resources. (48)
The expansion of the Nansha reefs will abide rigorously by the concept of "Green Construction, Eco-Friendly Reefs" in protecting the ecosystems. (SOA, 18 June 2015)	It is not clear what the concept of "Green Construction, Eco-Friendly Reefs" means. (55)
China embraces "nature simulation" as its "comprehensive technical concept" in the Nansha reef expansion project. This method simulates the displacement of bioclasts such as corals and sands during wind storms and high waves; this biological detritus settles on the combined equilibrium points of the shallow reef flats to form stable supratidal zones which then evolve into oceanic oases. (SOA, 18 June 2015)	The description of the "nature simulation" method neglects the importance of biogenic sediment production (i.e., sediment production stemming from the activities of living organisms) in reef island formation and maintenance. (54–55)

(continued)

Table 5.3 (continued)

Chinese authorities	Independent experts (Ferse report)
Big cutter suction dredgers are used to collect the loose coral fragments and sands in the lagoon and deposit them on bank-inset reefs to form supratidal platform foundation on which certain kinds of facilities can be built. Through the natural functions of the air, the rain, and the sun, paving it with some quick man-made material, the land reclamation area will produce the ecological effects by going from desalination, solidification, efflorescence, to a green coral reef ecological environment. (SOA, 18 June 2015)	China's construction activities utilized carbonate fragments and sands that were produced biologically but they simultaneously affected adversely the very biota (mostly corals, foraminifera, and calcareous algae) that are the basis for sediment generation in the first place. Rather than simulating the natural process of island development, the construction process increases the erosion of the reefs by shifting the balance between carbonate accretion and erosion, and thus increases the risk of drowning the reef as sea levels continue to rise. (54–55)
Measures to minimize impact on Coral Reefs	
The extent of the reclamation and dredging areas was minimized. (SOA, 10 June 2015)	On the largest reefs (Mischief and Subi Reef), about half and 60% of the reef, respectively, have been lost due to land reclamation. Dredging-related sediment plumes have affected nearly all of the lagoon areas of these two reefs. (50)
Trash collecting screens were set. (SOA, 10 June 2015)	It is not entirely clear what kind of trash this refers to. No indication of sediment screens is visible on the available aerial and satellite imagery, and sediment plumes extend far beyond the immediate areas of dredging and land-filling. (50)
The construction was planned on sites containing dead coral. (SOA, 18 June 2015) Most of the construction sites selected are located in reef flats with the lowest hermatypic coral coverage or where hermatypic corals are mostly dead. (SOA, 10 June 2015)	While it is correct that the reef flat environments of the impacted reefs appeared to have been degraded to some extent before construction began, Subi Reef reportedly had higher coral cover on the reef flat than on the outer slope, and the lagoonal reefs of Mischief and Subi Reefs had significant coral communities which have been severely impacted, and likely killed for the most part, by construction-related dredging and sedimentation. (50)
The timing of island building tried to avoid the spawning season of red snappers (mid-April), tuna (July and August) and bonito (March and April). (SOA, 10 June 2015)	Construction occurred in April on all reefs except Gaven Reef, between June and August on Cuarteron, Mischief and Gaven Reef, and between March and August on Cuarteron and Mischief Reef. Construction activities on several reefs thus coincided with the spawning periods of red snapper, tuna and bonito. (51)

(continued)

Table 5.3 (continued)

Chinese authorities	Independent experts (Ferse report)
The change of grain size of sand sediments was monitored regularly, ensuring that the area where sands are taken always consists of grits and fine sands were prevented from going into reclamation areas to maintain the water quality of coral reef areas. (SOA, 10 June 2015)	Satellite and aerial imagery shows clearly that the water quality in the vicinity of each construction site was affected by increased sediment load and turbidity from dredging. (51)
Construction intensity during the peak of growth of coral reefs was reduced, the turbidity change of waters was monitored dynamically and dredging intervals in light of the biological characteristics of coral reefs were adjusted. (SOA, 10 June 2015)	Coral growth is light- and temperature-dependent, which in the Spratly Islands are at their maximum in spring and early summer. Construction took place throughout this time. To what extent the environment was monitored throughout the construction activities, and dredging intervals adjusted, cannot be ascertained based on the available information. (52)
The growth and health of coral reefs in construction areas and indicators such as species, population, and diversity of swimming animals and plankton in coral reef areas were monitored. (SOA, 10 June 2015)	The accuracy of these three statements cannot be verified based on the available information. (52)
The wastewater and solid waste produced from life and construction to be sent for treatment at land facilities of harbours were centrally collected. (SOA, 10 June 2015)	
Newer vessels to ensure no oil spill happen were used Weather and marine condition forecasts were regularly listened to. (SOA, 10 June 2015)	
Prior preparation for typhoons and strong waves to avoid the loss of sands from structures was done. (SOA, 10 June 2015)	
Construction projects were planned on bank-inset reefs made of basically dead corals. (SOA, 18 June 2015)	While the deep lagoon basins contained less live corals than other reef habitats, they nonetheless constituted a vital habitat for benthic macrofauna, such as mollusks, echinoderms, and crustaceans. (56)
A cutter suction dredger was used to collect loose coral fragments and sands from flat lagoon basins, which do not constitute hospitable environment for corals, to fill the land reclamation areas. (SOA, 18 June 2015)	On the reefs lacking a deep lagoon (Hughes, Cuarteron, Gaven and Johnson Reef), material for land reclamation was gathered by excavating parts of the shallow reef flat habitat with the use of cutter suction dredgers. (56) In all cases, land reclamation targeted shallow reef flat habitat, which contained higher amounts of live coral, is the habitat with the highest primary productivity and likely constitutes an important nursery for juvenile fishes. Furthermore, sediment plumes from dredging have affected both lagoon and outer reef slope habitats. Thus, the dredging impacts were not confined to the deep lagoon basins. (56)

(continued)

Table 5.3 (continued)

Chinese authorities	Independent experts (Ferse report)
Impact on Coral Reefs	
The ecological environment (sic) of the South China Sea will not be damaged. (MOFA, 9 April 2015) Such projects will not damage the ecological environment (sic) of the South China Sea. (MOFA, 28 April 2015)	The statement that construction activity does not damage the environment on the reefs is contradicted by the facts underlining certain damage to reefs detailed in the Ferse Report. (48)
Due to the strong currents and waves in Spratly waters, the water bodies are updated fairly fast so that little suspended sands are produced from the constructions, leaving the photosynthesis of corals largely unaffected. Because the sites are located in areas where coverage of coral reefs is low, the overall community structure of coral reefs is not changed. In addition, since oceanographic and sediment status. (SOA, 10 June 2015) The changes are only limited to areas near the construction sites, the physical and chemical living environment of coral reefs are not fundamentally changed, therefore, their health was not significantly harmed by the construction activities. (SOA, 10 June 2015)	The available imagery of the construction process shows significant sediment plumes at all impacted reefs. At Subi, Mischief, and Fiery Cross Reef, the sediment plumes are seen to envelop large sections of the outer reef slope, indicating that at least within the nearer vicinity of land reclamation, reef communities were subject to intense sedimentation impacts. Within the lagoons of Mischief and Subi Reef, water exchange is limited, and the sediment plumes generated by dredging will thus have remained in the water column for several weeks. As coral species differ in their susceptibility to sedimentation and turbidity based on their morphology, the sedimentation plumes are very likely to have altered the community structure of the affected coral reefs. The construction activities have permanently altered the hydrodynamics of the affected reefs, and the resuspension of sediments generated by the dredging activities is likely to maintain an elevated level of sediments, from months to years in those parts of the reefs that are well-flushed by open ocean waters, and from years to decades in areas with less flushing, such as the lagoons of Mischief and Subi Reef. (53)
The ecological impact was "partial, temporary, controllable and recoverable." (SOA, 18 June 2015)	Ecological impacts from the construction activities affected large parts of the reefs and include permanent (for reclaimed reef flats and excavated channels) and long-lasting (for sediment resuspension in lagoons) effects. The extensive sediment plumes that remained near the construction areas for several weeks to months render the amount of control over potential impacts doubtful. For large areas of reef affected by the construction activities, recovery is unlikely or may take decades to centuries. (57)

(continued)

Table 5.3 (continued)

Chinese authorities	Independent experts (Ferse report)
The construction activities neither affected the health of the ecosystems of the Spratly Islands nor harmed the coral reef ecosystems. (SOA, 10 June 2015)	The available evidence leaves little doubt that the coral reef ecosystem of the seven affected reefs have suffered significant and extensive harm as a result of construction activities. (53)
Prospects for Recovery The South China Sea is not a body of closed waters, therefore, nutrients and food organisms can be replenished constantly from surrounding waters. (SOA, 10 June 2015)	The connectivity of reefs particularly in the western Spratly Islands is limited. Thus, the available information provides very limited support for the potential for replenishment of the impacted reefs from waters outside the Spratly Islands, let alone beyond the South China Sea. (50)
Domestic and international experts have experimented several ways to restoring coral reef ecosystems and designed multiple structures of artificial coral reefs, which proved that the restoration of coral reef communities could be realized should effective measures be taken. (SOA, 10 June 2015)	Restoration is not likely to succeed if stressors (such as sedimentation and destructive fishing) persist, and if ecological connectivity and larval supply are disturbed. In the latter case, transplantation of coral fragments if required to re-introduce corals, and this is generally not recommended for large-scale impacts, as it simply displaces the impact over a larger area from which fragments are sourced. In general, active reef restoration is recommended mostly for impacts of limited spatial extent—it has been carried out with some success only at scales of up to a few hectares. Furthermore, transplantation is not likely to succeed as long as sedimentation persists, which will be the case in large parts of the area affected by the construction activities. (52)
It is important to enhance monitoring of regional ecosystems and implement measures including release, coral restoration, and transplantation in order to better protect the coral reefs. (SOA, 10 June 2015)	Coral transplantation is unlikely to be a suitable approach for restoration of the impacted reefs for a number of reasons. First, the resuspension of residual sediments will reduce the suitability of the affected reef areas for coral survival and growth for a long time (weeks up to decades, depending on specific local environmental conditions. Second, a large amount of transplants will have to be sourced from elsewhere. If they are transplanted from other Spratly reefs, this means considerable additional impact on the region's reefs, with uncertain prospects of success. If they are meant to be sourced from coral farms or reefs outside the region, there is a risk of introducing non-native species and genotypes, reduced genetic diversity, and non-compatibility in terms of environmental requirements. (54)

(continued)

Table 5.3 (continued)

Chinese authorities	Independent experts (Ferse report)
Coral reefs that have been severely damaged by natural factors or human activities can be restored initially in 5–10 years provided that effective measures are taken, and complex and complete ecosystems can be fully restored in 50–100 years. (SOA, 10 June 2015)	Under the best possible conditions, recovery of coral communities takes upwards of ten years, and in that case will consist only of fast-growing species. Large parts of the seven reefs have been permanently destroyed by construction, and for the remaining areas, recovery is uncertain and, if it occurs, it will take more than a century until the large massive coral colonies have regrown. It not very clear what is meant by "effective measures"—this could refer either to measures to improve the environmental conditions by attempting to remove lingering stressors, or to active restoration activities. Restorative activities are extremely expensive and have only ever been attempted on small scales, far smaller than the scale of reclamation impacts. (54)

Source Prepared by the author, based on the Ferse Report, Ministry of Foreign Affairs ("MOFA") Statement (9 April 2015, Annex 624), MOFA Statement (28 April 2015, Annex 625), Interview, Embassy of the People's Republic of China in Canada (27 May 2015, Annex 820), State Oceanic Administration ("SOA") Report (10 June 2015, Annex 872), and SOA Statement (18 June 2015, Annex 821)

the three independent experts showed that construction had occurred during these spawning periods.[213] China further claimed that it had selected sites containing dead coral. The Ferse Report noted that even deep lagoon basins containing less live corals than other reef habitats constitute a vital habitat for mollusks, echinoderms, and crustaceans. At Cuarteron Reef, Hughes Reef, Gaven Reef, and Johnson Reef, which lacked a deep lagoon, material for land reclamation was gathered from the shallow reef habitat. Sediment plumes affected both lagoon and outer reef slope habitats.[214] China asserted that it prevented fine sands from going into reclamation areas to maintain the water quality of coral reef areas. Satellite and aerial imagery examined by the experts clearly showed that water quality in the vicinity of each construction site was affected by increased sediment and turbidity from dredging.[215]

A third set of doubtful Chinese claims concerned the impact of China's construction activities on reef structure and reef health. For the SOA, due to strong currents and waves, the photosynthesis of corals was left largely unaffected. Moreover, the sites were located in areas where coverage of corals was low; therefore, the overall structure of coral reefs and their physical and chemical living environment had not fundamentally changed. In contrast, the independent experts pointed out that the sediment plumes generated by dredging altered the structure of the reefs and a high level of sediments would be likely for months to years in those parts of the reefs that are well-flushed by open ocean waters and from years to decades in areas with less flushing.[216]

As regards the impact on reef health, China's SOA claimed that construction neither affected the health of the Spratly Islands ecosystems nor harmed the coral reef ecosystems. The independent experts had little doubt that the coral reef ecosystems of the seven reefs have suffered significant and extensive harm as a result of China's construction activities.[217] The last set of Chinese claims that the Ferse Report impugned concerned the prospects for recovery of the coral reefs. China's SOA believed that in the South China Sea nutrients and organisms can be

---

[213] Ibid., 391–92, para. 982(b).

[214] Ibid., 392, para. 982(g).

[215] Ibid., para. 982(c).

[216] Ibid., 392–93, para. 982(e).

[217] Ibid., 393, para. 982(f).

replenished constantly from surrounding waters. The Ferse Report found very limited support for the potential for replenishment from outside the Spratly Islands in light of larval connectivity patterns within the South China Sea.[218] China's SOA believed that severely damaged coral reefs can be restored in 5–10 years and complete ecosystems in 50–100 years. It also suggested transplantation to better protect the coral reefs. The Ferse Report explained that restoration would not succeed if stressors persisted and if ecological connectivity and larval supply were disturbed. Large parts of the seven reefs have been permanently destroyed; for the remaining areas, recovery is uncertain and, if it occurs, it will take more than a century. Moreover, restoration is extremely expensive and has only been attempted on far smaller scales. Transplantation is unlikely to be suitable on the scale of the impacts from construction, as it could risk impacting other reefs in the region, and involves prohibitive labor and costs.[219] Overall, China's SOA believed that the ecological impact on the coral reefs is partial, temporary, controllable, and reversible.[220] In contrast, the independent experts were of the view that the ecological impact would be permanent (for reclaimed reef flats and excavated channels) and long-lasting (for sediment suspension in lagoons). For large parts of the reefs affected by construction, recovery is unlikely or may take decades to centuries.[221] The Ferse Report concluded that Chinese assessments of the environmental impact of China's activities contradicted the available knowledge and underestimated the kind and extent of damage and long-lasting environmental impact of island-building.[222]

The Tribunal adopted the independent experts' threefold classification of the impact of construction and dredging activities on reef systems.[223] The first impact is direct destruction of reef habitat through burial

---

[218] Ibid., 391, para. 982(a).

[219] Ibid., 392, para. 982(d).

[220] Nilufer Oral describes this statement as an example of "Orwellian doublespeak." Oral, "The South China Sea Arbitral Award," 238.

[221] Award of 12 July 2016, 393–94, para. 982(i).

[222] Ferse Report, 60. This summary and Table 5.3 contradict the statements that the SOA failed to explain the impact of the big cutter section dredger on the marine environment and that the Tribunal did not assess contents of the SOA Statement of 18 June 2015 (Annex 821) or the SOA Report of 10 June 2015 (Annex 872). Oral, "The South China Sea Arbitral Award," 238, 241.

[223] Ferse Report, 21–29; Award of 12 July 2016, 331, para. 857.

under sand, gravel, and rubble. A second category of effects consists of indirect impacts on benthic organisms such as corals and seagrasses via altered hydrodynamics, increased sedimentation, turbidity, and nutrient enrichment.[224] The third category of impacts comprises indirect impacts on organisms in the water column, such as fishes and larvae, from sediments, chemical and nutrient release, and noise.[225] To understand the three types of impacts, one must read the Ferse Report itself.

The Ferse Report explains that destruction of reef flats will reduce the most productive environment, since the algae that grow at the greatest rates on the reef flats are consumed mostly by herbivores (fishes and invertebrates) and form a major part of the coral reef food web that supports fisheries. The loss of coral habitat through direct dredging and acute sedimentation will also diminish the suitability to provide shelter for fish.[226] The indirect impacts could occur through sedimentation, leading to burial and choking or, at the very least, affecting the ecology of benthic and pelagic coral reef organisms, including seagrass, corals, and fishes.[227] Sedimentation and turbidity impair the photosynthesis of seagrass and corals.[228] Sedimentation can diminish the reproductive output, wound healing, and growth of corals; it also negatively affects the settlement and survival of coral larvae.[229] Sedimentation can modify predator–prey interactions, impair the feeding of planktivorous fishes,

---

[224] "Benthon" refers collectively to the sedentary plant and animal life living on the sea bottom. The adjective "benthic" means "living at the soil-water interface at the bottom of a sea or lake." *Larousse Dictionary of Science and Technology*, 100.

[225] The water column is "a vertical section of water from the surface to the bottom of the sea, a lake, a river, etc. The water column is a way of describing the different features found in seawater at different depths." *Cambridge English Dictionary* (Cambridge: Cambridge University Press, 2019), http://dictionary.cambridge.org/dictionary/english/water-column, accessed 18 April 2019.

[226] Ferse Report, 24–25.

[227] "Pelagic" is defined as "living in the middle depths and surface of the sea." *Larousse Dictionary of Science and Technology*, 805.

[228] The Second Carpenter Report explains that sediment particles clog the feeding and respiration valves of corals and shade them, preventing them from getting enough light to photosynthesize. Second Carpenter Report, 263.

[229] Larvae do not settle on sand of any sort or on substrates coated with bacterial slime. John E.N. Veron, "Corals: Biology, Skeletal Deposition and Reef Building," in David Hopley (ed.), *Encyclopedia of Modern Coral Reefs: Structure, Form and Process* (Dordrecht: Springer, 2011), 279.

and deter fish herbivory.[230] The indirect impact on organisms in the water column may be illustrated by several examples. Parrotfish reduces grazing when coarse sediment is present and when organic loads in sediments are low. Scallops may have higher mortality when buried under fine sediment than under coarse sediment and with longer burial time. As for corals, their early-life history stages can be negatively affected by sediment release and turbidity. Once eggs and sperm are developed and ready to spawn, they are grouped together into sperm bundles ready for release. These float to the surface of the water, where they break open and fertilization can begin. Sediment concentrations of 35 mg.L^{-1} can prevent the ascent of coral egg–sperm bundles by intercepting them and ballasting them back to the bottom.[231] Higher sediment concentrations remove the sperm from the water surface during coral spawning events due to the entanglement of sperm within the sediment, causing the sperm to sink. The removal of sperm from the surface reduces fertilization rates considerably. Coral larvae do not like to settle on surfaces with too much sediment. They are easily smothered for a long period following sedimentation due to their small size and the lack of ability to repel sediments. Meanwhile, the release of nutrients can lead to the proliferation of bacteria, reducing oxygen, and of bacteria that compete with corals for space.[232]

The Tribunal also adopted the independent experts' conclusion concerning the duration and the extent of harm to the marine environment. The duration of harm to the areas affected by dredging and dredging-related release of sediments and nutrients and the prospects and likely

---

[230]Ferse Report, 26–27; Second Carpenter Report, pp. 264, 267, 272, Herbivory is defined as "the consumption of plants." *Collins English Dictionary* (Glasgow: HarperCollins Publishers, 2019), https://www.collinsdictionary.com/dictionary/english/herbivory, accessed 19 April 2019. For an overview of the impacts of sediment on coral reefs, see Michael J. Risk and Evan Edinger, "Impacts of Sediment on Coral Reefs," in David Hopley (ed.), *Encyclopedia of Modern Coral Reefs: Structure, Form and Process* (Dordrecht: Springer, 2011), 575–86. The Second Carpenter Report also devoted some attention to the impact of island-building on seagrass meadows and on reefs adjacent to those where construction took place, but these considerations were not incorporated in the Award. Second Carpenter Report, 263, 271.

[231]0.1 mg·L^{-1} is the same as 0.1 mg/L (milligram per liter), https://onlineconversion.vbulletin.net/forum/main-forums/convert-and-calculate/9350-0-1-mg%C2%B7l-1, accessed 19 April 2019.

[232]Ferse Report, 28; Second Carpenter Report, 272.

rates for rejuvenation differ depending on the environmental setting of each habitat. The harm to areas affected by dredging for navigable channels and basins will likely be near-permanent. The prospects for rejuvenation are low, especially as long as maintenance dredging for the use of artificial islands continues. Where the major geomorphological structures have been removed, such as accumulations of corals that stand several meters above the substrate, there is little prospect for recovery on ecological time scales. The harm to areas affected by the smothering of sediments and increased turbidity, which includes most of the lagoons at Mischief Reef and Subi Reef, and parts of the outer reef slopes of all seven reefs, is likely to endure for years to decades within the lagoon (due to limited water exchange) and for weeks to months on the outer reef slope.[233]

The Ferse Report's explanation of the six reasons for the long process of recovery is highly instructive. The first reason is low population size. Due to extensive damage from dredging and sedimentation, coral recovery will be limited by a paucity of replenishing larvae, particularly in the western part of the Spratly Islands, which are more isolated from one another. Other reefs in the area that could be a source of larvae are degraded. Second, deeper reefs do not necessarily provide a refuge for shallow-water populations. The connectivity between deep-water and shallow-water corals of the same species is limited. Hence, even if deep-water corals remain healthy, this does not necessarily help the shallow-water corals to recover. The third reason is poor habitat quality. The areas of reef dominated by rubble from dredging provide an unstable substratum that is continually rolled by the turbulence of incident waves. Settling corals are periodically buried as the substrate to which they are attached moves around. Fourth, recovery rates for massive corals are slower. They can resist modest levels of sedimentation, but they will experience mortality when they are directly affected by dredging. Recovery of these corals will take a minimum of decades and likely centuries. The fifth reason is a fundamental change in reef attractiveness. Larval fish can literally hear reefs and will swim toward noisy reefs. The noise is created by snapping shrimps, populations of which will have been devastated in the dredged areas. If the reefs become quiet, the recovery of reef fish will be delayed. The sixth and final reason is the loss

---

[233]Award of 12 July 2016, 380, para. 979.

of fish nursery habitats. The habitats of commercially important coral reef fishes, such as groupers and snappers, were subjected to dredging. The loss of nursery habitats will increase the mortality of juvenile reef fish and reduce fishery productivity.[234]

As for the extent of the harm, the Tribunal accepted the independent experts' finding that China's construction has led to reduced productivity and complexity of the reefs, with significant reductions of nursery habitat for a number of fish species. The reefs affected by construction will have a greatly reduced capacity to sustain local fisheries. Their ability to help replenish the fisheries of the neighboring countries will be reduced to a third. The construction will have a broader impact on the marine ecosystem in and around the South China Sea and on fisheries resources. The magnitude of the impact will depend on the role of the seven reefs as critical habitat and source of larvae for fisheries resources compared to that of other Spratly Islands reefs. The magnitude is difficult to quantify due to a lack of empirical studies, but cascading effects cannot be ruled out.[235]

The Ferse Report's unambiguous conclusion is that China's island-building has caused and will cause environmental harm to coral reefs at Cuarteron Reef, Fiery Cross Reef, Gaven Reef, Johnson Reef, Hughes Reef, Mischief Reef, and Subi Reef beyond the pre-existing damage to reefs that resulted from such phenomena as destructive fishing and collection of corals and clams, storm damage, and human presence on the small garrisons on the reefs. The scale of these previous impacts cannot be compared with the scale and duration of the environmental harm caused by the recent island-building.[236] Professor McManus may have been exaggerating only a little when he asserted that the loss of reef area due to burial was the most nearly permanent loss of coral area in human history.[237]

There was no doubt on the part of the Tribunal that it was the Chinese State as such that was responsible for the damage to the marine environment. It is a basic principle of international responsibility that the conduct of an organ of the State is attributable to that State. This principle described by the UN International Law Commission as the first

---

[234] Ferse Report, 40–43.

[235] Award of 12 July 2016, 390, para. 979.

[236] Ferse Report, 59.

[237] McManus, "Offshore Coral Reef Damage," 220.

principle of attribution for the purposes of State responsibility and by the ICJ as a customary rule. Attribution to the State is justified "when conduct consisting of an act or omission or a series of acts or omissions is to be considered as the conduct of the State."[238] In the South China Sea, the identity of the actor responsible for island-building was no mystery. As the Tribunal put it, "[t]here is no question that the artificial island-building program is part of an official Chinese policy and program implemented by organs of the Chinese State."[239]

The Tribunal, basing itself on "compelling evidence, expert reports and a critical assessment of China's claims," had no doubt that China's artificial island-building on seven reefs in the Spratly Islands had caused "devastating and long-lasting damage to the marine environment."[240] In relation to Philippine Submissions No. 11 and 12(b), the Tribunal concluded that

> ...through its construction activities, China has breached its obligation under Article 192 to protect and preserve the marine environment, has conducted dredging in such a way as to pollute the marine environment with sediment in breach of Article 194(1), and has violated its duty under Article 194(5) to take measures necessary to protect and preserve rare or fragile ecosystems as well as the habitat of depleted, threatened or endangered species and other forms of marine life.[241]

The Tribunal's conclusions regarding Philippine Submission No. 14, which claimed that China had aggravated and extended the dispute by, among others, conducting dredging, artificial island-building, and construction activities at Mischief Reef, Cuarteron Reef, Fiery Cross Reef, Gaven Reef, Johnson Reef, Hughes Reef, and Subi Reef, were expressed in even harsher terms, describing the harm to the coral reef habitat as "permanent" and "irreparable." The Tribunal added that in practical terms, "neither this decision nor any action that either Party may take in response can undo the permanent damage that has been done to the coral reef habitats of the South China Sea."[242]

---

[238] United Nations International Law Commission, "Commentaries," 80, 86, 84.

[239] Award of 12 July 2016, 388, para. 976.

[240] Ibid., 394, para. 983.

[241] Ibid.

[242] Ibid., 462, para. 1178.

The CSIL criticized the Tribunal's findings for allegedly committing errors in fact-finding and for treating the due diligence duty provided for in Articles 192, 194(1), and 194(5) as obligations of result rather than obligations of conduct. The alleged errors in fact-finding were of two sorts. The Ferse Report was said to be "dubious" and its reliability "doubtful," because it was prepared in only 17 days, the time that elapsed between the appointment of the second and third coral reef experts on 12 April 2016 and the submission of the report on 29 April 2016.[243] The Report itself was criticized for lacking first-hand, empirical data, for citing research having no direct relevance to China's construction in the Spratly Islands, and for repeatedly using, without examination, aerial and satellite images provided by the Philippines. Among these images were those provided by a US think tank, the Center for Strategic and International Studies (CSIS), which were said to be of questionable reliability.[244] The second error in fact-finding allegedly arose from the Tribunal's disregarding of the environmental protection measures adopted by China in the course of its construction activities. The CSIL referred to the "Green Construction Eco-Friendly Reefs" methods described in the 500-word SOA statement of 18 June 2015. These measures were said to be sufficient to establish that China had fulfilled its due diligence obligation to protect the environment of the Spratly Islands.[245] The second major criticism of the CSIS was that the formulation of the Tribunal's conclusion in paragraphs 982 and 983 in effect imposed an obligation of result on China. The Tribunal is said to have failed to determine whether China had used the best practicable means at its disposal and in accordance with its capabilities to protect and preserve fragile marine ecosystems and the habitats of depleted, threatened, or endangered species.[246]

Most of the CSIL's criticisms of the Tribunal's fact-finding are disingenuous. The characterization of the Ferse Report as dubious because of the length of time it took to produce it makes one wonder whether the CSIL believed that the experts had to start from scratch, without any knowledge whatsoever of coral reefs, and that for this reason they would have needed an inordinate length of time to produce a report.

---

[243] CSIL, 585, para. 809.

[244] Ibid.

[245] Ibid., 586–87, paras. 810–11. It should be recalled here that this document was found by the Philippines between the first and second rounds of the Hearing on the Merits.

[246] Ibid., 585, paras. 806–7.

A quick perusal of the experts' curriculum vitae reveals that all three are specialists on the recovery of coral reefs. Dr. Ferse's research has focused largely on coral reef restoration and ecological functioning, in particular the roles of reef fishes and their link to the reef habitat as well as on the impact of environmental and anthropogenic factors on coral reef benthic communities. Professor Mumby's research focuses on tropical coastal ecosystems and the development of ecosystem models to investigate conservation measures in mitigating disturbances on reefs. He is an expert on remote sensing of coral reefs, on which he did his Ph.D. dissertation. Dr. Ward has conducted research into the responses of corals to environmental stress including elevated nutrients, mechanical damage, and elements of climate change such as ocean acidification and temperature elevations. Given their training and experience, it is hardly surprising that they were able to prepare a report in 17 days.[247]

Other alleged criticisms of the Tribunal's fact-finding are in reality consequences of China's refusal to appear before and cooperate with the Tribunal. The reason that the independent experts, and for that matter, the experts presented by the Philippines were unable to make a site visit and for that reason had to rely in part on analogies with data drawn from other regions of the world was that China had expressed its "firm opposition" to site visits in a letter addressed in February 2015 to individual members of the Tribunal.[248] Contrary to an impression that the CSIL's critique might create, the Tribunal had in fact envisaged the possibility of site visits, as indicated in the Rules of Procedure in 2013, and asked the Philippines and China whether one would be possible.[249] The reason that the Tribunal had to rely on images provided by the Philippines was that China had not made any submissions to the Tribunal. It is churlish, to say the least, to criticize the Tribunal for the consequences of China's own conduct.

[247] Ferse Report, 4–6, 64–110.

[248] Award of 12 July 2016, 15–16, para. 42.

[249] Ibid., 15, para. 40. Article 2, para. 2 of the Rules of Procedure provides:

> The Arbitral Tribunal may take all appropriate measures in order to establish the facts including, when necessary, the conduct of a visit to the localities to which the case relates. The Parties shall afford the Arbitral Tribunal all reasonable facilities in the event of such a visit.

> *South China Sea Arbitration*, Rules of Procedure, PCA Doc. 101991 (27 August 2013), https://pcacases.com/web/sendAttach/233, accessed 7 April 2019.

The allegation that the Tribunal repeatedly used the aerial and satellite imagery without analyzing them is false, as demonstrated by Table 5.2. The further allegation that the Tribunal had disregarded the environmental protection measures that China had taken is patently untrue. As we have seen above and in Table 5.1, the Tribunal examined them on the basis of the Ferse Report and found them to be inadequate. The charge that the Tribunal's conclusion converted an obligation of conduct into an obligation of result misunderstands the Award of 12 July 2016. As we have seen above, the Tribunal did examine, with the help of independent experts, whether China had taken all necessary measures to protect and preserve fragile ecosystems, as China had claimed. This examination concluded that China had not taken all the measures required to enable it to comply with its obligation to protect and preserve the marine environment, for several reasons. In one instance, China did not actually take the measures that it claimed to have taken—it had not timed the construction to avoid the spawning periods of three species of fish. In another instance, the measure taken was not adequate—carrying out construction projects on bank-inset reefs made of dead corals ignored the fact that deep lagoon basins constituted a vital habitat for benthic macrofauna. In yet another instance, there was absolutely no evidence that China had taken the measure that it claimed to have taken—no trash-collecting screens that China had allegedly set were visible on the aerial and satellite imagery.

China's conduct as its construction activities was ongoing, particularly its wholesale rejection of Philippine protests, not to mention those of other littoral States, made it almost inevitable to conclude that China had not complied with its obligation to cooperate with other States bordering the South China Sea to protect and preserve the marine environment.

B. The Protection and Preservation of the Marine Environment and the Obligation to Cooperate

Articles 197 and 123 of the Convention are among the numerous provisions of the latter that impose on States a duty to cooperate.[250]

---

[250] Tim Stephens, "Article 197. Cooperation on a Global or Regional Basis," in Alexander Proelss (ed.), *The United Nations Convention on the Law of the Sea. A Commentary* (München: Verlag C.H. Beck oHG, 2017), 1329. Stephens gives as examples

The Arbitral Tribunal, implicitly confirming that States in a region and States bordering a semi-enclosed sea are under an obligation to cooperate, found no evidence that China had cooperated with the other littoral States of the South China Sea on the protection and preservation of the marine environment in relation to its construction activities.

1. The Obligation to Cooperate Among States in a Region and Among States Bordering an Enclosed or Semi-Enclosed Sea

Two provisions of the Convention are potentially applicable to cooperation among the littoral States of the South China Sea: Articles 197 and 123. Article 197 (Cooperation on a Global or Regional Basis) stipulates that:

> States shall cooperate on a global basis and, as appropriate, on a regional basis, directly or through competent international organizations, in formulating and elaborating international rules, standards and recommended practices and procedures consistent with this Convention, for the protection and preservation of the marine environment, taking into account characteristic regional features.

The use of the word "shall" undoubtedly means that a legal obligation is intended. Lacking in the Article is a definition of the concept of "region."[251] In contrast, Article 123, entitled "Cooperation of States Bordering Enclosed or Semi-Enclosed Seas," is preceded by an Article that provides a definition of an enclosed or semi-enclosed sea.

According to Article 122 of the Convention, an enclosed or semi-enclosed sea is

> a gulf, basin or sea surrounded by two or more States and connected to another sea or the ocean by a narrow outlet or consisting entirely or primarily of the territorial seas and exclusive economic zones of two or more coastal States.

Article 100, relating to the suppression of piracy, Article 118, on the conservation of the living resources of the high seas, and Article 143(a), concerning the conduct of marine scientific research in the Area.

[251] Nilufer Oral, *Legal Aspects of Sustainable Development. Regional Co-Operation and Protection of the Marine Environment Under International Law* (Leiden: Martinus Nijhoff Publishers, 2013), 33 ("Oral, *Legal Aspects*").

Relying on this definition, the Tribunal had no trouble in characterizing the South China Sea as a "semi-enclosed sea" in the Western Pacific.[252] Yet reliance on Article 123 might have posed a problem of a different order, relating to the difference in the formulations of the duty to cooperate in the two articles. Many scholars doubt that Article 123 of the Convention imposes on States bordering an enclosed or semi-enclosed a duty "to coordinate the implementation of their rights and duties with respect to the protection and preservation of the marine environment." The Tribunal did not make a distinction between Articles 123 and 197, tacitly accepting that such an obligation did exist, as did the Philippines and China.

The doubt concerning the existence of an obligation to cooperate under Article 123 arises from the formulation of the introductory text preceding the enumeration of the areas of cooperation. The introductory text stipulates that States bordering a semi-enclosed or enclosed sea "shall endeavour, directly or through an appropriate regional organization...." It seems that only Fleischer is of the view that Article 123 represents a legal obligation to cooperate.[253] Most scholarly commentators are of the view that the language of Article 123 is that of exhortation and not of a binding obligation.[254] Its drafting history indicates that

---

[252] Award on Jurisdiction, 1, para. 3; Award of 12 July 2016, 1, para. 3.

[253] Carl-August Fleischer, "La pêche [Fisheries]," in René-Jean Dupuy and Daniel Vignes (eds.), *Traité du Nouveau Droit de la mer* [Treatise on the New Law of the Sea] (Paris: Éditions Economica, 1985), 877.

[254] Erick Franckx, "Regional Marine Environment Protection Regimes in the Context of UNCLOS," *The International Journal of Marine and Coastal Law* 13 (1998): 315; Erik Franckx and Marco Benatar, "The 'Duty' to Co-operate for States Bordering Enclosed or Semi-Enclosed Seas," *Chinese (Taiwan) Yearbook of International Law and Affairs* 31 (2013): 69–70; Myron H. Nordquist et al. (eds.), *United Nations Convention on the Law of the Sea 1982 Commentary*, vol. III, *Second Committee: High Seas, Regime of Islands, Enclosed or Semi-Enclosed Seas, and Right of Access of Land-Locked States to and from the Sea and Freedom of Transit* (Leiden: Martinus Nijhoff Publishers, 1995), 39; Nordquist et al., vol. IV, 78; Nilufer Oral, *Legal Aspects of Sustainable Development. Regional Co-Operation and Protection of the Marine Environment Under International Law* (Leiden: Martinus Nijhoff Publishers, 2013), 40, 43; Budislav Vukas, "United Nations Convention on the Law of the Sea and the Polar Marine Environment," in Davor Vidas (ed.), *Protecting the Polar Marine Environment: Law and Policy for Pollution Prevention* (Cambridge: Cambridge University Press, 2000), 41; and Budislav Vukas, "Enclosed and Semi-Enclosed Seas," *Max Planck Encyclopedia of Public International Law* (Oxford: Oxford University Press, 2015), para. 15.

it was not intended to create an international legal obligation of regional cooperation.[255] The doubt as to the binding character of Article 123 is strengthened by the contrast between the language of Article 123 and that of Article 197.

Understandably, the difference in the formulation of the two articles has been described as confusing. After all, States have a general duty to cooperate at the global and regional levels.[256] Vukas interprets Article 123 to mean that a State bordering an enclosed or semi-enclosed sea violates Article 123 if it refuses to enter into meaningful negotiations requested by other States bordering the sea.[257] The most sensible approach seems to be that of Bernard Oxman, who cautions against placing too much emphasis on the use of the word "should" in place of "shall." In his view, where cooperation and agreement are required, it is not clear how much difference (if any) it makes whether one uses "should" or "shall."[258]

This probably represents the general attitude of those who have had to interpret the Convention provisions on cooperation. The Philippines pointed out during the Hearings on the Merits that the existence of a duty to cooperate, described as a fundamental principle in the preservation and the protection of the marine environment, had been reaffirmed on several occasions by the ITLOS.[259] In only one of these cases did a party specifically invoke Article 123, on which the ITLOS chose not

---

[255] Ingo Winkelmann, "Article 123. Cooperation of States Bordering Enclosed or Semi-Enclosed Seas," in Alexander Proelss (ed.), *The United Nations Convention on the Law of the Sea. A Commentary* (München: Verlag C.H. Beck oHG, 2017), 887.

[256] Oral, *Legal Aspects*, 39.

[257] Vukas, "United Nations Convention on the Law of the Sea," 47; Vukas, "Enclosed or Semi-Enclosed Seas," para. 19.

[258] Bernard Oxman, "Observations on the Interpretation and Application of Article 43 of UNCLOS With Particular Reference to the Straits of Malacca and Singapore," *Singapore Journal of International and Comparative Law* 2 (1998): 409,

https://repository.law.miami.edu/cgi/viewcontent.cgi?referer=https://www.google.com/&httpsredir=1&article=1415&context=fac_articles, accessed 25 February 2019. Oxman was a delegate of the US to the Third UNCLOS III, which drafted the Convention between 1973 and 1982, and one of the legal counsels of the Philippines in the *South China Sea Arbitration*.

[259] Hearing on the Merits, Transcript, Day 3, 40.

to make a pronouncement.[260] In the *South China Sea Arbitration*, the Arbitral Tribunal attached great importance to the fact that in the last two decades, the case law of the ICJ and of the ITLOS have all stressed the importance of the duty to cooperate.[261] The Tribunal also found it significant that the duty to cooperate is highlighted in the Preamble of the Convention, which declared that the States Parties to the Convention were "prompted by the desire to settle, in a spirit of mutual understanding and cooperation, all issues relating to the law of the sea." It seems therefore that the Tribunal considered that there existed an obligation to cooperate under both Articles 197 and 123.

It is interesting to note that the CSIL's critique of the Award of 12 July 2016 did not allude to a distinction between Article 197 and Article 123 on the basis of the binding force of the provision. The CSIL's criticism was directed at the Tribunal's findings that China's construction activities had breached Article 197 and that China had failed to promote cooperation under Article 123.

2. China's Compliance with Its Obligation to Cooperate
In their reasoning, the Philippines and the Tribunal relied on both Articles 197 and 123. The CSIL critique of the Award of 12 July 2016 was founded on a distinction relating to the scope of the obligations under the two articles.

In its arguments, the Philippines invoked both Articles 197 and 123. For the Philippines, Article 197 was valuable, in that it specified that cooperation at the regional level may take into account "characteristic regional features." According to the Philippines, in the South China Sea these "characteristic regional features" included "the fundamental biological and ecological importance and fragile nature of coral reef ecosystems." There was very little evidence of Chinese cooperation on environmental protection in the South China Sea. There were only cursory references to such cooperation in the 2002 Declaration on the Conduct of the

---

[260] *MOX Plant (Ireland v. United Kingdom), Provisional Measures, Order of 3 December 2001, ITLOS Reports 2001*, 100, para. 26(3),
https://www.itlos.org/fileadmin/itlos/documents/cases/case_no_10/published/C10-O-3_dec_01.pdf, accessed 5 April 2019; Oral, *Legal Aspects*, 40, 42.

[261] Hearing on the Merits, Transcript, Day 4, 40–41; Award of 12 July 2016, 394–95, para. 985.

Parties in the South China Sea (DOC).[262] The FAO Asia-Pacific Fisheries Commission had adopted no measures on conservation. No regional seas agreement for the South China Sea existed. The United Nations Environment Programme (UNEP) Regional Seas Programme for East Asia, in which both China and the Philippines participated, had adopted a 1994 Action Plan, covering the rehabilitation of vital ecosystems, such as mangroves, seagrasses, and coral reefs; the restoration of ecologically or economically important species and communities; the establishment of a viable network of Marine Protected Areas; and EIA. It seemed obvious to the Philippines that China's construction activities in the South China Sea did not correspond with these environmental priorities. In the disputed areas of the South China Sea, there were no Marine Protected Areas (MPAs) or areas designated as vulnerable marine ecosystems; there was no evidence of serious restraints on illegal fishing and little evidence of goodwill. China had done nothing to give effect to the obligation to cooperation in the protection and preservation of the marine environment. Indeed, China's behavior toward the Philippines and other States bordering the South China Sea had been aggressive. On the contrary, China had sought to exclude others, rather than to cooperate with them.[263]

The Tribunal, like the Philippines, found no convincing evidence that China had attempted to cooperate or coordinate with other States bordering the South China Sea: "China has not cooperated or coordinated with the other States bordering the South China Sea concerning the protection and preservation of the marine environment concerning such activities."[264] The rationale for its conclusion differed from that presented by the Philippines. For the Tribunal, what was significant was the fact that China's construction activities had been met with protest from the Philippines and other neighboring States.[265] As evidence of

---

[262] *MP*, Annex 144, Association of Southeast Asian Nations, Declaration on the Conduct of Parties in the South China Sea (4 November 2002), vol. V, 321–25, https://files.pca-cpa.org/pcadocs/The%20Philippines%27%20Memorial%20-%20Volume%20V%20%28Annexes%20103-157%29.pdf, accessed 16 April 2019. The DOC was signed by the 10 members of ASEAN and China. According to paragraph 6 of the DOC "[p]ending a comprehensive and durable settlement of the disputes, the Parties concerned may explore or undertake cooperative activities. These may include the following: a. marine environmental protection...."

[263] Hearing on the Merits, Transcript, Day 3, 40–44.

[264] Award of 12 July 2016, 475, para. 1202, (13)(b).

[265] Ibid., 394–95, paras. 984, 986.

the protests that China had disregarded, the Tribunal cited one of the Philippines' protests to China as well as the Joint Communiqué of the 48th ASEAN Foreign Ministers Meeting, held in Kuala Lumpur and published on 4 August 2015. The Foreign Ministers had

> discussed extensively the matters relating to the South China Sea and remained seriously concerned over recent and ongoing developments in the area. We took note of the serious concerns expressed by some Ministers on the land reclamations in the South China Sea, which have eroded trust and confidence, increased tensions and may undermine peace, security and stability in the South China Sea.[266]

We might add that there are several forums for dialogue between ASEAN and China. Following the ASEAN Foreign Ministers' Meeting is a dialogue with external partners, which include China. The 2015 ASEAN Foreign Ministers' Meeting was followed by the 16th ASEAN Plus Three Foreign Ministers' Meeting, China being one of the "Plus Three," and the 5th East Asia Summit Foreign Ministers' Meeting. If China had been serious about consultation and cooperation, it would have taken advantage of these opportunities to engage in a dialogue with the Philippines and other ASEAN Member States on its construction activities in the South China Sea. Heedless of the ASEAN Member States' protests, China had proceeded with the island-building. Other than this, the Tribunal did not provide guidance in the application of Articles 197 and 123, as Nilufer Oral has pointed out.[267]

The CSIL criticized the Award of 12 July 2016 for finding that China's construction activities had breached the provisions of the Convention concerning the obligation to cooperate and for disregarding China's efforts to promote cooperation on the protection and preservation of the marine environment in the South China Sea. It argued that Article 197 mandated cooperation in rule-making, "in formulating

---

[266] *Joint Communiqué 48th ASEAN Foreign Ministers Meeting Kuala Lumpur, Malaysia 4th August 2015*, 25, para. 150, https://www.asean.org/wp-content/uploads/images/2015/August/48th_amm/JOINT%20COMMUNIQUE%20OF%20THE%2048TH%20AMM-FINAL.pdf, accessed 22 March 2019. The ten members of ASEAN are Brunei Darussalam, Cambodia, Indonesia, Lao People's Democratic Republic, Malaysia, Myanmar, the Philippines, Singapore, Thailand, and Vietnam.

[267] Oral, "The South China Sea Arbitral Award," 242.

and elaborating international rules, standards and recommended practices and procedures." This formulation meant that China's construction activities did not fall within the matters regulated by Article 197.[268] The CSIL also argued that since the 1990s, China has actively championed regional cooperation on the protection and preservation of the marine environment among littoral States in the South China Sea. As evidence of this cooperation, the CSIL mentioned the signature of the DOC in 2002; a Chinese proposal in 2011 to establish a special technical committee on marine scientific research and environmental protection within the DOC framework; the establishment of a China-ASEAN Maritime Cooperation Fund to promote cooperation on, among others, marine environmental protection; the approval in 2012 by China's State Council of a Cooperation Framework Plan on the South China Sea and Surrounding Waters (2011–2015); and China's bilateral cooperation with littoral States, such as Indonesia, Cambodia, and Thailand.[269]

There is no disagreement that Article 197 refers to rule-making. Disagreement might arise over the content of "international rules, standards and recommended practices and procedures." It can be plausibly argued that the notion of "standards and recommended practices and procedures" encompasses the measures that China claimed to have taken (and in reality failed to take) in the course of its construction activities in the South China Sea, such as the timing of the works so as not to coincide with the spawning of certain species of fish, the setting up of trash screens, the minimization of the area of construction, the planning of construction on sites with dead corals, and so on. None of these measures was ever alluded to in response to the numerous protests made by the Philippines, all of which were summarily rejected by China. Once it is accepted that these measures fall within the category of standards, practices, and procedures, the CSIL's criticism relating to Article 197 loses credibility. Moreover, once we consider that the duty of cooperation referred to China's construction activities and not to other matters, the information provided to refute the Tribunal's interpretation of Article 123 becomes irrelevant.

The Tribunal observed that China's lack of coordination is "not unrelated to its lack of communication."[270] Underlying this observation is

---

[268] CSIL, 587–88, paras. 811–12.

[269] Ibid., 588–89, paras. 813–14.

[270] Award of 12 July 2016, 394, para. 986.

China's failure to communicate the results of its assessment of the environmental impact of its construction activities in the South China Sea.

C. The Protection and Preservation of the Marine Environment and the Obligation to Monitor and Assess

The most widespread instrument used to make decision-makers aware of the environmental consequences of proposed activities is the Environmental Impact Assessment ("EIA"). The Convention, though not a multilateral environmental agreement, was one of the first international instruments to impose an obligation on States to carry out an assessment of the impact of planned activities on the marine environment, under certain conditions. China's failure to carry out an EIA prior to the initiation of its construction activities would contradict any claim by China that its activities took into account environmental protection. The Tribunal, preferring to err on the side of caution, confined itself to the finding that China had failed to communicate the results of any such assessment.

1. The Obligation to Conduct an EIA and to Communicate Its Results

Underlying the obligation to conduct an EIA, which was first introduced in the United States in 1969, is the idea that decisions affecting the environment should be made in light of a comprehensive understanding of their effects.[271] The first international instruments incorporating EIA requirements seem to have been adopted at the regional level in the early 1980s.[272] It thus appears that the Convention is one of the first binding

---

[271] Neil Craik, *The International Law of Environmental Impact Assessment: Process, Substance and Integration* (Cambridge: Cambridge University Press, 2008), 4. The National Environmental Policy Act of 1969, as amended (Pub. L. 91-190, 42 U.S.C. 4321-4347, January 1, 1970, as amended by Pub. L. 94-52, July 3, 1975, Pub. L. 94-83, August 9, 1975, and Pub. L. 97-258, § 4(b), September 13, 1982), https://ceq.doe.gov/laws-regulations/laws.html, accessed 22 April 2019.

[272] These were the *Convention for Co-Operation in the Protection and Development of the Marine and Coastal Environment of the West and Central African Region*, done at Abidjan on 23 March 1981 and entered into force 5 August 1981, http://sedac.ciesin.columbia.edu/entri/texts/marine.coastal.west.central.africa.1981.html, accessed 21 April 2019, and the *Convention for the Protection of the Marine Environment and Coastal Areas of the South-East Pacific*, done at Lima on 12 November 1981 and entered into force 19 May 1986, http://sedac.ciesin.org/entri/texts/marine.environment.coastal.south.east.pacific.1981.html, accessed 21 April 2019; cited in Craik, 286–87.

multilateral instruments that contain provisions on EIA. This might be one circumstance that explains why the Convention requirements are relatively modest, compared to those of later instruments, and in particular, the Convention on Environmental Impact Assessment in a Transboundary Context ("Espoo Convention"), which was concluded in 1991 under the auspices of the UN Economic Commission for Europe.[273]

Under Article 206 of the Convention, the threshold for initiating an EIA is that a State has "reasonable grounds for believing that planned activities under ...[its] jurisdiction or control may cause substantial pollution of or significant and harmful changes to the marine environment."[274] If these conditions are fulfilled, the State party to the Convention shall carry out an assessment, "as far as practicable." The expressions "reasonable grounds for believing" and "as far as practicable" appear to leave room for subjective assessment. The scope for subjective assessment seems to be widened by the provision that it is the State carrying out the EIA that has the right to determine whether it is practicable to do so.[275] The concepts "substantial pollution of or significant and harmful changes to the marine environment" are left undefined. In contrast, under the Espoo Convention, the obligation to carry out an EIA is an absolute obligation. Article 2(3) of the Espoo Convention requires a State proposing to undertake an activity that is listed in its Appendix I and that is likely to cause a significant adverse transboundary impact to ensure that an EIA is undertaken prior to a decision to authorize or undertake the proposed activity. Appendix I lists 17 activities that are likely to cause significant adverse transboundary impact, including crude oil refineries, thermal power stations, and installations

---

[273]The fact that the Convention was concluded among a group of developed States at the regional level might also explain why the signatories were willing to adopt stricter requirements. *Convention on Environmental Impact Assessment in a Transboundary Context*, done at Espoo (Finland), 25 February 1991, entered into force on 10 September 1997, https://www.unece.org/fileadmin/DAM/env/eia/documents/legaltexts/Espoo_Convention_authentic_ENG.pdf, accessed 21 April 2019.

[274]Award of 12 July 2016, 395, para. 987.

[275]Lingjie Kong, "Environmental Impact Assessment under the United Nations Convention on the Law of the Sea," *Chinese Journal of International Law* 10 (2011): 659; Robin Churchill, "The LOS Regime for Protection of the Marine Environment— Fit for the Twenty-First Century?" in Rosemary Rayfuse (ed.), *Research Handbook of International Marine Environmental Law* (Cheltenham, Gloucestershire: Edward Elgar, 2015), 7.

for the production or enrichment of nuclear fuels, for the reprocessing of irradiated nuclear fuels or for the storage, disposal, and processing of radioactive waste. All of these activities could undoubtedly cause substantial pollution of or significant and harmful changes to the marine environment. Article 206 of the Convention provides for an EIA only for planned activities, while Article 2(7) of the Espoo Convention urges States to carry out EIAs, if possible, for policies, plans, and programs.

Article 206 of the Convention stipulates that a State Party to the Convention that has carried out the EIA must publish the results of the assessment according to the procedure laid down in Article 205, whereby States are required to publish the reports or "to provide such reports at appropriate intervals to the competent international organizations, which should make them available to all States." The obligation is an obligation of result, but the obligation may be somewhat weakened by the silence of Article 205 on the content of reports and on the absence of definition of competent international organizations. In effect, States are free to choose the way they wish to disseminate the results of an EIA.[276]

The Espoo Convention specifies the contents of the EIA in its Appendix II. The EIA must describe the proposed activity and its purposes or reasonable alternatives to the proposed activity, the environment likely to be significantly affected by the proposed activity and its alternatives, the potential environmental impact of the proposed activity and its alternatives, and mitigation measures; indicate environmental data, predictive methods, and underlying assumptions; identify gaps in knowledge and uncertainties encountered in compiling the required information; outline monitoring and management programs and any plans for post-project analysis; and present a non-technical summary.

The Espoo Convention's provisions go further than Articles 205 and 206 of the Convention in other respects. It recognizes the roles of actors other than the State initiating the EIA in the assessment process. Under Article 3(1) of the Espoo Convention, the State that proposes to undertake the activity must notify States that it considers may be affected by the activity. According to Article 3(7), if a State that may be significantly

---

[276]Eike Blitza, "Article 205. Publication of Reports," in Alexander Proelss (ed.), *The United Nations Convention on the Law of the Sea. A Commentary* (München: Verlag C.H. Beck oHG, 2017), 1367–68.

affected by adverse transboundary impact has not been notified of the activity, it may request information from the State proposing to undertake the activity with a view to holding discussions on the possible impact of the activity. Articles 2(6) and 3(8) stipulate that the public in the areas likely to be affected by the activity should participate in relevant environmental impact assessment procedures, should be informed of the activity and be provided with possibilities for making comments or objections on, the proposed activity.

The Espoo Convention also lays down rules applicable to the post-EIA phase. Once the EIA has been completed, the State proposing to undertake or authorize an activity must, under the terms of Article 5, enter into consultations with the affected Party concerning the potential transboundary impact of the proposed activity and measures to reduce or eliminate its impact. When a final decision is to be taken, the EIA must be taken into account, and the decision must be transmitted to the States affected by the activity, along with the reasons for the decision.

Whatever the lacunae of Articles 205 and 206 of the Convention, they nonetheless remain important. It is noteworthy that the two previous cases involving the interpretation and application of Part XII of the Convention that have been brought before the ITLOS and/or an arbitral tribunal stemmed from the alleged failure of the respondent State to conduct an EIA. In both cases, States argued about whether it took place, not whether it was compulsory.[277] The ITLOS has declared the duty to carry out an EIA to be a customary rule, as has the ICJ, putting an end to doctrinal debates on its nature.[278]

The Tribunal in the *South China Sea Arbitration* had to determine whether China undertook an EIA in circumstances that made the Tribunal's task much more difficult.

---

[277] Birnie, Boyle and Redgewell, 169. The two cases are *The MOX Plant Case (Ireland v. United Kingdom)*, *Provisional Measures, Order of 3 December 2001, ITLOS Reports 2001*, 110, para. 82, https://www.itlos.org/cases/list-of-cases/case-no-10/, accessed 23 April 2019, and *Case Concerning Land Reclamation by Singapore in and Around the Straits of Johor (Malaysia v. Singapore)*, *Provisional Measures, Order of 8 October 2003, ITLOS Reports 2003*, 25, para. 92, https://www.itlos.org/cases/list-of-cases/case-no-12/, 23 April 2019.

[278] Laura Pineschi, "The Duty of Environmental Impact Assessment in the First ITLOS Chamber Advisory Opinion: Toward the Supremacy of the General Rule to Protect and Preserve the Marine Environment as a Common Value?" in Nerina Boschiero et al. *International Courts and the Development of International Law. Essays in Honour of Tullio Treves* (The Hague: T.M.C. Asser Press, 2013), 427, 431.

2. China's Compliance with Its Obligation to Monitor and Assess

The Philippines argued that China was under an obligation to carry out an EIA yet it had failed to do so. The Tribunal agreed that the threshold for undertaking an EIA had been reached but since it was unable to determine with certainty whether China had indeed carried out an EIA, it limited itself to declaring that China had breached the obligation to communicate the results of an EIA.

The Philippines contended that China was "fairly and squarely" required to carry out an EIA.[279] In the absence of any indication in Article 206 of the contents of an EIA, the Philippines referred to the UN International Law Commission's commentary to its 2001 Articles on the Prevention of Transboundary Harm, which explained that the specific content of an EIA is left to domestic law but added that at an EIA must include at a minimum, an assessment of the possible effects of the activity on people and property and environment of other States.[280] In the view of the Philippines, an EIA should have assessed the possible effects of island-building on the South China Sea's marine ecosystem, the coral reefs, biodiversity and sustainability of the living resources, and the endangered species.[281] There was no evidence that China had carried out an EIA. No science-based evaluation had been made public to the Philippines or to the competent international organization, as required by Articles 205 and 206 of the Convention.[282]

During the Hearing on the Merits, the Philippines was asked to comment on the May 2015 statement by the Director General of Department of Boundary and Ocean Affairs of China's Ministry of Foreign Affairs,[283] which the Tribunal had found and made available to

[279] Hearing on the Merits, Transcript, Day 3, 34–36; Award of 12 July 2016, 362, para. 911.

[280] Hearing on the Merits, Transcript, Day 3, 38–39; Award of 12 July 2016, 362, para. 910. See United Nations International Law Commission, "International Liability for Injurious Consequences Arising Out of Acts Not Prohibited by International Law (Prevention of Transboundary Harm From Hazardous Activities)", in *Report of the International Law Commission on the Work of Its Fifty-Third Session (23 April–1 June and 2 July–10 August 2001)*, UN Doc. GAOR A/56/10 (2001), 405, http://www.un.org/documents/ga/docs/56/a5610.pdf, accessed 23 April 2019.

[281] Hearing on the Merits, Transcript, Day 3, 38–39; Award of 12 July 2016, 362, para. 911.

[282] Hearing on the Merits, Transcript, Day 4, 183.

[283] *The Philippines' Annexes Cited During the Merits Hearing*, Annex 820, 5–10.

it. The statement claimed, among others, that the construction projects had gone through science-based evaluation and assessment, with equal importance given to construction and environmental protection; that full account had been taken of ecological protection and fishery conservation; and that China would step up ecological monitoring of the reefs, waters, and islands. It was easy for the Philippines to point out that the statement was unaccompanied by any supporting evidence. [284] At the Hearing, the Philippines, which had been impelled by a Tribunal question to research Chinese statements, had found a statement published by China's SOA on 18 June 2015, but it was only 500 words long. Its length—or rather its brevity—prompted the Philippines to describe it as a pseudo-evaluation.[285] After the conclusion of the Hearings on the Merits, the Philippines, commenting on further Chinese documents that the Tribunal had found and made available to it, observed that SOA communiqués published in 2012, 2013, and 2014 focused largely on waters close to the Chinese mainland and Hainan and did not address the ecological conditions of the areas of the South China Sea where China had been carrying out island-building. None referred to risk assessment having been carried out to evaluate the possible impacts of the island-building or to evaluate the actual impacts.[286] Contrary to the

[284] Hearing on the Merits, Transcript, Day 4, 182–83.

[285] Ibid., Day 4, 185; *The Philippines' Annexes Cited During the Merits Hearing*, Annex 821, 11–16.

[286] *Responses of the Philippines to the Tribunal's 5 February 2016 Request for Comments*, 7–9, paras. 17–24. The following were the documents found by the Tribunal and transmitted to the Philippines for comments:

1. Annex 864, China State Oceanic Administration, The Guidance for the Assessment of Coastal Marine Ecosystem Health, Marine Industry Standards of the People's Republic of China, No. HY/T 087-2005 (2005);
2. Annex 865, China State Oceanic Administration, "2012 Communique on Marine Environment of China, Part 2: Marine Biodiversity and Ecological Conditions" (1 April 2013);
3. Annex 866, China State Oceanic Administration, Code of Practice for Marine Monitoring Technology, Part 5: Marine Ecology, Marine Industry Standards of the People's Republic of China, No. HY/T 147.5-2013 (25 April 2013);
4. Annex 867, China State Oceanic Administration, "2013 Communique on Marine Environment of China, Part 2: Conditions of Marine Ecology" (25 March 2014);

assertions of a 2500-word SOA report published on 10 June 2015,[287] there was no evidence of the existence of scientific studies by experts in civil engineering, marine engineering, marine ecological, environmental protection and hydrology.[288] It comes as no surprise that the Philippines concluded that China was in breach of Article 206.[289]

Even assuming that an EIA had been conducted, the Philippines argued that China was under an obligation to communicate it to the Philippines or to the "competent international organisations," as required by Articles 205 and 206 of the Convention. The Philippines believed that China's failure to communicate an EIA was conclusive evidence of the violation of the two articles. On other matters raised in the arbitration, China had not been slow to publicize its views, through various means. If there had been an evaluation, China could easily have made it available to the Tribunal directly or in some other way, even without appearing before it. The question therefore was why China had not produced an evaluation.[290] It could only mean that China had not carried out an EIA.

The Tribunal identified three Articles that were relevant for the determination of China's compliance with the Convention: Articles 204, 205, and 206.[291] Article 204, which stipulates that "States shall, endeavor,

5. Annex 868, China State Oceanic Administration, Technical Guidelines for Environmental Impact Assessment of Marine Engineering, National Standards of the People's Republic of China, No. GB/T 19485-2014 (1 April 2014);
6. Annex 870, China State Oceanic Administration, "2014 Communique on Marine Environment of China, Part 2: Conditions of Marine Ecology" (16 March 2015);
7. Annex 871, China State Oceanic Administration, South China Sea Branch, "Communique on the Oceanic Conditions of the South China Sea Region in 2014" (28 May 2015);
8. Annex 872, China State Oceanic Administration, "Construction Activities at Nansha Reefs Did Not Affect the Coral Reef Ecosystem" (10 June 2015).
    All of these documents were published in https://files.pca-cpa.org/pcadocs/ The%20Philippines%27%20Written%20Responses%20%2811%20March%20 2016%29%20%28Annexes%20864-892%29.pdf, accessed 19 April 2019.

[287] *The Philippines' Written Responses (Annexes 864–892)*, Annex 872.

[288] *Responses of the Philippines to the Tribunal's 5 February 2016 Request for Comments.*, 14–15, para. 36.

[289] Hearing on the Merits, Transcript, Day 3, 35–39.

[290] Ibid., Transcript, Day 4, 183; Award of 12 July 2016, 363, para. 911.

[291] Award of 12 July 2016, 377, para. 947.

as far as practicable, to observe, measure, evaluate and analyse the risks or effects of pollution of the marine environment," had not been raised by the Philippines, as pointed out by Nilufer Oral; in so doing the Tribunal is said to have expanded the application of Article 204 and its monitoring obligation beyond cases of pollution.[292] In this particular instance, the reference to Article 204 did not seem to have had much direct impact on the Tribunal's reasoning and was not repeated in the Tribunal's findings. The Tribunal explained further that Article 206 was particularly important for ensuring that planned activities with potentially damaging effects may be effectively controlled and Article 205 for ensuring that other States are kept informed of their essential risks. The Tribunal quoted an Advisory Opinion of the ITLO that had declared that the obligation to conduct an EIA is a direct obligation under the Convention and a general obligation under customary international law.[293]

In the Tribunal's view, the threshold for initiating an EIA had been met by China's island-building in the South China Sea. As the Tribunal put it, the scale and impact of the island-building were such that "China could not have reasonably held any belief other than that the constructions 'may cause significant and harmful changes to the marine environment'."[294] China was undoubtedly required to prepare an EIA and to communicate the assessment results.

The Tribunal acknowledged that Chinese officials and institutions had made statements that could have been interpreted to mean that China had carried out an EIA. Typical of such statements was one made by a Foreign Ministry spokesperson in April 2015 regarding Mischief Reef, claiming that China's construction projects on islands and reefs had gone through scientific assessment and rigorous tests.[295] A 500-word SOA statement published on 18 June 2015 and located by the Philippines between the two rounds of the Hearing on the Merits claimed that in order to ascertain the effects of the construction work on oceanic systems, scientific studies had been conducted by experts in civil engineering, marine engineering, marine ecology, environmental protection,

---

[292] Oral, "The South China Sea Arbitral Award," 239.

[293] Award of 12 July 2016, 377, para. 947.

[294] Ibid., 395, para. 988.

[295] Ibid., 365, para. 917.

and hydrogeology. Many protection measures were said to have been adopted in the planning, design, and construction stages.[296] Yet simple assertions that an EIA existed "did not equate to having 'adduced any evidence that it actually carried out such a preliminary assessment'," as both the Philippines and the Tribunal recalled.[297]

The Tribunal did not find credible "China's repeated assertions by officials at different levels, that it has undertaken thorough environmental studies," for the simple reason that neither the Tribunal, the Tribunal-appointed experts, the Philippines, nor the Philippines' experts, were able to identify any report that would meet the requirements of Article 206 or even the requirements of China's own EIA Law, adopted in 2002.[298] The reference to domestic law is significant, in view of the absence of any indication in Article 206 of the contents of an EIA and of the position of the UN International Law Commission that the specific content of an EIA is left to domestic law. The Tribunal went out of its way to demonstrate that China's EIA law set very high standards, going beyond those of the Convention itself. China's EIA law requires that the EIA should be "objective, open and impartial, comprehensively consider impacts on various environmental factors and the ecosystem they form after the implementation of the plan or construction project, and thus provide scientific basis for the decision-making."[299] It is striking that Article 5 of the Law appeared to encourage widespread participation in EIAs: "the state shall encourage all relevant units, experts and the public to participate in the EIA in proper ways." The Law devoted an entire Chapter, Chapter 3, to construction projects, the EIA for which must include "analysis, projection and evaluation on the potential environmental impacts of the construction project" and "Suggestions

---

[296] Ibid., 366, para. 922.

[297] Ibid., 395, para. 989. The passage quoted is from *Certain Activities Carried Out by Nicaragua in the Border Area (Costa Rica v. Nicaragua) and Construction of a Road in Costa Rica Along the San Juan River (Nicaragua v. Costa Rica), Merits Judgment, I.C.J. Reports 2015*, 57, para. 154, https://www.icj-cij.org/files/case-related/152/152-20151216-JUD-01-00-BI.pdf, accessed 23 April 2019.

[298] Award of 12 July 2016, 396, para. 989.

[299] *SDP*, Annex 615, People's Republic of China, *Law of the People's Republic of China on Evaluation of Environmental Effects* (28 October 2002), Article 4, 114–23, https://files.pca-cpa.org/pcadocs/The%20Philippines%27%20Supplemental%20Documents%20-%20Volume%20I%20%28Annexes%20607-667%29.pdf, accessed 9 April 2019; Award of 12 July 2016, 396, para. 990.

on implementation of environmental monitoring for the construction project."[300] One commentator has found fault with the Tribunal for not resorting to "the international law of environmental impact assessment."[301] Quite apart from the fact that there is no multilateral treaty on EIAs with universal scope,[302] the quotations from China's law leave the observer with the impression that the high standards set in China's Law would have rendered the resort to any putative international standards superfluous.

As mentioned in Chapter 3, China's national reports on the implementation of the CBD claimed that since 1979, China had been conducting EIAs. Marine engineering projects were said to be under an obligation to conduct an EIA, which should take full consideration of the negative impact of construction project on biodiversity and formulate measures to minimize such impact.[303] In 2005, China duly noted the adoption of the 2002 law. The Law was said to prescribe "that analysis, forecast and assessment should be performed for the environmental impacts, including impacts upon biodiversity, that may arise out of the planning and construction projects, and countermeasures and measures to prevent or ease the negative environmental impacts."[304]

Whatever may have been China's record in carrying out EIA prior to 2013, the Tribunal observed that the SOA Report published on 10 June 2015 and the SOA Statement published on 18 June 2015 fell far short of the criteria set by Chinese law itself and were far less comprehensive than EIAs reviewed by other international courts and tribunals or

[300] Award of 12 July 2016, 396, para. 990.

[301] Ilias Plakokefalos, "Environmental Law Aspects of the South China Sea Arbitration Award," Symposium on the South China Sea Award, Netherlands Institute for the Law of the Sea (NILOS), Utrecht Centre for Oceans Water and Sustainability Law, School of Law Utrecht University, 7 December 2016, 16, https://papers.ssrn.com/sol3/papers.cfm?abstract_id=2880624, accessed 23 April 2019.

[302] Pineschi, 425.

[303] People's Republic of China, State Environmental Protection Administration of China, *China's 2nd National Report on Implementation of the Convention on Biological Diversity* (Beijing: China Environmental Science Press, 2001), 55, 78, 121, https://www.cbd.int/reports, accessed 23 April 2019.

[304] Ibid., *China's Third National Report on Implementation of the Convention on Biological Diversity* (Beijing: State Environmental Protection Administration of China, 2005), 136, https://www.cbd.int/reports, accessed 23 April 2019.

the EIAs filed in foreign construction projects to which the SOA scientists referred.[305] The Tribunal's allusion to the existence of international standards as regards the content of EIAs must be a first step in filling the lacunae of Article 206.[306] The absence of definition of comprehensiveness, to which one commentator drew attention,[307] is perhaps not a major lacuna in the Award, given the extreme brevity of the SOA publications in question—500 words for the Statement and 2500 words for the Report.

With all of the circumstances pointing to the conclusion that China had not undertaken an EIA, the Tribunal still preferred to be cautious and refrained from making a definitive finding that China had (or had not) carried out an EIA. In its view, a finding that China had or had not prepared an EIA was not necessary for it to find a breach of the obligation to communicate the results of the EIA under Article 206.[308] The latter required not only that a State prepare an EIA but also that it communicates it to the competent international organization. It is worth recalling at this point that none of the Chinese statements and publications that were examined by the Tribunal had been transmitted by China to the Tribunal. The SOA Statement published on 18 June 2015 had been found by the Philippines between the two rounds of the Hearing on the Merits.[309] All the other Chinese documents had been retrieved as a result of research by the Tribunal itself as the proceedings were ongoing. Leaving no stone unturned, the Tribunal wrote directly to China in February and March 2016, three months after the conclusion of the Hearing on the Merits, to ask whether it had conducted an EIA. If it had done so, the Tribunal requested that China provide a copy to it. China did not provide any. The Tribunal agreed with the Philippines that even if China did not participate in the proceedings, it had found "occasions and means to communicate statements by its own officials, or by others

[305] Award of 12 July 2016, 396, para. 990.

[306] Chie Kojima, "South China Sea Arbitration and the Protection of the Marine Environment: Evolution of UNCLOS Part XII Through Interpretation and the Duty to Cooperate," *Asian Yearbook of International Law* 21 (2015): 177.

[307] Mbengue, 287.

[308] Nilufer Oral sees this move of the Tribunal's as a "clever twist," rather than as a sign of prudence. Oral, "The South China Sea Arbitral Award," 240.

[309] Hearing on the Merits, Transcript, Day 4, 181–82.

writing in line with China's interests." The Tribunal agreed with the Philippines that "had it (China) wished to draw attention to the existence and content of an EIA, the Tribunal has no doubt it could have done so."[310]

The Tribunal was of the view that the formulation of Article 206, particularly the word "reasonable" and the phrase "as far as practicable," left a State a degree of discretion in deciding whether to undertake an EIA or not. On the contrary, the obligation to communicate the reports of the results of EIAs was absolute.[311] Even assuming for the sake of argument that China had prepared an EIA, it had breached Article 206 by failing to communicate the results of the assessment: "China has failed to communicate an assessment of the potential effects of such activities on the marine environment, within the meaning of Article 206 of the Convention."[312] The Tribunal's focus on the fulfillment of a procedural requirement, according to Tanaka, is useful in circumstances when it may be difficult to determine whether a State has actually carried out an EIA.[313]

Nearly two years after the issuance of the Award on the merits, the CSIL confirmed that China had not prepared an EIA.[314] The belated admission was not tantamount to an avowal of China's breach of its obligations under the Convention, for the CSIL contended that the Tribunal had disregarded the fact that China had conducted real-time monitoring of the construction and its environmental impacts as well as the regular scientific assessments of monitoring. Such assessments allegedly showed that the construction would not cause "substantial pollution of or significant and harmful changes to the marine environment."[315] In the CSIL's view, China had reasonable grounds to believe that the construction would not cause "substantial pollution of or significant and harmful changes to the marine environment."[316] The grounds for China's

---

[310] Award of 12 July 2016, 397, para. 991.

[311] Ibid., 378, para. 948.

[312] Ibid., 475, para. 1202, (13)(c).

[313] Yoshifumi Tanaka, "The South China Sea Arbitration: Environmental Obligations Under the Law of the Sea Convention," *Review of European, Comparative and International Environmental Law* 27 (2018): 96.

[314] CSIL, 592, para. 820.

[315] Ibid., 590, para. 817.

[316] Ibid., 591, para. 817.

conclusion were to be found in the statements of the Foreign Ministry spokesperson on 9 and 28 April 2015, the SOA Report of 10 June 2015, and the SOA Statement of 18 June 2015.[317] The CSIL alleged that the Tribunal disregarded China's discretion in determining whether an EIA is required or not. Since China had no reason to conduct an EIA—and in fact it did not conduct any—there was no result to communicate. The CSIL concluded that the Tribunal had erred in finding that China had breached the obligation to communicate the results of an assessment.[318]

One cannot help but be amazed by the fact that two years after the issuance of the Award on the merits, the CSIL could find no other supporting evidence for the conclusion that China's construction activities would not harm the marine environment than two statements by a spokesperson of the Ministry of Foreign Affairs, the 500-word Statement published by the SOA on 18 June 2015, and the 2500-word Report published by the SOA on 10 June 2015, all of which had already been produced, scrutinized, and found wanting during the arbitral proceedings. The first two statements provided no evidence at all that any scientific studies had been conducted, while the last two statements, assuming they are not to be considered as EIAs but as pre-EIA assessments for the purpose of determining whether a full-fledged EIA was required, had been analyzed in detail by the Philippines, the experts appointed by the Philippines, and the independent experts appointed by the Tribunal. All had concluded that China's construction activities would cause harm to the marine environment, either because there was no evidence that China would take or had taken the measures that it claimed it would take or because the evidence actually contradicted China's claim that measures had been taken. The assertion that China had "reasonable grounds" for believing that its construction activities would not cause "substantial pollution of or significant and harmful changes to the marine environment" beggars belief.

This rebuttal of the CSIL's clumsy attempt to defend China closes the sad chapter on the harm to endangered species and fragile marine ecosystems caused by China's toleration of the illegal harvesting of endangered species by its nationals and by its construction activities in the South China Sea.

---

[317] Ibid., 591–92, para. 820.
[318] Ibid., 591, para. 819.

REFERENCES

Birnie, Patricia, Alan Boyle, and Catherine Redgewell. *International Law and the Environment*, 3rd ed. Oxford: Oxford University Press, 2009.

Blitza, Eike. "Article 205. Publication of Reports." *The United Nations Convention on the Law of the Sea. A Commentary*, 1364–69. Eds. Alexander Proelss et al. München: Verlag C.H. Beck oHG, 2017.

———. "Article 206. Assessment of Potential Effects of Activities." *The United Nations Convention on the Law of the Sea. A Commentary*, 1369–78 Eds. Alexander Proelss et al. München: Verlag C.H. Beck oHG, 2017.

Cambridge English Dictionary. Cambridge: Cambridge University Press, 2019, https://dictionary.cambridge.org/dictionary/, accessed 18 April 2019.

Case Concerning Land Reclamation by Singapore In and Around the Straits of Johor (Malaysia v. Singapore), Provisional Measures, Order of 8 October 2003, ITLOS Reports 2003, https://www.itlos.org/cases/list-of-cases/case-no-12/, 23 April 2019.

Cassese, Antonio. *International Law*, 2nd ed. Oxford: Oxford University Press, 2005.

Certain Activities Carried Out by Nicaragua in the Border Area (Costa Rica v. Nicaragua) and Construction of a Road in Costa Rica Along the San Juan River (Nicaragua v. Costa Rica), Merits Judgment, I.C.J. Reports 2015, 665, https://www.icj-cij.org/files/case-related/152/152-20151216-JUD-01-00-BI.pdf, accessed 23 April 2019.

Cesar, Herman S.J. "Coral Reefs: Their Threats, Functions and Economic Value." *Collected Essays on the Economics of Coral Reefs*, 14–39. Ed. Herman S.J. Cesar. Kalmar: Linnaeus University, 2002, http://www.reefbase.org/resource_center/publication/pub_12370.aspx, accessed 31 March 2019.

Chagos Marine Protected Area Arbitration (Mauritius v. United Kingdom), Award of 18 March 2015, https://files.pca-cpa.org/pcadocs/MU-UK%20 20150318%20Award.pdf, accessed 5 April 2019.

Chinese Society of International Law. "The South China Sea Arbitration Awards: A Critical Study." *Chinese Journal of International Law* 17 (2018): 207–748.

Churchill, Robin. "The LOS Regime for Protection of the Marine Environment—Fit for the Twenty-First Century?" *Research Handbook for International Marine Environmental Law*, 3–30. Ed. Rosemary Rayfuse. Cheltenham, Gloucestershire: Edward Elgar, 2015.

Collins English Dictionary. Glasgow: HarperCollins Publishers, 2019, https://www.collinsdictionary.com/dictionary/english/, accessed 18 April 2019.

Combacau, Jean. "Obligation de résultat et obligation de comportement: Quelques questions et pas de réponse [Obligation of Result and Obligation of Conduct: Some Questions and No Response]." *Mélanges offerts à Paul Reuter. Le droit international: Unité et diversité* [Essays in Honor of Paul

Reuter. International Law: Unity and Diversity], 181–204. Eds. Daniel Bardonnet et al. Paris: Editions A. Pedone, 1981.

Convention for Co-Operation in the Protection and Development of the Marine and Coastal Environment of the West and Central African Region, done at Abidjan on 23 March 1981 and entered into force 5 August 1981, http://sedac. ciesin.columbia.edu/entri/texts/marine.coastal.west.central.africa.1981.html, accessed 21 April 2019.

Convention for the Protection of the Marine Environment and Coastal Areas of the South-East Pacific, done at Lima on 12 November 1981 and entered into force 19 May 1986, http://sedac.ciesin.org/entri/texts/marine.environment.coastal.south.east.pacific.1981.html, accessed 21 April 2019.

Convention on Biological Diversity, signed at Rio de Janeiro on 5 June 1992, entered into force on 29 December 1993, https://www.cbd.int/doc/legal/cbd-en.pdf, accessed 26 March 2019.

———. Conference of the Parties (COP) 9. Decision IX/20. Marine and Coastal Biodiversity. Doc. UNEP/CBD/COP/DEC/IX/20 (9 October 2008), https://www.cbd.int/doc/decisions/cop-09/cop-09-dec-20-en.pdf, accessed 6 May 2019.

———. COP 10. Decision X/2. X/2.Strategic Plan for Biodiversity 2011–2020 and the Aichi Biodiversity Targets. Doc. UNEP/CBD/COP/DEC/X/2 (29 October 2010), https://www.cbd.int/doc/decisions/cop-10/cop-10-dec-02-en.pdf, accessed 6 May 2019.

Convention on Environmental Impact Assessment in a Transboundary Context, done at Espoo (Finland) 25 February 1991 and entered into force on 10 September 1997, https://www.unece.org/fileadmin/DAM/env/eia/documents/legaltexts/Espoo_Convention_authentic_ENG.pdf, accessed 21 April 2019.

Convention on International Trade in Endangered Species of Wild Fauna and Flora, signed at Washington, DC, on 3 March 1973, amended at Bonn, on 22 June 1979, amended at Gaborone, on 30 April 1983, https://www.cites.org/eng/disc/text.php, accessed 26 March 2019.

Craik, Neil. *The International Law of Environmental Impact Assessment: Process, Substance and Integration.* Cambridge: Cambridge University Press, 2008.

Czybulka, Detlef. "Article 192. General Obligation." *The United Nations Convention on the Law of the Sea. A Commentary*, 1277–87. Ed, Alexander Proelss. München: Verlag C.H. Beck oHG, 2017.

Dupuy, Pierre-Marie. "Reviewing the Difficulties of Codification: On Ago's Classification of Obligations of Means and Obligations of Result in Relation to State Responsibility." *European Journal of International Law* 10 (1999): 371–85.

——— and Martine Rémond-Guilloud. "La préservation du milieu marin [The Preservation of the Marine Environment]." *Traité du Nouveau Droit de la mer* [Treatise on the New Law of the Sea], 819–956. Eds. René-Jean Dupuy and Daniel Vignes. Paris: Éditions Economica, 1985.

Fitzmaurice, Malgosia A. "International Protection of the Environment." *Recueil des Cours de l'Académie de Droit International de La Haye* [Collected Courses of the Hague Academy of International Law]. Vol. 293 (2001-VI), 9–488.

Fleischer, Carl-August. "La pêche [Fisheries]." *Traité du Nouveau Droit de la mer* [Treatise on the New Law of the Sea], 819–956. Eds. René-Jean Dupuy and Daniel Vignes. Paris: Éditions Economica, 1985.

Franckx, Erick. "Regional Marine Environment Protection Regimes in the Context of UNCLOS." *The International Journal of Marine and Coastal Law* 13 (1998): 307–24.

———— and Marco Benatar. "The 'Duty' to Co-Operate for States Bordering Enclosed or Semi-Enclosed Seas." *Chinese (Taiwan) Yearbook of International Law and Affairs* 31 (2013): 66–81.

French, Duncan, and Tim Stephens. *ILA Study Group on Due Diligence in International Law. First Report.* 7 March 2014, https://olympereseauinternational.files.wordpress.com/2015/07/due_diligence_-_first_report_2014.pdf, accessed 11 April 2019.

Indus Waters Kishenganga Arbitration (Pakistan v. India). Partial Award, 18 February 2013, https://pcacases.com/web/sendAttach/1681, accessed 17 April 2019.

International Union for Conservation of Nature. *About.* Gland: IUCN, 2019, https://www.iucn.org/about, accessed 12 April 2019.

Iron Rhine Arbitration (Belgium/Netherlands), Award of 24 May 2005, https://pcacases.com/web/sendAttach/478, accessed 17 April 2019.

Jiménez de Aréchaga, Eduardo. "International Responsibility." *Manual of Public International Law*, 531–603. Ed. Max Sørensen. London: Macmillan, 1968.

———— and Atilla Tanzi. "International State Responsibility." *International Law: Achievements and Prospects*, 347–80. Ed. Mohammed Bedjaoui. Paris: UNESCO, 1991.

Kojima, Chie. "South China Sea Arbitration and the Protection of the Marine Environment: Evolution of UNCLOS Part XII Through Interpretation and the Duty to Cooperate." *Asian Yearbook of International Law* 21 (2015): 166–80.

König, Doris. "The Elaboration of Due Diligence Obligations as a Mechanism to Ensure Compliance with International Legal Obligations by Private Actors." *La contribution du Tribunal international du droit de la mer à l'état de droit: 1996–2016* [The Contribution of the International Tribunal for the Law of the Sea to the Rule of Law: 1996–2016], 83–95. Leiden: Martinus Nijhoff Publishers, 2017.

Kong, Lingjie. "Environmental Impact Assessment Under the United Nations Convention on the Law of the Sea." *Chinese Journal of International Law* 10 (2011): 651–59.

Larousse Dictionary of Science and Technology. Edinburgh: Larousse plc, 1995.

London Convention on the Prevention of Marine Pollution by Dumping of Wastes and Other Matter, done at London on 13 November 1972 and entered into force on 30 August 1975, http://www.imo.org/en/OurWork/Environment/LCLP/Documents/LC1972.pdf, accessed 6 May 2019.

McManus, John W. "Offshore Coral Reef Damage, Overfishing and Paths to Peace in the South China Sea." *The International Journal of Marine and Coastal Law* 32 (2017): 199–237.

The MOX Plant Case (Ireland v. United Kingdom), Provisional Measures, Order of 3 December 2001, ITLOS Reports 2001, 95, https://www.itlos.org/cases/list-of-cases/case-no-10/, accessed 23 April 2019.

Nordquist, Myron H., et al. (eds.). *United Nations Convention on the Law of the Sea 1982 Commentary*. Vol. III. *Second Committee: High Seas, Regime of Islands, Enclosed or Semi-Enclosed Seas, and Right of Access of Land-Locked States to and from the Sea and Freedom of Transit*. Leiden: Martinus Nijhoff Publishers, 1995.

———. *United Nations Convention on the Law of the Sea 1982 Commentary*. Vol. IV. *Third Committee: Protection and Preservation of the Marine Environment, Marine Scientific Research, and Development and Transfer of Marine Technology*. Leiden: Martinus Nijhoff Publishers, 1991.

Oliveira, Thiago Braz Jardim. "La diligence due dans la prévention des dommages à l'environnement [Due Diligence in the Prevention of Damage to the Environment]." *Anuário Brasileiro de Direito Internacional/Annuaire Brésilien de Droit International* [Brazilian Yearbook of International Law] 7 (2012): 205–42, https://papers.ssrn.com/sol3/papers.cfm?abstract_id=2253408, accessed 11 April 2019.

Oral, Nilufer. "Implementing Part XII of the 1982 UN Law of the Sea Convention and the Role of International Courts." *International Courts and the Development of International Law. Essays in Honour of Tullio Treves*, 403–24. Eds. Nerina Boschiero et al. The Hague: T.M.C. Asser Press, 2013.

———. *Legal Aspects of Sustainable Development. Regional Co-Operation and Protection of the Marine Environment Under International Law*. Leiden: Martinus Nijhoff Publishers, 2013.

———. "The South China Sea Arbitral Award, Part XII of UNCLOS and the Protection and Preservation of the Marine Environment." *The South China Sea Arbitration: The Legal Dimension*, 223–46. Eds. S. Jayakumar et al. Cheltenham, Gloucestershire: Edward Elgar, 2018.

Ouedraogo, Awalou. "La due diligence en droit international: de la règle de la neutralité au principe général [Due Diligence in International Law: From the Rule of Neutrality to the General Principle]." *Revue générale de droit* [General Journal of Law] 42 (2012): 642–83, https://www.erudit.org/fr/revues/rgd/2012-v42-n2-rgd01542/1026909ar.pdf, accessed 11 April 2019.

Oxman, Bernard. "Observations on the Interpretation and Application of Article 43 of UNCLOS With Particular Reference to the Straits of Malacca and Singapore." *Singapore Journal of International and Comparative Law* 2 (1998): 408–26, https://repository.law.miami.edu/cgi/viewcontent.cgi?referer=https://www.google.com/&httpsredir=1&article=1415&context=⤢ fac_articles, accessed 25 February 2019.

People's Republic of China. State Environmental Protection Administration of China. *China's 2nd National Report on Implementation of the Convention on Biological Diversity.* Beijing: China Environmental Science Press, 2001, https://www.cbd.int/reports, accessed 23 April 2019.

———. *China's Third National Report on Implementation of the Convention on Biological Diversity.* Beijing: State Environmental Protection Administration of China, 2005, https://www.cbd.int/reports, accessed 23 April 2019.

Permanent Court of Arbitration. *Indus Waters Kishenganga Arbitration (Pakistan v. India). Court of Arbitration Issues Partial Award in First Arbitration Under the Indus Waters Treaty 1960.* Press Release. PCA Doc. 87220 (19 February 2013), https://pcacases.com/web/sendAttach/1684, accessed 17 April 2019.

Pineschi, Laura. "The Duty of Environmental Impact Assessment in the First ITLOS Chamber Advisory Opinion: Toward the Supremacy of the General Rule to Protect and Preserve the Marine Environment as a Common Value?" *International Courts and the Development of International law. Essays in Honour of Tullio Treves,* 425–40. Eds. Nerina Boschiero et al. The Hague: T.M.C. Asser Press, 2013.

Plakokefalos, Ilias. "Environmental Law Aspects of the South China Sea Arbitration Award." Symposium on the South China Sea Award, Netherlands Institute for the Law of the Sea (NILOS), Utrecht Centre for Oceans Water and Sustainability Law, School of Law Utrecht University (7 December 2016), https://papers.ssrn.com/sol3/papers.cfm?abstract_id=2880624, accessed 23 April 2019.

Protocol to the Convention on the Prevention of Marine Pollution by Dumping of Wastes and Other Matter, 1972, done at London, 2 November 1996 (as amended in 2006), http://www.imo.org/en/OurWork/Environment/LCLP/Documents/PROTOCOLAmended2006.pdf, accessed 11 April 2019.

Risk, Michael J., and Evan Edinger. "Impacts of Sediment on Coral Reefs." *Encyclopedia of Modern Coral Reefs: Structure, Form and Process,* 575–86. Ed. David Hopley. Dordrecht: Springer, 2011.

South China Sea Arbitration. Award of 12 July 2016, https://pcacases.com/web/sendAttach/2086, accessed 26 March 2019.

———. Award on Jurisdiction and Admissibility, 29 October 2015, https://pcacases.com/web/sendAttach/1506, accessed 26 March 2019.

————. Hearing on Jurisdiction and Admissibility. Transcript, Day 2 (8 July 2015), https://pcacases.com/web/sendAttach/1400, accessed 3 April 2019.

————. Hearing on the Merits and Remaining Issues of Admissibility. Transcript. Day 3 (26 November 2015), https://pcacases.com/web/sendAttach/1549, accessed 26 March 2019.

————. Transcript, Day 4 (30 November 2015), https://pcacases.com/web/sendAttach/1550, accessed 10 April 2019.

————. Independent Expert Report. Assessment of the Potential Environmental Consequences of Construction Activities on Seven Reefs in the Spratly Islands in the South China Sea, by Sebastian C.A. Ferse, Peter Mumby, and Selina Ward, 26 April 2016, https://pcacases.com/web/sendAttach/1809, accessed 24 March 2019.

————. *Memorial of the Philippines* (30 March 2014). Vol. I, https://files.pca-cpa.org/pcadocs/Memorial%20of%20the%20Philippines%20Volume%20I.pdf, accessed 26 March 2019.

————. Annex 16. Memorandum to the Assistant Secretary, Office of Asian and Pacific Affairs, Department of Foreign Affairs, Republic of the Philippines (23 March 1992). Vol. III, 201–204, https://files.pca-cpa.org/pcadocs/The%20Philippines%27%20Memorial%20-%20Volume%20III%20%28Annexes%201-60%29.pdf, accessed 26 March 2019.

————. Annex 19. Memorandum from Erlinda F. Basilio, Acting Assistant Secretary, Office of Asian and Pacific Affairs, Department of Foreign Affairs, Republic of the Philippines, to the Secretary of Foreign Affairs of the Republic of the Philippines (29 March 1995). Vol. III, 213–17, https://files.pca-cpa.org/pcadocs/The%20Philippines%27%20Memorial%20-%20Volume%20III%20%28Annexes%201-60%29.pdf, accessed 26 March 2019.

————. Annex 20. Memorandum from Lauro L. Baja, Jr., Assistant Secretary, Office of Asian and Pacific Affairs, Department of Foreign Affairs, Republic of the Philippines, to the Secretary of Foreign Affairs of the Republic of the Philippines (7 April 1995). Vol. III, 219–24, https://files.pca-cpa.org/pcadocs/The%20Philippines%27%20Memorial%20-%20Volume%20III%20%28Annexes%201-60%29.pdf, accessed 26 March 2019.

————. Annex 21. Memorandum from the Ambassador of the Republic of the Philippines in Beijing to the Undersecretary of Foreign Affairs of the Republic of the Philippines (10 April 1995). Vol. III, 225–31, https://files.pca-cpa.org/pcadocs/The%20Philippines%27%20Memorial%20-%20Volume%20III%20%28Annexes%201-60%29.pdf, accessed 26 March 2019.

————. Annex 29. Memorandum from the Assistant Secretary for Asian and Pacific Affairs, Department of Foreign Affairs, Republic of the Philippines, to the Secretary of Foreign Affairs, Republic of the Philippines (23 March 1998). Vol. III, 277–82, https://files.pca-cpa.org/pcadocs/The%20Philippines%27%20Memorial%20-%20Volume%20III%20%28Annexes%201-60%29.pdf, accessed 26 March 2019.

————. Annex 41. Situation Report the Philippine Navy to the Chief of Staff, Armed Forces of the Philippines, No. 004-18074 (18 April 2000). Vol. III, 341–44, https://files.pca-cpa.org/pcadocs/The%20Philippines%27%20 Memorial%20-%20Volume%20III%20%28Annexes%201-60%29.pdf, accessed 26 March 2019.

————. Annex 42. Letter from Vice Admiral, Armed Forces of the Philippines, to Secretary of National Defense of the Republic of the Philippines (27 May 2000). Vol. III, 345–50, https://files.pca-cpa.org/pcadocs/The%20 Philippines%27%20Memorial%20-%20Volume%20III%20%28Annexes%20 1-60%29.pdf, accessed 26 March 2019.

————. Annex 43. Memorandum from the Embassy of the Republic of the Philippines in Beijing to the Secretary of Foreign Affairs of the Republic of the Philippines, No. ZPE-06-2001-S (13 February 2001). Vol. III, 351–55, https://files.pca-cpa.org/pcadocs/The%20Philippines%27%20Memorial%20 -%20Volume%20III%20%28Annexes%201-60%29.pdf, accessed 26 March 2019.

————. Annex 44. Memorandum from Acting Secretary of Foreign Affairs of the Republic of the Philippines to the President of the Republic of the Philippines (5 February 2001). Vol. III, 357–62, https://files.pca-cpa.org/ pcadocs/The%20Philippines%27%20Memorial%20-%20Volume%20III%20 %28Annexes%201-60%29.pdf, accessed 26 March 2019.

————. Annex 45. Memorandum from the Assistant Secretary for Asian and Pacific Affairs, Department of Foreign Affairs, Republic of the Philippines, to the Secretary of Foreign Affairs, Republic of the Philippines (14 February 2001). Vol. III, 363–68, https://files.pca-cpa.org/pcadocs/The%20 Philippines%27%20Memorial%20-%20Volume%20III%20%28Annexes%20 1-60%29.pdf, accessed 26 March 2019.

————. Annex 46. Office of Asian and Pacific Affairs, Department of Foreign Affairs, Republic of the Philippines, Apprehension of Four Chinese Fishing Vessels in the Scarborough Shoal (23 February 2001). Vol. III, 369–74, https://files.pca-cpa.org/pcadocs/The%20Philippines%27%20Memorial%20 -%20Volume%20III%20%28Annexes%201-60%29.pdf, accessed 26 March 2019.

————. Annex 47. Memorandum from the Embassy of the Republic of the Philippines in Beijing to the Secretary of Foreign Affairs, Republic of the Philippines, No. ZPE-09-2001-S (17 March 2001). Vol. III, 375–78, https://files.pca-cpa.org/pcadocs/The%20Philippines%27%20Memorial%20 -%20Volume%20III%20%28Annexes%201-60%29.pdf, accessed 26 March 2019.

————. Annex 49. Memorandum from Perfecto C. Pascual, Director, Naval Operation Center, Philippine Navy, to The Flag Officer in Command, Philippine Navy (11 February 2002). Vol. III, 395–97, https://files.pca-cpa.

org/pcadocs/The%20Philippines%27%20Memorial%20-%20Volume%20
III%20%28Annexes%201-60%29.pdf, accessed 26 March 2019.

———. Annex 50. Letter from Vice Admiral, Philippine Navy, to the Assistant
Secretary for Asian and Pacific Affairs, Department of Foreign Affairs,
Republic of the Philippines (26 March 2002). Vol. III, 399–404, https://
files.pca-cpa.org/pcadocs/The%20Philippines%27%20Memorial%20-%20
Volume%20III%20%28Annexes%201-60%29.pdf, accessed 26 March 2019.

———. Annex 51. Memorandum from Josue L. Villa, Embassy of the Republic
of the Philippines in Beijing, to the Secretary of Foreign Affairs of the
Republic of the Philippines (19 August 2002). Vol. III, 405–9, https://
files.pca-cpa.org/pcadocs/The%20Philippines%27%20Memorial%20-%20
Volume%20III%20%28Annexes%201-60%29.pdf, accessed 26 March 2019.

———. Annex 52. Report from CNS to Flag Officer in Command, Philippine
Navy, File No. N2D-0802-401 (1 September 2002). Vol. III, 411–14,
https://files.pca-cpa.org/pcadocs/The%20Philippines%27%20Memorial%20
-%20Volume%20III%20%28Annexes%201-60%29.pdf, accessed 26 March
2019.

———. Annex 57. Letter from Rear Admiral, Armed Forces of the Philippines,
to the Assistant Secretary for Asian and Pacific Affairs, Department of Foreign
Affairs, Republic of the Philippines (2006). Vol. III, 461–80, https://
files.pca-cpa.org/pcadocs/The%20Philippines%27%20Memorial%20-%20
Volume%20III%20%28Annexes%201-60%29.pdf, accessed 26 March 2019.

———. Annex 59. Report from the Commanding Officer, NAVSOU-2,
Philippine Navy, to the Acting Commander, Naval Task Force 21, Philippine
Navy, No. NTF21-0406-011/NTF21 OPLAN (BANTAY AMIANAN)
01-05 (9 April 2006). Vol. III, 485–502, https://files.pca-cpa.org/pca-
docs/The%20Philippines%27%20Memorial%20-%20Volume%20III%20
%28Annexes%201-60%29.pdf, accessed 26 March 2019.

———. Annex 75. Memorandum from the Embassy of the Republic of the
Philippines in Beijing to the Secretary of Foreign Affairs of the Republic of
the Philippines, No. ZPE-121-2011-S (2 December 2011). Vol. IV, 121–24,
https://files.pca-cpa.org/pcadocs/The%20Philippines%27%20Memorial%20
-%20Volume%20IV%20%28Annexes%2061-102%29.pdf, accessed 26 March
2019.

———. Annex 77. Memorandum from Colonel, Philippine Navy, to Chief
of Staff, Armed Forces of the Philippines, No. N2E-0412-008 (11 April
2012). Vol. IV, 131–47, https://files.pca-cpa.org/pcadocs/The%20
Philippines%27%20Memorial%20-%20Volume%20IV%20%28Annexes%20
61-102%29.pdf, accessed 26 March 2019.

———. Annex 78. Report from the Commanding Officer, SARV-003,
Philippine Coast Guard, to Commander, Coast Guard District Northwestern
Luzon, Philippine Coast Guard (28 April 2012). Vol. IV, 147–59, https://

files.pca-cpa.org/pcadocs/The%20Philippines%27%20Memorial%20-%20 Volume%20IV%20%28Annexes%2061-102%29.pdf, accessed 26 March 2019.

———. Annex 94. Armed Forces of the Philippines, Near-Occupation of Chinese Vessels of Second Thomas (Ayungin) Shoal in the Early Weeks of May 2012 (May 2013). Vol. IV, 333–46, https://files.pca-cpa.org/pca-docs/The%20Philippines%27%20Memorial%20-%20Volume%20IV%20 %28Annexes%2061-102%29.pdf, accessed 26 March 2019.

———. Annex 117. Ministry of Foreign Affairs of the People's Republic of China, Foreign Ministry Spokesperson Liu Weimin's Regular Press Conference on April 12, 2012 (12 April 2012). Vol. V, 109–14, https:// files.pca-cpa.org/pcadocs/The%20Philippines%27%20Memorial%20-%20 Volume%20V%20%28Annexes%20103-157%29.pdf, accessed 16 April 2019.

———. Annex 144. Association of Southeast Asian Nations, Declaration on the Conduct of Parties in South China Sea (4 November 2002). Vol. V, 321–25, https://files.pca-cpa.org/pcadocs/The%20Philippines%27%20Memorial%20 -%20Volume%20V%20%28Annexes%20103-157%29.pdf, accessed 16 April 2019.

———. Annex 211. Note Verbale from the Embassy of the People's Republic of China in Manila to the Department of Foreign Affairs of the Republic of the Philippines, No. (12) PG-239 (25 May 2012). Vol. VI, 401–4, https:// files.pca-cpa.org/pcadocs/The%20Philippines%27%20Memorial%20-%20 Volume%20VI%20%28Annexes%20158-221%29.pdf, accessed 26 March 2019.

———. Annex 240. Eastern South China Sea Environmental Disturbances and Irresponsible Fishing Practices and their Effects on Coral Reefs and Fisheries (22 March 2014), by Kent E. Carpenter, Ph.D. Vol. VII, 389–437, https:// files.pca-cpa.org/pcadocs/The%20Philippines%27%20Memorial%20-%20 Volume%20VII%20%28Annexes%20222-255%29.pdf, accessed 7 April 2019.

———. *The Philippines' Annexes Cited During Hearing on Jurisdiction (Annexes 574–583).* Annex 579. Ministry of Foreign Affairs, People's Republic of China, Foreign Ministry Spokesperson Lu Kang's Remarks on Issues Relating to China's Construction Activities on the Nansha Islands and Reefs (16 June 2015), 158–59, https://files.pca-cpa.org/pcadocs/Annexes%20cited%20 during%20Hearing%20on%20Jurisdiction%20%28Annexes%20574-583%29. pdf, accessed 17 April 2019;.

———. *The Philippines Annexes Cited During the Merits Hearing (Annexes 820–59)* (30 November 2015). Annex 850. "Offshore Coral Reef Damage, Overfishing and Paths to Peace in the South China Sea," by John W. McManus, 578–608, https://pcacases.com/web/view/7, accessed 26 March 2019.

———. Annex 820. Embassy of the People's Republic of China in Canada, An Interview on China's Construction Activities on the Nansha Islands and Reefs

27 May 2015, 5–10, https://pcacases.com/web/view/7, accessed 26 March 2019.

———. Annex 821. China State Oceanic Administration, "Construction Work at Nansha Reefs Will Not Harm Oceanic Ecosystems" (18 June 2015), 11–16, https://files.pca-cpa.org/pcadocs/The%20Philippines%27%20Annexes%20cited%20during%20Merits%20Hearing%20%28Annexes%20820-859%29.pdf, accessed 7 April 2019.

———. *The Philippines' Written Responses (Annexes 864–892)* (11 March 2016). Annex 864. China State Oceanic Administration, The Guidance for the Assessment of Coastal Marine Ecosystem Health, Marine Industry Standards of the People's Republic of China, No. HY/T 087-2005 (2005), 5–7, https://files.pca-cpa.org/pcadocs/The%20Philippines%27%20Written%20Responses%20%2811%20March%202016%29%20%28Annexes%20864-892%29.pdf, accessed 19 April 2019.

———. Annex 865. China State Oceanic Administration. "2012 Communique on Marine Environment of China, Part 2: Marine Biodiversity and Ecological Conditions" (1 April 2013), 8–21, https://files.pca-cpa.org/pcadocs/The%20Philippines%27%20Written%20Responses%20%2811%20March%202016%29%20%28Annexes%20864-892%29.pdf, accessed 19 April 2019.

———. Annex 866. China State Oceanic Administration. Code of Practice for Marine Monitoring Technology, Part 5: Marine Ecology, Marine Industry Standards of the People's Republic of China, No. HY/T 147.5-2013 (25 April 2013), 22–24, https://files.pca-cpa.org/pcadocs/The%20Philippines%27%20Written%20Responses%20%2811%20March%202016%29%20%28Annexes%20864-892%29.pdf, accessed 19 April 2019.

———. Annex 867. China State Oceanic Administration, "2013 Communique on Marine Environment of China, Part 2: Conditions of Marine Ecology" (25 March 2014), 25–36, https://files.pca-cpa.org/pcadocs/The%20Philippines%27%20Written%20Responses%20%2811%20March%202016%29%20%28Annexes%20864-892%29.pdf, accessed 19 April 2019.

———. Annex 868. China State Oceanic Administration. Technical Guidelines for Environmental Impact Assessment of Marine Engineering, National Standards of the People's Republic of China, No. GB/T 19485 -2014 (1 April 2014), 37–128, https://files.pca-cpa.org/pcadocs/The%20Philippines%27%20Written%20Responses%20%2811%20March%202016%29%20%28Annexes%20864-892%29.pdf, accessed 19 April 2019.

———. Annex 869. China State Oceanic Administration, South China Sea Branch, "Communique on the Oceanic Conditions of the South China Seas Region in 2013" (14 August 2014), 129–54, https://files.pca-cpa.org/pcadocs/The%20Philippines%27%20Written%20Responses%20%2811%20March%202016%29%20%28Annexes%20864-892%29.pdf, accessed 19 April 2019.

————. Annex 870. China State Oceanic Administration, "2014 Communique on Marine Environment of China, Part 2: Conditions of Marine Ecology" (16 March 2015), https://files.pca-cpa.org/pcadocs/The%20 Philippines%27%20Written%20Responses%20%2811%20March%20 2016%29%20%28Annexes%20864-892%29.pdf, accessed 19 April 2019.

————. Annex 871. China State Oceanic Administration, South China Sea Branch, "Communique on the Oceanic Conditions of the South China Sea Region in 2014" (28 May 2015), 160–70, https://files.pca-cpa.org/ pcadocs/The%20Philippines%27%20Written%20Responses%20%2811%20 March%202016%29%20%28Annexes%20864-892%29.pdf, accessed 19 April 2019.

————. Annex 872. China State Oceanic Administration, "Construction Activities at Nansha Reefs Did Not Affect the Coral Reef Ecosystem" (10 June 2015), 201–208, https://files.pca-cpa.org/pcadocs/The%20 Philippines%27%20Written%20Responses%20%2811%20March%20 2016%29%20%28Annexes%20864-892%29.pdf, accessed 19 April 2019.

————. *Philippines Written Responses to the Tribunal's November 2015 Questions (Annexes 860–63)* (18 December 2015). Annex 862. "Why Are Chinese Fishermen Destroying Coral Reefs in the South China Sea?," by Rupert Wingfield-Hayes, BBC (15 December 2015), 14–24, http://www.pca-cases.com/pcadocs/The%20Philippines%27%20Written%20Responses%20 to%20the%20Tribunal%27s%20November%202015%20Question%20 %28Annexes%20860-863%29.pdf, accessed 11 April 2019.

————. *Responses of the Philippines to the Tribunal's 5 February 2016 Request for Comments* (11 March 2016), https://pcacases.com/web/sendAttach/1849, accessed 19 April 2019.

————. Rules of Procedure. PCA Doc. 101991 (27 August 2013), https://pca-cases.com/web/sendAttach/233, accessed 7 April 2019.

————. *Supplemental Documents of the Philippines* (19 November 2015). Annex 615. People's Republic of China. *Law of the People's Republic of China on Evaluation of Environmental Effects* (28 October 2002), 114–23, https:// files.pca-cpa.org/pcadocs/The%20Philippines%27%20Supplemental%20 Documents%20-%20Volume%20I%20%28Annexes%20607-667%29.pdf, accessed 9 April 2019.

————. Annex 624. Ministry of Foreign Affairs, People's Republic of China, Foreign Ministry Spokesperson Hua Chunying's Regular Press Conference (9 April 2015). Vol. I, 165–69, https://files.pca-cpa.org/pcadocs/The%20 Philippines%27%20Supplemental%20Documents%20-%20Volume%20I%20 %28Annexes%20607-667%29.pdf, accessed 9 April 2019.

————. Annex 608, Department of Foreign Affairs of the Republic of the Philippines, Statement on China's Reclamation Activities and Their Impact on the Region's Marine Environment (13 April 2015). Vol. I, 7–9, https://

files.pca-cpa.org/pcadocs/The%20Philippines%27%20Supplemental%20
Documents%20-%20Volume%20I%20%28Annexes%20607-667%29.pdf,
accessed 17 April 2019.

———. Annex 609. Republic of the Philippines, Bureau of Fisheries and Aquatic
Resources, Press Release: DA-BFAR, National Scientist Condemns the
Destruction of Marine Resources in the West Philippine Sea (23 April 2015).
Vol. I, 11–15, https://files.pca-cpa.org/pcadocs/The%20Philippines%27%20
Supplemental%20Documents%20-%20Volume%20I%20%28Annexes%20607-
667%29.pdf, accessed 17 April 2019.

———. Annex 670. Note Verbale from the Department of Foreign Affairs of the
Republic of the Philippines to the Embassy of the People's Republic of China
in Manila, No. 14-1180 (4 April 2014). Vol. II, 9–12, https://files.pca-cpa.
org/pcadocs/The%20Philippines%27%20Supplemental%20Documents%20
-%20Volume%20II%20%28Annexes%20608-709%29.pdf, accessed 7 April
2019.

———. Annex 671. Verbatim Text of Response by Deputy Chief of Mission,
Embassy of the People's Republic of China in Manila, to Philippine
Note Verbale No. 14-1180 dated 4 April 2014 (11 April 2014). Vol. II,
13–16, https://files.pca-cpa.org/pcadocs/The%20Philippines%27%20
Supplemental%20Documents%20-%20Volume%20II%20%28Annexes%20608-
709%29.pdf, accessed 7 April 2019.

———. Annex 672. Note Verbale from the Department of Foreign Affairs of the
Republic of the Philippines to the Embassy of the People's Republic of China
in Manila, No. 14-2093 (6 June 2014). Vol. II, 17–20, https://files.pca-cpa.
org/pcadocs/The%20Philippines%27%20Supplemental%20Documents%20
-%20Volume%20II%20%28Annexes%20608-709%29.pdf, accessed 7 April
2019.

———. Annex 673. Note Verbale from the Department of Foreign Affairs of
the Republic of the Philippines to the Embassy of the People's Republic of
China in Manila, No. 14-2276 (23 June 2014). Vol. II, 21–24, https://
files.pca-cpa.org/pcadocs/The%20Philippines%27%20Supplemental%20
Documents%20-%20Volume%20II%20%28Annexes%20608-709%29.pdf,
accessed 7 April 2019.

———. Annex 675. Note Verbale from the Embassy of the People's Republic
of China in Manila to the Department of Foreign Affairs, Republic of the
Philippines, No. 14 (PG)-195 (30 June 2014). Vol. II, 29–32, https://
files.pca-cpa.org/pcadocs/The%20Philippines%27%20Supplemental%20
Documents%20-%20Volume%20II%20%28Annexes%20608-709%29.pdf,
accessed 7 April 2019.

———. Annex 676. Note Verbale from the Embassy of the People's Republic
of China in Manila to the Department of Foreign Affairs, Republic of the
Philippines, No. 14 (PG)-197 (4 July 2014). Vol. II, 35–39, https://
files.pca-cpa.org/pcadocs/The%20Philippines%27%20Supplemental%20

Documents%20-%20Volume%20II%20%28Annexes%20608-709%29.pdf, accessed 7 April 2019.

———. Annex 678. Note Verbale from the Embassy of the People's Republic of China in Manila to the Department of Foreign Affairs, Republic of the Philippines, No. 14 (PG)-264 (2 September 2014). Vol. II, 45–50, https://files.pca-cpa.org/pcadocs/The%20Philippines%27%20Supplemental%20Documents%20-%20Volume%20II%20%28Annexes%20608-709%29.pdf, accessed 7 April 2019.

———. Annex 680. Note Verbale from the Embassy of the People's Republic of China in Manila to the Department of Foreign Affairs, Republic of the Philippines, No. 14 (PG)-336 (28 October 2014). Vol. II, 55–59, https://files.pca-cpa.org/pcadocs/The%20Philippines%27%20Supplemental%20Documents%20-%20Volume%20II%20%28Annexes%20608-709%29.pdf, accessed 7 April 2019.

———. Annex 681. Note Verbale from the Ministry of Foreign Affairs, People's Republic of China to the Embassy of the Republic of the Philippines in Beijing, No. (2015) Bu Bian Zi No. 5 (20 January 2015). Vol. II, 61–66, https://files.pca-cpa.org/pcadocs/The%20Philippines%27%20Supplemental%20Documents%20-%20Volume%20II%20%28Annexes%20608-709%29.pdf, accessed 7 April 2019.

———. Annex 685. Note Verbale from the Embassy of the People's Republic of China to the Department of Foreign Affairs, Republic of the Philippines, No. 15 (PG)-068 (4 March 2015). Vol. II, 83–87, https://files.pca-cpa.org/pcadocs/The%20Philippines%27%20Supplemental%20Documents%20-%20Volume%20II%20%28Annexes%20608-709%29.pdf, accessed 7 April 2019.

———. Annex 686. Note Verbale from the Department of Boundary and Ocean Affairs, Ministry of Foreign Affairs, People's Republic of China, to the Embassy of the Republic of the Philippines in Beijing, No (2015) Bu Bian Zi No. 22 (30 March 2015). Vol. II, 89–93, https://files.pca-cpa.org/pcadocs/The%20Philippines%27%20Supplemental%20Documents%20-%20Volume%20II%20%28Annexes%20608-709%29.pdf, accessed 7 April 2019.

———. Annex 687. Note Verbale from the Department of Boundary and Ocean Affairs, Ministry of Foreign Affairs, People's Republic of China, to the Embassy of the Republic of the Philippines in Beijing, No. (2015) Bu Bian Zi No. 23 (30 March 2015). Vol. II, 97–101, https://files.pca-cpa.org/pcadocs/The%20Philippines%27%20Supplemental%20Documents%20-%20Volume%20II%20%28Annexes%20608-709%29.pdf, accessed 7 April 2019.

———. Annex 699. Environmental Consequences of Land Reclamation Activities on Various Reefs in the south China Sea (14 November 2015), by K. E. Carpenter and L. M. Chou, Vol. II, 245–94, http://www.pcacases.com/pcadocs/The%20Philippines%27%20Supplemental%20Documents%20-%20Volume%20II%20%28Annexes%20608-709%29.pdf, accessed 7 April 2019.

————. *Supplemental Written Submission of the Philippines* (16 March 2015). Vol. I, http://www.pcacases.com/pcadocs/Supplemental%20Written%20 Submission%20Volume%20I.pdf, accessed 5 April 2019.

Stephens, Tim. "Article 197. Cooperation on a Global or Regional Basis." *The United Nations Convention on the Law of the Sea. A Commentary*, 1328–33. Ed. Alexander Proelss München: Verlag C.H. Beck oHG, 2017.

———— and Duncan French. *ILA Study Group on Due Diligence in International Law. Second Report* (July 2016), https://ila.vettoreweb.com/Storage/ Download.aspx?DbStorageId=1427…4796, accessed 11 April 2019.

Stockholm Declaration of the United Nations Conference on the Human Environment, 16 June 1972, https://www.jus.uio.no/english/services/ library/treaties/06/6-01/stockholm_decl.xml, accessed 22 April 2019.

Tanaka, Yoshifumi. "The South China Sea Arbitration: Environmental Obligations Under the Law of the Sea Convention." *Review of European, Comparative and International Environmental Law* 27 (2018): 90–96.

United Nations Conference on Environment and Development. *Agenda 21.* Rio de Janeiro, Brazil, 3 to 14 June 1992, https://sustainabledevelopment. un.org/content/documents/Agenda21.pdf, accessed 11 April 2019.

United Nations Convention on the Law of the Sea, concluded at Montego Bay on 10 December, entered into force on 16 November 1994, http://www. un.org/Depts/los/convention_agreements/texts/unclos/closindx.htm, accessed 21 March 2019.

United Nations International Law Commission. "Commentaries to the Draft Articles on *Responsibility of States for Internationally Wrongful acts* adopted by the International Law Commission at its Fifty-Third Session (2001)." *Report of the International Law Commission on the Work of its Fifty-Third Session. Official Records of the General Assembly, Fifty-Sixth Session, Supplement No. 10* (A/56/10) (2001), http://www.un.org/documents/ga/docs/56/a5610. pdf, accessed 19 April 2019.

————. *Draft Articles on Responsibility of States for Internationally Wrongful Acts, with Commentaries 2001. Report of the International Law Commission on the Work of its Fifty-Third Session, 23 April–1 June and 2 July–10 August 2001. Official Records of the General Assembly, Fifty-Sixth Session*, Supplement No. 10, Doc. A/56/10, http://legal.un.org/ilc/texts/instruments/english/ commentaries/9_6_2001.pdf, accessed 12 April 2019.

————. "International Liability for Injurious Consequences Arising Out of Acts Not Prohibited by International Law (Prevention of Transboundary Harm from Hazardous Activities)." *Report of the International Law Commission on the Work of its Fifty-Third Session (23 April–1 June and 2 July–10 August 2001).* UN Doc. GAOR A/56/10 (2001), http://www.un.org/documents/ ga/docs/56/a5610.pdf, accessed 23 April 2019.

United States. National Environmental Policy Act of 1969, as amended ʿPub. L. 91-190, 42 U.S.C. 4321-4347, January 1, 1970, as amended by Pub. L. 94-52, July 3, 1975, Pub. L. 94-83, August 9, 1975, and Pub. L. 97-258, § 4(b), Sept. 13, 1982), https://ceq.doe.gov/laws-regulations/laws.html, accessed 22 April 2019.

Veron, John E. N. "Corals: Biology, Skeletal Deposition and Reef Building." *Encyclopedia of Modern Coral Reefs: Structure, Form and Process*, 275–81. Ed. David Hopley. Dordrecht: Springer, 2011.

Vienna Convention on the Law of Treaties, done at Vienna on 23 May 1969, entered into force on 27 January 1980, http://legal.un.org/ilc/texts/instruments/english/conventions/1_1_1969.pdf, accessed 9 April 2019.

Vukas, Budislav. "Enclosed or Semi-Enclosed Seas," *Max Planck Encyclopedia of Public International Law*. Oxford: Oxford University Press, 2015.

———. "United Nations Convention on the Law of the Sea and the Polar Marine Environment." *Protecting the Polar Marine Environment: Law and Policy for Pollution Prevention*, 24–56. Ed. Davor Vidas. Cambridge: Cambridge University Press, 2000.

Winkelmann, Ingo. "Article 123. Cooperation of States Bordering Enclosed or Semi-Enclosed Seas." *The United Nations Convention on the Law of the Sea. A Commentary*, 886–92. Ed. Alexander Proelss. München: Verlag C.H. Beck oHG, 2017.

CHAPTER 6

Summary and Conclusions

The *South China Sea Arbitration* will be remembered as much for its interpretation and application of Part XII of the United Nations Convention on the Law of the Sea, 1982 ("the Convention"), on the protection and preservation of the marine environment as for its decisions on the survival of historic rights following the Convention's entry into force or the distinctions among low-tide elevations, islands, and rocks. An impartial third party has ruled that China has violated its obligations under the Convention to protect and preserve the marine environment through its toleration of the harvesting of endangered species by its nationals as well as through its construction activities in the South China Sea; its failure to cooperate with other littoral States of the South China Sea; and its failure to communicate the results of its assessment of the environmental impact of its construction activities.

The South China Sea, a large marine ecosystem characterized by a high degree of biodiversity, has been over-exploited for over thirty years now. The role of Chinese nationals in this overexploitation is unique, in that they are for the most part harvesting not fish but high-value endangered species, such as corals, giant clams, and sea turtles, and they use dynamite and cyanide for this purpose. From the mid-1990s onward, their activities triggered incidents between the Philippines and China. Further tension in bilateral relations was created by China's construction activities on Mischief Reef, a low-tide elevation on the Philippine continental shelf that China occupied in 1995. These activities were expanded to other maritime features in the South China Sea from 2013 onward.

© The Author(s) 2020
A. C. Robles Jr., *Endangered Species
and Fragile Ecosystems in the South China Sea*,
https://doi.org/10.1007/978-981-13-9813-1_6

The activities of Chinese fishermen and of China itself prompted repeated protests from the Philippines. Initially, Philippine arguments stressed the importance of the conservation of marine living resources but as time went on, the Philippines increasingly based its arguments on considerations of biodiversity. The Philippine concern with biodiversity is understandable, for the endangered species, such as sea turtles, and fragile ecosystems, mainly coral reefs, play significant ecological roles in the Philippines and in the South China Sea. The Philippine concern is all the more remarkable, given that their economic value for the country differed markedly: Coral reefs are of vital importance for many coastal communities in the country, but sea turtles are an important economic resource only for marginalized and indigenous communities.

While the Philippines did not fail to charge Chinese fishermen with violations of Philippine laws, it also invoked China's obligation to protect and preserve the marine environment under the Convention. Such a move enabled the Philippines to circumvent Chinese rejection of Philippine protests on the grounds that China had sovereignty over the waters in which Chinese nationals were carrying out their activities. At the same time, the Philippines also invoked other international environmental agreements, notably the Convention on Biological Diversity, 1992 ("CBD"), for it undoubtedly found the concepts associated with the CBD useful for describing the conduct of Chinese fishermen and of China.

Nevertheless, the Philippines did not submit a claim under the CBD. Unlike other multilateral environmental agreements, the CBD lacks a fully developed non-compliance mechanism. Like other multilateral environmental agreements, the CBD lacks provision for compulsory dispute settlement entailing binding decisions. The only compulsory procedure that is available in the CBD framework is conciliation, which by definition, does not entail binding decisions. While the Convention did not set in place a non-compliance mechanism, it provides for compulsory dispute settlement (unilateral submission of the dispute to a judicial body) entailing binding decisions, which would be attractive to a country that had attempted over the years to resolve its differences with China by bilateral channels and failed. Understandably, the Philippines formulated its allegations of damage to biodiversity in terms of the general obligation under Article 192 of the Convention to protect and preserve the marine environment and the obligation to take all necessary measures to preserve rare and fragile marine ecosystems under Article 194(5) of the Convention. The CBD, together with the Convention on the ¯rade in Endangered Species of Wildlife and Fauna, 1973 ("CITES"), w̄as to be valuable in informing the interpretation of the Convention provisions relating to the conservation of fragile marine ecosystems.

Before the Tribunal constituted under Annex VII of the Convention could hear the merits of the Philippine claims, the Tribunal had to satisfy itself that it had jurisdiction over them. In the Position Paper that laid out its reasons for rejecting the Tribunal's jurisdiction, China had not expressed any specific objection to the Tribunal's jurisdiction with respect to the environmental claims of the Philippines, but the Tribunal decided to examine two jurisdictional issues: Whether certain preconditions to compulsory dispute settlement had been fulfilled; and whether automatic limitations and optional exceptions to jurisdiction applied.

The preconditions refer to the absence of other agreements (such as the CBD) between the Philippines and China to settle disputes regarding the interpretation and application of the Convention and to the absence of other general, bilateral or regional agreements (such as the CBD) between the Philippines and China that provided for compulsory mechanisms entailing binding decisions to settle disputes relating to the interpretation and application of the Convention. A further precondition was the exchange of views between the Philippines and China on the settlement of the disputes. The Tribunal arrived at the conclusion that both preconditions had been fulfilled. As the CBD was not an agreement to resolve disputes relating to the interpretation and application of the UN Convention on the Law of the Sea and it did not involve compulsory dispute settlement binding decisions, it could not preclude the jurisdiction of the Tribunal. The voluminous diplomatic documents submitted by the Philippines to the Tribunal proved that exchanges of views had taken place.

The Philippines contended that the automatic limitation to jurisdiction, which specified that disputes relating to the coastal State's implementation of international rules and standards for the protection and preservation of the marine environment would be subject to compulsory dispute settlement entailing binding decisions, did not in fact constitute a limitation. The optional exceptions, specifically the military activities and law enforcement exceptions, were not applicable either. The Tribunal accepted that China had repeatedly declared that its construction activities and island-building at Mischief Reef and other reefs were for civilian purposes only, rendering the military activities exception inapplicable. The Tribunal also agreed with the Philippines that the law enforcement activities exception was not a blanket exception for law enforcement activities; it only referred to certain disputes arising over marine scientific research.

The affirmation of the Tribunal's jurisdiction with respect to the environmental claims enabled it to examine the merits of the Philippine claims. The Tribunal agreed with the Philippines that the interpretation of the general obligation to preserve and protect the marine environment laid down in Article 192 of the Convention could be guided by reference to other provisions of Part XII of the Convention, particularly Article 194(5), and other multilateral environmental agreements, which included CITES and the CBD. The latter provided an internationally accepted definition of ecosystem, a concept that was not defined by the Convention.

The evidence submitted by the Philippines was sufficient to convince the Tribunal that China had tolerated illegal harvesting of endangered species (sea turtles, giant clams, and certain corals) by its nationals. The Tribunal agreed with the Philippines that CITES, whose annexes list sea turtles, giant clams, and corals as endangered species, forms part of the general corpus of international law that informed the content of Article 192. In this context, harvesting of sea turtles constitutes harm to the environment as such, while the harvesting of corals and giant clams had a harmful effect on a fragile marine ecosystem, the coral reefs. The Tribunal also accepted the Philippine argument that China's obligation under Article 192 was a due diligence obligation to take measures necessary to prevent such acts and to maintain a certain level of vigilance in the enforcement of these measures. The Tribunal found no evidence that China had taken any steps to enforce its rules against its fishermen. China, which was well aware of these practices, not only turned a blind eye to them but provided armed government vessels to protect the fishing vessels engaged in the harmful activities. China thus breached its obligation under Articles 192 and 194(5) of the Convention to take necessary measures to protect and preserve the marine environment, with respect to the harvesting of endangered species from fragile ecosystems at Scarborough Shoal and Second Thomas Shoal.

The Tribunal arrived at a different conclusion concerning dynamite and cyanide fishing by Chinese nationals in the South China Sea. The absence of Philippine protests since the turn of the century and the fact that China updated its laws and prohibited the use of explosives, poisons, electricity, and other measures that impaired fisheries resources suggested to the Tribunal that China may have taken measures to prevent this practice in the Spratly Islands. These considerations led the Tribunal to refrain from making any conclusion concerning China's responsibility for dynamite and cyanide fishing.

There was no question in the Tribunal's mind that the island-building program on seven reefs—Cuarteron Reef, Fiery Cross Reef, Gaven Reef (North), Johnson Reef, Hughes Reef, Subi Reef, and Mischief Reef—in the South China Sea was part of an official Chinese policy and program adopted by organs of the Chinese State. In this regard, the Tribunal agreed with the Philippines that China had three sets of obligations. The first was a general obligation to preserve and protect the marine environment under Article 192 of the Convention and particularly to protect and preserve fragile marine ecosystems as well as the habitat of depleted, threatened or endangered species and other forms of marine life in conformity with Article 194(5). The second was an obligation to cooperate with regional States under Article 197 and with States bordering a semi-enclosed or enclosed sea under Article 123 of the Convention. The third was to monitor and assess the impact of its construction activities on the marine environment by carrying out an Environmental Impact Assessment ("EIA") and communicating the results of the assessment, in conformity with Articles 206 and 205 of the Convention.

The Tribunal had no doubt that the coral reef systems in the South China Sea were fragile ecosystems and that China's construction activities had caused devastating and long-lasting damage to the marine environment. Moreover, the Tribunal found no convincing evidence of China attempting to cooperate or coordinate with other littoral States. On the contrary, China had proceeded with island-building, heedless of the protests of its neighbors. Finally, neither the Tribunal, the Tribunal-appointed experts, the Philippines nor the experts appointed by the Philippines were able to identify any Chinese report that would resemble an EIA. The Tribunal did not rule out the possibility that China had carried out an EIA, as required by Article 206. What was certain was that China had failed to communicate the results of any such assessment as stipulated by Article 205. These were the reasons that led the Tribunal to declare that China had breached the Convention.

The non-appearance of a party to the dispute does not warrant a default judgment. Annex VII of the Convention on arbitration requires that the arbitral tribunal ensure that its decisions would be well founded in fact and in law. The steps that the Tribunal in the *South China Sea Arbitration* took for this purpose are all the more noteworthy given the scientific and technical issues raised by the Philippine claims. The Tribunal set in place a procedure that enabled the Philippines to address issues that in the Tribunal's view it may not have adequately addressed or not at all addressed. The Tribunal posed a large number of questions to the Philippines prior to, during, and after the Hearings on Jurisdiction and on the Merits. Similarly, the Tribunal posed questions to the expert presented by the Philippines during the Hearing on the Merits. The Tribunal did not rely for scientific evidence only on the expert and the expert reports presented by the Philippines but appointed its own experts, after the conclusion of the Hearing on the Merits. These steps do not justify claims that the Tribunal inaugurated a new form of fact-finding called "hard fact-finding," nor do they support the view that expert evidence was the basis of the conclusion that China had taken measures to combat cyanide and dynamite fishing. The Tribunal instructed the Philippines to search for Chinese EIAs, while undertaking its own research to determine whether China had conducted an EIA. Finally, the Tribunal wrote directly to China to ask whether it had indeed carried out an EIA and if so, to request that China provide a copy to the Tribunal. The Tribunal was only prevented from undertaking a site visit by Chinese opposition to the idea. The steps taken by the Tribunal to ensure that its decisions were well founded in fact and in law are of such a nature as to create confidence in the decisions of the Tribunal relating to Submissions No. 11 and 12(b). The post-arbitration critique of the Awards prepared by the Chinese Society for International Law is not likely to counter this impression; indeed, one is tempted to argue that even had China appeared before the Tribunal or otherwise made submissions to it, China's arguments regarding Submissions No. 11 and 12(b) would not have prospered.

The Tribunal's decisions on the merits of the disputes are remarkable for several reasons. First, they affirmed that the Tribunal had jurisdiction to entertain the Philippine claims despite the fact that the maritime features and surrounding waters in which the activities were carried out were the object of competing claims of sovereignty. Second, they broadened the interpretation of the general obligation under Article 192 of the Convention to protect and preserve the marine environment to encompass the protection of marine biodiversity, notably endangered species as well as rare or fragile marine ecosystems and the habitats of depleted, threatened or endangered species, and other forms of marine life. Third, the Tribunal interpreted the Convention's provisions in the light of the corpus of general international law, which included some of the most important international environmental agreements.

The research demonstrates that these decisions of the Tribunal had their roots in the diplomatic practice of the Philippines, which were subsequently embodied in the submissions to the Tribunal. The confidential diplomatic documents that the Philippines submitted to the Tribunal showed that the Philippines, without abandoning its claims to sovereignty, found it essential to formulate its claims in terms of the general obligation to protect and preserve the marine environment under the Convention, as a means of preempting China's rejections of its protests on the grounds that China had sovereignty over the waters in which Chinese nationals carried out their activities. In its communications with China, the Philippines also emphasized the need to preserve endangered species and fragile marine ecosystems, and not merely the conservation of marine living resources, an attitude that arguably reflects the significance of biodiversity in general for the Philippines. In the search for legal bases for its protests to China, the Philippines invoked the CBD and the CITES alongside the Convention, because the concepts associated with these two Conventions seemed to characterize more accurately the conduct of Chinese nationals and of China.

Examination of the diplomatic documents submitted by the Philippines to the Tribunal retrospectively provides validation, if it were at all necessary, of the decision to submit the disputes over the activities of Chinese fishermen and of China itself to arbitration. The documents show that over nearly twenty years the Philippines protested these activities, requested that China instruct its fishermen to refrain from their illegal activities, explained to China the harm to the marine environment caused by the activities of Chinese fishermen and of China itself, changed its practice from arresting Chinese fishermen to escorting them away from the disputed waters, and sought agreements with China. China for the most part, rejected the protests, denied that the Philippines even had the right to protest, reasserted its sovereignty over the disputed waters, declared against all the evidence that the activities of the Chinese fishermen were legitimate, denied that its construction activities caused harm to the marine environment, and warned the Philippines that the Philippines would be responsible for the consequences of Philippine actions. Occasionally, China displayed creativity when responding to the

Philippines—witness the time when China explained construction on Mischief Reef as the installation of a TV aerial, or the time when China stated that the shelters being built on Mischief Reef could even be used by Filipino fishermen. This last statement did not go very far, for as soon as the Philippines expressed interest in the proposal, China made any such use conditional on an agreement between the two governments.

The irreversible damage caused to the marine environment by China's construction activities and China's refusal to comply with the Awards, not to mention the Philippine government's decision to set them aside in the interest of better relations with China, might tempt the observer to dismiss the Awards. But the task of the Tribunal was to interpret the law—and that it did. The decisions by an impartial third party have given the lie to Chinese claims that its activities and that of its nationals do not cause harm to the marine environment. This is the reason why Chinese scholars, notwithstanding the passivity of the Philippine government, are unrelenting in their efforts to undermine the credibility of the Awards.

If the Awards have been powerless to prevent damage to the marine environment, the denials by the Chinese scholars will probably not be able to prevent the looming collapse of fisheries resulting from the unchecked harvesting of endangered species and the unprecedented destruction by China of the coral reef ecosystems on which much of the fisheries depends. Even a dramatic—and improbable—reversal of policy on the part of the Chinese government will not repair the damage that the marine environment has already suffered. If and when fisheries stocks collapse, there can be little doubt about where much of the responsibility will lie. Of course, it is not just the State that is responsible and its population that will have to bear the consequences—all of the littoral States of the South China Sea and their populations will be adversely affected to varying degrees. It is not certain whether smaller littoral States will be able to call a great power to account for any environmental disasters that might occur. But the State that is responsible will have to face the risk—one that is greatly feared by authoritarian governments—that the devastating and long-lasting damage that it has inflicted on the marine environment will attract the opprobrium of its own population.

INDEX